THE WINE MANUAL

Jacques Marie

Shari Darling **Konrad Ejbich**

gage EDUCATIONAL PUBLISHING COMPANY
A DIVISION OF CANADA PUBLISHING CORPORATION
Vancouver · Calgary · Toronto · London · Halifax

Book design and maps: Visutronx

Technical art: Jane Whitney

Cover photograph: image copyright PhotoDisc, Inc., 1997

Acknowledgments

We would like to thank the following wine lovers for their invaluable help in the making of this book:

Tony Aspler
Editor of *Wine Tidings* magazine, and wine columnist for the *Toronto Star*.

Laszlo Buzas
Director of Food and Beverage, Inter-Continental Hotel, Toronto.

Karen Richardson-Norris
Instructor, Algonquin College.

Ivan Stephen
Consultant, Snowdon Public Relations.

Mark W. Waldron, Ph.D.
Professor of Rural Extension Studies, Ontario Agricultural College.

Thanks also to the following wineries for allowing us to use their labels:
Freixenet, Inniskillin, Kendall-Jackson, Mateus, Ruffino, Stoneleigh, St. Stephan's Crown, Taylor Fladgate, Veuve Clicquot Ponsardin, Vineland Estates, Wyndham Estate

Canadian Cataloguing in Publication Data

Marie, Jacques

The wine manual

Includes index.

ISBN 0-7715-5118-5

1. Wine and wine making. I. Darling, Shari.

II. Ejbich, Konrad. III. Title.

TP548.M37 1997 641.2'2 C96-932230-5

ISBN 0-7715-**5118-5**

1 2 3 4 5 PC 01 00 99 98 97

Written, printed, and bound in Canada.

Table of Contents

PREFACE

As every experienced restaurateur knows, serving the public is a far cry from serving the public well. This book hopes to provide you, the aspiring wine professional, with the tools to advance in your career.

The Wine Manual is the first book of its kind in North America, because it is written specifically for a person entering the hospitality field of wine service. This book is intended to give you a basic and thorough understanding of the business of making, buying, storing, selling, and serving wine. And as a collection of international facts and references, it will help you to put the world of wine into perspective.

To get the most from your study of wine, use this book together with contemporary magazines and any other publications written for people interested in wine. That is, read what wine professionals read, but also know what your customers are reading.

Viticulture and Viniculture

The chapters contained in Part I explore viticulture (how grapes are grown) and viniculture (how wines are made). Both are subjects that combine craft, art, science, geography, and history.

CHAPTER I

A BRIEF HISTORY OF WINE

Wine has fascinated us since our prehistoric ancestors chanced upon the natural process that turns grapes into wine.

ANCIENT TIMES (TO A.D.500)

Paleontologists have discovered fossilized leaves that date back at least sixty million years, near Sezanne in the French region of Champagne. They are of the earliest known variety, Vitis sezannensis, which flourished during a time when much of France was still covered by tropical forest.

Wild vines were available to ancient tribes in abundance, and played an important role in their lives. Since the grape tolerates various conditions—arid soils, prolonged drought, and even flooding along water-saturated river banks—it would have been valued by early nomadic people.

The plant and its fruit were found to have many uses. The ripened fruit could be eaten as food or crushed and drunk as fluid. The strong and flexible branches could be used to make ropes and lines. With time, people learned to dry the grapes for use during the winter months. The grape was also found to be useful in flavouring, tenderizing, and preserving food, and to have antiseptic and medicinal qualities. Skins of red grapes could be used to dye fabrics and faces.

Early Winemakers

At some point, early humans found that grapes ferment naturally, and the result had flavour and "power." But it was not until nomadic tribes began to settle down that conscious attempts to improve and refine this completely natural process were initiated. The precise period in which people started to cultivate the vine is not clear. Archaeologists and theologians alike suggest that cultivation was practised in Asia Minor, south of the Black and Caspian seas. The oldest known pips of cultivated vines were discovered on Mount Ararat, where the Bible tells us Noah's Ark set down after the flood. Carbon-dated to between 7000 B.C. and 5000 B.C., this find led to further excavations along the southern slopes of the Caucasus Mountains in the areas of Armenia, Georgia, and northern Iran. Here, archaeologists unearthed large clay jars used to make and store wine.

Vitis vinifera sylvestris (a wild woodland variety) became Vitis vinifera sativa (a cultivar) when nomads settled down and selected this specimen to cultivate.

As early as 3500 B.C., Sumerians developed efficient irrigation systems in the barren land known as Mesopotamia, between the Tigris and Euphrates rivers. Successive civilizations showed equally great interest in making and trading wine—Assyrians and Babylonians (3300 B.C.–2000 B.C.), Egyptians (2800 B.C.–550 B.C.), Phoenicians (1400 B.C.–200 B.C.), Greeks (2000 B.C.–150 B.C.), Etruscans (800 B.C.–350 B.C.), and Romans (750 B.C.– A.D. 450) all praised wine and believed it to be a special gift from the gods.

Egyptians drank beer for the most part, but wine was celebrated in wealthy and royal circles. The insides of pyramids in the Valley of the Kings and the Valley of the Nobles are decorated with friezes and tablets depicting the harvesting of grapes and the making of wine. During the Pre-dynastic period, Egyptians had some knowledge of pruning, and had developed a crude wine press. Records from 1792 B.C.–1750 B.C. show they understood the negative effects of air on wine, since they began to seal their wine vessels with pitch and grease. They were also the first to mark these containers with details about the origin of grapes (they had identified six regions), the date of the harvest, and even the name of the winemaker, thus establishing the first record of wine labelling.

The Phoenicians are credited with propagating the vine throughout the Mediterranean basin. As seafaring merchants, they introduced vines to lands along their trade routes—along the coast of Africa to present-day Morocco, from the Dardanelles to the southern coast of Spain, and around the Balaeric Islands, the eastern coast of Sardinia and Corsica, Sicily, Crete, Rhodes, and Cyprus.

The achievements of the Greeks include: identifying Vitis vinifera varieties, recording winemaking techniques, developing clay amphorae in which to store and age wine, and inventing the pruning knife. They found the soil and climatic conditions of Italy so conducive to growing vines that they gave to southern Italy the name Oenotria ("land of wine"). According to Greek legend, Dionysus, the son of Zeus, created wine. In Roman mythology, the god of wine was known as Bacchus.

The Romans made many great strides in winemaking. They brought ingenuity and reverence to viticulture and to viniculture. Learning as much as they could from the Greeks and the Etruscans, the Romans significantly improved the quality of wine by reducing grape yields through pruning and fertilization. They conducted many experiments, such as introducing sea water to improve wine's brightness, adding alkaline substances to reduce its acidity, boiling grape juice to increase its sugar content, and mixing in herbs and spices (the first vermouths!). Pliny the Elder, himself a grower, classified grapes as to colour, time of ripening, and soil preferences. Not all Roman improvements were well founded, however—they stored grape concentrates in lead containers, and drank wine from pewter cups.

Viticulture is the management of the vine to the point of the healthy ripening of the grapes. Viniculture is the management of the winemaking process once the grapes have been picked.

Their most significant contribution was to spread the practice of grape growing and the culture of winemaking throughout their empire. Bordeaux, Burgundy, Champagne, Alsace, the Rhône, the Rhine, the Moselle, Switzerland, Austria, Hungary, Spain, Portugal, and even Britain were all Roman wine domains.

Perhaps they overdid it. Near the end of the first century A.D., wine production in most of their territories was brought to a halt by an edict from Rome. In an attempt to keep the wines of Spain and the vast lands of Gaul from competing with Roman exports, the Emperor Domitian prohibited grape growing and ordered all vines north and west of the Alps to be uprooted. By the time the Roman Empire was beginning to fall (A.D. 476), Europe's wine regions had already deteriorated.

THE MIDDLE AGES (A.D. 500–1400)

There was no significant improvement in the fields of grape growing and winemaking in the centuries that followed the fragmentation of the Roman Empire. As demand collapsed, so did the supply. Vineyards were uprooted, and agriculture focussed on other crops.

The Christian church, in need of a steady supply of wine for sacramental purposes, preserved some of the know-how of the winemaking process and undertook a program of acquisition of the lands it needed for grape growing. By the end of the first millenium, it was the largest holder of European vineyards. As part of a stable, tax-exempt organization with an ample supply of free labour to work the vineyards, monasteries and their educated clergy continued to advance the art of making wine.

Whereas quantity had been the foremost objective in the bacchanalian past, the Church was more interested in quality—what was not used in the celebration of Mass could be sold to offset the cost of caring for the poor and the weak. Devoted Christian winemakers meticulously recorded their results and endeavoured to classify wine by variety, by vineyard, and by vintage. Knowledge was freely shared with other monasteries. (The Cistercians of Burgundy, for example, communicated regularly with their brethren at Kloster Eberbach in the Rheingau.) Their continued efforts and achievements in preserving the craft of winemaking through the Dark Ages are of profound significance in the history of wine.

FROM THE RENAISSANCE TO MODERN TIMES (A.D. 1400–1900)

For centuries, wine was anything but a luxury. The water in most towns and cities was impure, and in some instances disease-bearing. Wine (because of its antiseptic properties) was a much safer beverage. People of all ages regularly consumed it on its own, or added wine to purify the water. For the most part, production focussed on high potency, with little thought given to taste. Increasing quantities were produced for export to countries where wine could not be produced (Norway, Sweden) or where production was too small to satisfy the needs of the population (Britain).

With the revival of interest in art and letters during the Renaissance, a new desire for wine as a refined, cultured beverage was born among the ruling classes. The wine trade became a lucrative business. There followed vineyard expansion, experimentation with new techniques (barrel ageing, topping up ullage, lees contact, racking, sulphuring), and development of technological wonders such as the glass bottle and the cork stopper. For the first time it became feasible to keep wine for more than a year. By the beginning of the eighteenth century, it was common to buy speculatively before the wine was made (wine futures), and to lay down wine for following generations.

The Industrial Revolution brought growth and, with it, unimaginably awful social conditions. A new demand for cheap wine came with the hordes of thirsty workers flocking to the towns and cities.

The colonization and commercial exploitation of North America did little for Europe's wine industry at first. Settlers had far more important things to do than sit around drinking wine. But once a social and political infrastructure had been established, an interest emerged in the cultivation of European vines to produce quality wine.

Unfortunately, North America's earliest influence on European viticulture was tragic and devastating. In the middle of the nineteenth century—just as the industry was poised for explosive growth since governments were imposing heavier taxes and restrictions on spirits—native American vines were shipped to Europe for experimental purposes. These vines carried with them a tiny insect called phylloxera. The wild vines of North America had developed a resistance to the insect over the millenia, but Europe's vinifera vines had no resistance whatsoever. Within a decade, the louse had spread all across France and into Italy, leaving ugly scenes of withered and dying vineyards.

Phylloxera is an aphidlike insect which kills Vitis vinifera varieties by feeding on their roots, thereby starving the vine.

The solution to phylloxera brought with it new problems. Since American vines were immune, it was concluded that grafting vinifera vines onto American rootstock would solve the problem. It did. But the vines from the new land were more susceptible to other diseases, especially downy mildew and powdery mildew, further reducing output at a time when demand was still climbing. The search for a solution to the series of blights led also to the creation and study of hybrid vine varieties. The hope was to develop disease-resistant vines by genetically crossing vinifera varieties with North American species resistant to phylloxera.

French hybridizers carried out thousands of experiments, creating hundreds of hybrids. Only a few were found to be viable in commercial wine production. But improved rootstocks, insecticides, and fungicides permitted the return and popularity of vinifera varieties.

THE TWENTIETH CENTURY

Europe's problems got much worse before they got any better. World War I not only deprived Europe of its agricultural labour force, it also saw the destruction of its wineries and the trampling of its vineyards. After the war, growing populations wanted wine and, with insufficient supply, cheating soon sprang up: cheap plonk labelled as great Bordeaux and Burgundy, imported raisins fermented in water, dangerous chemicals added to darken juices. Something had to be done to protect an innocent, and increasingly angry, public. France was the first to respond, passing laws and regulations to guarantee authenticity by delimiting wine regions to control production volumes and to regulate allowable grape varieties and winemaking techniques. These policies formed the basis for the first Appellation d'Origine Contrôlée law, in 1935.

Meanwhile, the budding wine industries in the United States and Canada suffered the onslaught of temperance movements and ultimately, Prohibition. In 1919, the Eighteenth Amendment prohibited the sale of alcoholic beverages in the United States. Vineyards were left to decay; winemaking skills were forgotten; a generation lost its taste for wine. By 1933, the United States government terminated the experiment, but the harm was already done.

World War II had the dual effect of reducing the drinking population overall, but at the same time exposing North American soldiers to the taste of good wine. In the half century since the end of the war, world consumption of fine wine has increased, as has interest in growing and making it. Since 1970, wine production has increased dramatically in Australia, Canada, Chile, New Zealand, and the United States.

Today, with increasing medical evidence of the health-giving qualities of wine consumed in moderation, there is great hope and excitement in the wine industry for continued celebration of the ancient grapevine.

CHAPTER 2

CLIMATE AND SOIL

A wine's quality and its style are determined by decisions made long before any grapes are harvested, even before they are planted. It would be nice if grapes could grow anywhere on the planet. The vine is tolerant of many different types of inhospitable conditions, but it cannot ripen at all as it gets nearer the North or South Poles, and it bakes into raisins, or completely fails to fruit, if planted too close to the equator.

A wine producer's success depends on several interdependent factors: climate, soil, grape variety, viticulture, viniculture, and nature's mercy (i.e., weather). Climate, soil, and grape variety are pondered and decided before establishing the vineyard. The right choices will make or break the resulting product.

Factors relating to viticulture and viniculture can be changed at any time by the vineyard manager or by the winemaker. To spray or not to spray in the fields, to de-stem, or skin ferment, or age a wine on its lees; these are questions debated all the time by producers and consumers alike. There are many schools of thought as to how best to care for the grapes before harvest, and for the wine afterward.

The last factor, weather, explains why so many wine producers are anxious in the last days before the harvest. Despite making all the correct choices, and using all the best agricultural practices, it takes just one hailstorm to ruin an entire crop.

TERROIR

Many growers believe fiercely in the importance of terroir—the natural environment of a site, taking into account its climate, soil, and topography.

Every grower has terroir, like it or not. Climate takes into account temperature, rainfall, and wind. Soil includes the surface and subsoils, with all their chemical components. Topography means the lay of the land, the degree and direction of any slope: an east-facing slope gets sun in the morning whereas a west-facing hillside gets the evening sun.

Terroir reflects all the idiosyncrasies of a piece of land. One can speak of the entire Bordeaux region as a terroir in the general sense, or of Château Le Pin, the smallest property in Bordeaux's tiny Pomerol district, as a distinct terroir. Many wine-producing regions (especially within the European Union) use terroir as a basis for demarcating, delimiting, and regulating wine zones.

CLIMATE

Climate is the combination of temperature, precipitation, air movement, and average amount of sun/cloud cover of a locality over an extended period of time. It can be further defined in terms of macroclimate, mesoclimate, and microclimate.

A macroclimate describes the climate of a large area or region. For example, the macroclimates of France include (in general terms) the moderate maritime zone on the Atlantic and northern seaboards, the harsher continental inland zone, and the warm Mediterranean zone of the south.

A mesoclimate is the climate specific to a subregion, a group of vineyards, a single vineyard, or even a section of a vineyard. It is created by the surrounding natural features (bodies of water, mountains, forests, etc.) and their relationship to one another. Within the Niagara Peninsula in Ontario, the strip of land bordering Lake Ontario is a mesoclimate. It benefits from the moderating effect of the lake, which raises winter temperatures along the shoreline by several degrees, protecting vines from winterkill, and reducing the danger of late-spring frosts. It also cools the summer air so the growing season is extended and grapes ripen slowly.

Deeper inland, another mesoclimatic zone follows the bench of the Niagara Escarpment where air movement is at its highest. The plateau between the lake and the escarpment has yet another distinct mesoclimate, with good air circulation that keeps the vineyards relatively dry and free of disease.

A microclimate lies within a narrowly defined area—such as between rows of vines. A vineyard may have as many microclimates as it has plants.

Climate remains much the same year after year, since it reflects the average of many years of fluctuating weather conditions. Changes in the weather from vintage to vintage determine the size of the crop, the ripeness attained by the grapes (ratio of sugars, acids, and percentage of pulp), and the degree of vine sickness and parasites in a particular season.

Contemporary thought leans toward separating the world's wine regions into three climatic zones: cool, intermediate, and warm. The warmest areas are useful mainly for table grapes and raisins. Intermediate-climate wines are often full bodied and alcoholic, such as those of southern France, southern California, Italy, Greece, Spain, the Middle East, and North Africa. The most delicate and complex wines are produced in cool-climate regions such as Alsace, Beaujolais, Bordeaux, Burgundy, Chablis, Champagne, the Loire, Germany, Canada, the northern United States, southern Chile, southern New Zealand, and Tasmania.

Temperature and Sunshine Hours

Virtually all the planet's wine-growing zones lie between 30°N and 50°N and between 30°S and 50°S.

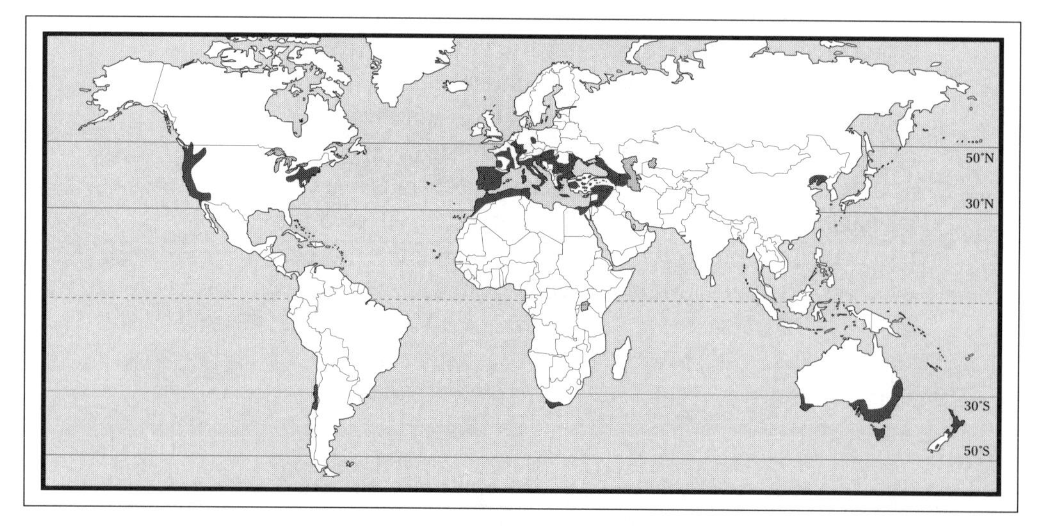

Although temperature and sunshine hours work together within a particular climate, they have distinctly different effects on the grape. Hours of sunshine are affected by latitude, altitude, vineyard orientation, obstacles (forests, buildings), humidity (haze, cloud cover), and pollution.

Sunshine is important to the process of photosynthesis, (which converts water and carbon dioxide into sugar), and to the concentration of pigments in the skin of the grape. Generally, the ideal amount of sunshine that wine grapes require for perfect maturation is as follows:

- 85–100 hours of sunshine per week during the ripening period.
- 1300 hours in total between budbreak and picking.
- 1500–1600 hours annually.

Budbreak is the stage at which shoots emerge from the vine bud.

DEGREE DAYS

Professors Maynard Amerine and A.J. Winkler of the University of California (Davis) calculated average daily temperatures in various California regions. They used temperatures expressed in degrees Fahrenheit over the threshhold of 50°F (equivalent to 10°C) but restricted their study to the period between April 1 to October 31, effectively, the growing season. They then classified five wine regions according to their "degree days." They proposed that their results demonstrated a measurable relationship between a region's mean growing temperatures and the quality of the wine produced.

To calculate degree days, the average monthly temperature over 50°F is multiplied by the number of days in the month. For example, if the mean temperature in the month of June is 66°F, the daily incremental heat over 50°F is 16 degrees. The number of days in June is 30. Therefore, there would be $16 \times 30 = 480$ degree days in June. The calculation for the growing season would combine the totals for each of the months from April to October.

Temperature affects the production of sugars, the level of acidity, and the development of moulds. That is not to say the hotter the better. In fact, grapes that take longer to ripen produce better balanced, more flavourful wines.

Generally speaking, grapes need a mean temperature during the ripening period of 19°C for white wines and 23°C for reds. But grape varieties react differently to temperature.

Riesling, for example, requires a cool, temperate climate in order to retain the high level of acidity which gives it a refreshing sensation on the palate. In contrast, Syrah prefers heat in order to develop its organoleptic character, i.e., its colour, aroma, taste, and alcoholic power.

Temperature as it affects the ripening of grapes can also be measured and expressed in terms of heat units, and degree days. The French researcher A.P. de Candolle reported (around the turn of the century) that the vine begins its growth phase after the surrounding temperatures reach 10°C. He concluded that the number of heat units beneficial to growth could be measured by summing up degrees of temperature over 10°C during the growing season.

Another way grape growers measure how favourable a site may be is by calculating the mean temperature of the warmest and coolest months of the year. The average temperature of the hottest summer month shows the heat potential for the growing season, while the average for the coldest winter month indicates how well the vine may survive during its period of dormancy. Growers estimate the average length of the growing season by studying weather data to identify the last possible date of spring frost and the earliest date of fall frost.

Frost

Wine producers fear frost more than any other climatic condition. It can ruin a harvest, or even kill off an entire vineyard. Cold air moves down the slopes of hills and settles on valley floors. To avoid the destructive effects of a harsh winter in those regions where cold is a serious problem, grape growers must avoid lowlands when planting vines. The only way to protect vines in these areas is by using the labour-intensive technique of burying the vine. Earth from between the rows is "hilled up" over the trunk and shoots before winter sets in, and must be removed each spring.

Where the danger is short lived, as in late-spring frosts that come after budbreak, some growers use oil- or gas-burning heaters to warm the air. Others may use wind machines throughout the vineyard, or even helicopters hovering, to force the cold air away from the plants. Yet another technique is to spray water over the vines so that the water itself will freeze into a protective ice coating around tender young buds, until the morning sun can warm the air.

Rainfall

Like all plants, the vine needs a certain amount of water during the growing season, as much as 2 L–4 L per vine per day. During flowering, some water is needed to ensure adequate fruit set. Water is needed again during veraison, as at this point the grapes are only half their full-grown size. Although irrigation is used in many wine regions, grape growers generally prefer the vines receive water naturally. Grapes usually prefer too little water rather than too much: insufficient rainfall can be supplemented by irrigation, but an excess is detrimental to healthy growth. Fungal and bacterial infections can set in and the fruit can swell, split, and oxidize on the vine. In addition, heavy rain just before harvest will swell the grape, reducing the percentage of sugar and pulp which, in turn, results in a watering down of flavours.

Veraison is the point at which the berry begins to enter the ripening phase, when the hard green fruit of the vine starts to soften and change colour.

Wind

Too much wind in the vineyard interferes with the flowering, can snap off shoots and buds, tears leaves, and bruises grape bunches. A gentle breeze helps to keep vines dry and free of disease, and allows them to ripen slowly and evenly.

SOIL

Soil provides a home for the vine, and all of its nutrition. It follows then that to grow healthy grapes, the soil itself must be healthy.

Soil is a work-in-progress combining three ingredients: solid matter (such as minerals and decaying organic substances), water, and air. As these change, so the soil changes.

In general, the best soils are:

- porous—to provide good water drainage.
- light textured—to allow easy penetration by roots.
- deep—to allow roots to reach mineral-laden subsoils and bedrock.
- fertile—but just barely. The soil should provide a variety of minerals and organic matter for a balanced and nutritious plant diet without "overfeeding" the vine. Overly fertile soil results in too much vegetative growth (stems and leaves) and interferes with the ripening of the grapes.

A well-balanced soil affects the strength of the vine, the time of its maturity, and the quality of the fruit. To ensure good results, grape growers study soil composition, structure, texture, depth, fertility, acidity, alkalinity, and colour.

Depth and Drainage

Although grape vines can survive in many soils, they flourish best in those that are well drained. The vine hates to have its roots sitting in wet or mucky dirt. To provide good drainage, the soil must be light textured, like loose gravel or sandy loam, so that the rootstock can extend deep into the ground. There it will absorb dissolved minerals that give the grapes more complexity in their flavour. Greater depth of roots is also important in protecting the health of the vine during severe drought: deep roots are better able to find water within the soil's depths.

Soil Composition

Soil derives from three types of rock:

- igneous—formed from the earth's hot magma (e.g., basalt, granite, porphyry).
- sedimentary—consolidated sediments (e.g., limestone, chalk, sandstone, shale).
- metamorphic—transformed rock (e.g., gneiss, slate, schist).

Most wine regions have soils that derive from a combination of rock types. Some have soils of only one type, such as the chalk soils of Champagne and Jerez de la Frontera.

Organic matter comes from decomposing animal and plant waste. To judge a vineyard's health you need look only as far as the worm population. No worms, no good. Organic matter plays a role in the nutrition of the soil, and (with its ability to loosen clay) can improve the soil's water-holding capacity.

Minerals are the inorganic substances such as rock, ores, crystals, metals, salts, and fossils that help the vine to build strength and the wine to gain structure.

Although most soils contain sufficient nutrients, grape growers monitor several of the more important elements on a regular basis. The following list gives some of the more important elements in soils.

- nitrogen—without it, plants cannot grow. It is a constituent of all proteins, amino acids, chlorophyl, and other organic compounds. How much nitrogen is needed depends on the grape variety, on the risk of winter injury, and the quality and yield expected. A vine deficient in nitrogen will have pale green or yellowish leaves. Nitrogen is applied to the soil in spring before the flowering; applied later, it can stimulate excessive shoot growth resulting in high yields, low sugar, and late maturity.

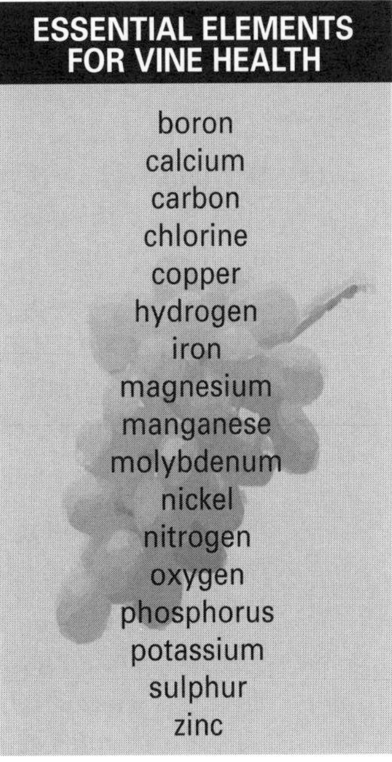

- phosphorus—photosynthesis and the conversion of starch to sugars depend on phosphorus in adequate concentrations. It is applied to the soil before the vineyard is planted. Dark green leaves, or thick purple veins on the leaf, indicate the soil may be deficient in phosphorus.

- potassium—activates enzymes during the formation of carbohydrates and in the synthesis of proteins and fats. Too much potassium can raise the pH level in the grapes to an undesirable level. Too little reduces winter hardiness, causing leaves to fall prematurely, and makes the berries small, tight, and unevenly ripe. It is required by the plant late in the season when the crop is maturing.

pH is a measure of the active acidity in any solution.

- magnesium—a key component of chlorophyl, applied to the vines in the form of salt sprays during the spring.

- calcium—essential for the plant's metabolism and a basic component of cell walls. Soils are rarely deficient in calcium but when they are (as in acidic quartz gravel), the deficiency can cause bunchstem failure in certain white varieties. Too much may cause lime-induced chlorosis—a yellowing of the leaves or green portion of young stems caused by interference in the assimilation of iron by the vine. (Deficiency in magnesium, manganese, zinc, and boron will also cause chlorosis.) Calcium in the soil is thought to add austerity and flintiness to a wine.

Soil Types

The most common types of soil for wine-grape growing are as follows:

- alluvium—fertile, finely grained, and made up of mud, sand, silt, and small stones; found around river beds and flood plains.
- clay—a very fine-grained sedimentary soil with a dense structure. Clay is beneficial in small doses in many soil types but, when the percentage is too high, the vineyard requires extensive tiling to provide drainage.
- ferruginous—containing a high percentage of iron.
- granite—the stuff of mountains. Granitic soil is low in fertility, but when broken down it offers excellent drainage.
- gravel—coarse, broken stones and rocks of any number of origins. Gravel retains daytime heat and radiates it back to the vines at night. Its most attractive feature, though, is that it provides excellent drainage, allowing water-stressed vine roots to stretch deep into the soil.
- humus—well-decomposed organic matter. Humus is added to soil to improve air circulation and water retention and to provide a wide variety of plant nutrients.
- limestone—calcium carbonate composed of the compressed, crushed and/or dissolved shells of marine life. It can be white (chalk), buff, grey, or red-brown (iron) in colour. Dolomitic limestone is rich in magnesium and manganese. Limestone soils are prized more in cool climates than in warmer ones.
- loam—easily crumbled mixture of clay, sand, and silt. Loam has the ability to store water, yet provides adequate water drainage. It is the best soil for agriculture, but because of its richness in nutritive elements, it is not always the best for grape growing, as it promotes too much vigour.
- loess—a mixture of fine clay and silt, combined with fossilized matter. Loess is pale beige in colour and offers adequate drainage.
- marl—a loose and crumbly alkaline mixture of sand, silt, and clay, with substantial amounts of calcium, chlorine, and phosphorus.
- rhyolite—a very acidic volcanic rock. It is the lava form of granite, and is less coarse.

- sand—a fine sedimentary soil, often with a high quartz content. Sand particles have speedy drainage and, therefore, poor ability to retain nutrients and organic matter. Irrigation is always necessary.
- shale—a sedimentary rock formed by clay, mud, or silt.
- slate—rock derived from clay or shale, packed in thin layers. It holds moisture and heat, so radiates warmth during cloudy days and at night. Slate is found in the soils of Mosel-Saar-Ruwer region, and is beneficial to Riesling grapes.
- stony—can contain any combination of igneous or metamorphic rocks including broken slate, basalt, gneiss, or porphyry. Stony soils are easily warmed by the sun and retain heat into the night. They often provide good drainage and significantly reduce erosion on slopes.
- silt—loose, sedimentary deposit with particles larger than clay but smaller than sand.
- tuff—fragmented rock composed of volcanic material. It can be as fine as sand or as coarse as gravel.

CHAPTER 3

*V*INE AND GRAPE

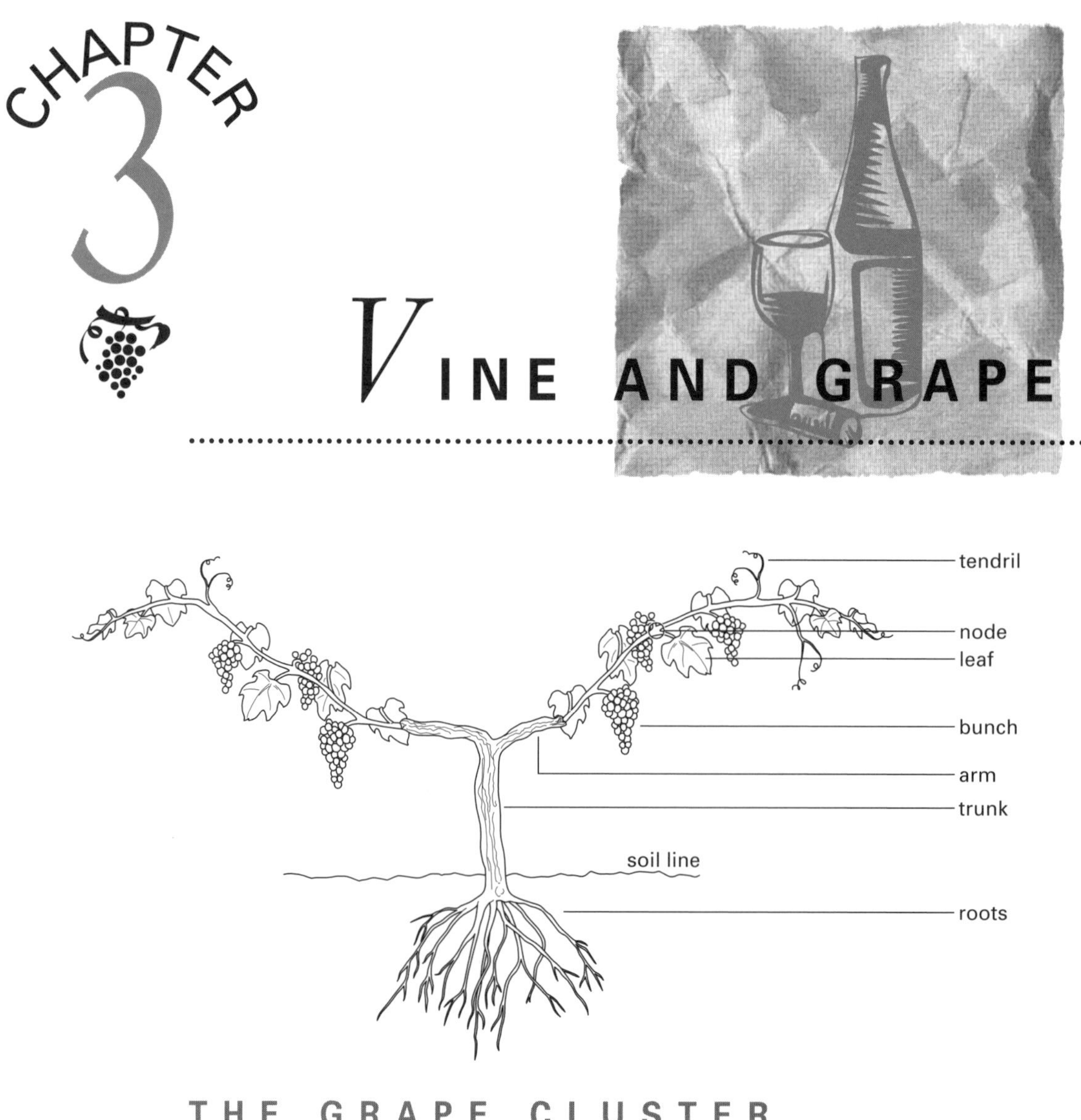

THE GRAPE CLUSTER

The grape cluster is composed of two distinct parts: the stalk and the berries. The berry itself consists of the bloom, skin, seeds, and—most important—the juicy pulp.

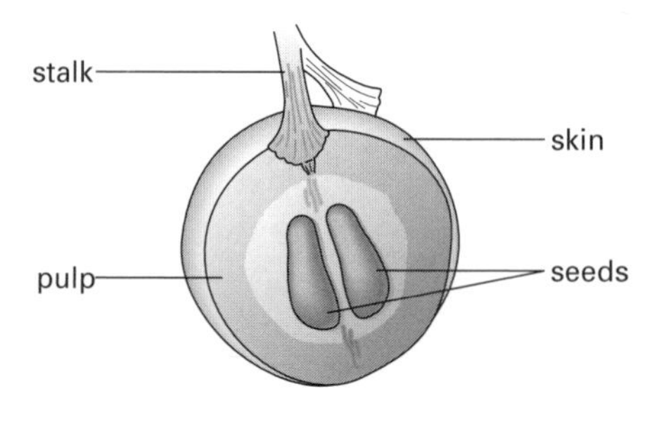

The Stalk

The stalk accounts for 3%–6% of the weight of the grape cluster. It is 78%–80% water. Traces of various sugars, organic acids, tannin, and minerals such as potassium are also found in the stalk.

Tannin is a natural preservative composed of various phenolic compounds. It is found in the stems, skins, and seeds of the grape.

The stalks generally are not included in the fermentation of white wines as they impart too much bitterness. With many reds, though, some stalks are included with the berries before or during fermentation, to give the wine more structure and a longer lifespan.

The Bloom

If you take a grape from the vine and rub one of your fingers across the skin, you will see you have rubbed something off. This is the bloom, a waxy, water-repellent substance that attaches itself to the grape skin. The bloom is covered with wild yeasts, wine yeasts, acetobacter, and other micro-organisms deposited by insects or by the wind.

Acetobacter is a group of bacteria which can turn wine into vinegar.

The Skin

The skin consists of an epidermis (a single layer of cells) and a hypodermis (the tissue beneath the epidermis). It accounts for 5%–12% of the weight of the cluster.

The hypodermis contains pigments and flavouring compounds. Anthocyanin is a blue-red pigment found in red grapes; flavone is a yellow pigment found in both white and red grapes. Pigments are highly soluble in alcohol. The flavouring compounds in each Vitis variety are distinctive. They give the wine its aromatic qualities. Muscat, Traminer, and Riesling, for example, are characterized by their signature aromas.

The Seeds

The average grape has four seeds, making up 3% of the grape's total weight. Seeds contain tannin (5%–8%) and oils (10%–12%). They should not be broken during pressing, as they release very bitter tannins and an enzyme (oxidase) that reacts with oxygen to darken the crushed grapes.

The Pulp

The pulp is the fleshy interior of the grape, made up of large cells separated by thin partitions. These cells are filled with a watery solution of sugars, acids, minerals, and other substances.

Within the grape, the pulp is divided into three zones: the area closest to the seeds, the area closest to the skin, and the area between these two. This middle zone is the richest in sugars, and the first to release its juice. So, the first pressing (vin de goutte) yields sweeter juice than the second pressing (vin de presse). This latter pressing releases the juice from the other zones, which have slightly less sugar but more tannin and acid. The vin de presse will have a harder taste, but some of it may be blended with the vin de goutte to give the wine more structure.

Most grapes have clear pulp, so even red grapes yield a clear juice when first pressed. It is only after extended soaking of the red skins in the juice that the pigments are released to colour the wine.

VARIETY SELECTION

A hybrid is a vine created by cross-breeding existing vine varieties. A clone is a vine propagated from a single selected parent vine.

A cultivar is any cultivated vine, not a wild vine.

After choosing a site, the most important decision a grape grower must make is to select the right cultivar to plant. Many hybrids and clones have been developed, and today there are approximately 10 000 identified cultivars. (Italy alone grows more than 600.) As a result, the choice of the best variety for a specific vineyard is rarely easy. About 250 grape varieties are commercially grown. A variety's fruit and vine characteristics, the soil and climatic conditions of the chosen site, perceived market demand, and economic factors have to be considered before making the selection.

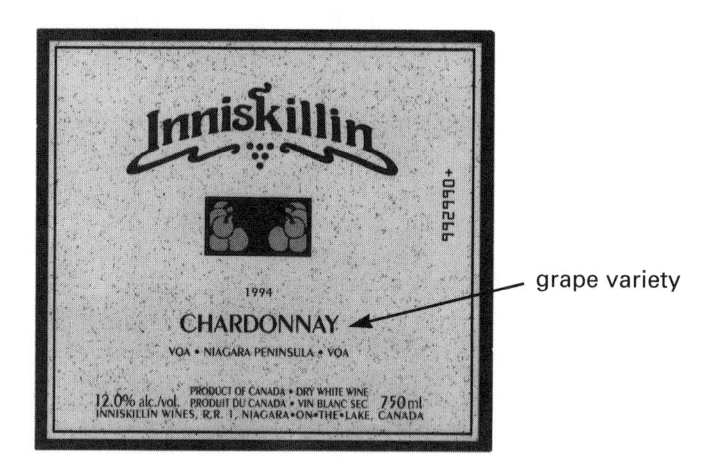

In older wine regions with long traditions, growers have adapted to their terroir. Over the centuries they have discovered which varieties do best in their district.

France's complex regulations govern where varieties can be planted, how the vines are to be cultivated, and even what will be the maximum yields in every district. In contrast, wine regions in Australia, Canada, Chile, New Zealand, and the United States are still in the early stages of discovery. Whereas France's entire Beaujolais region cultivates mainly Gamay, most newer vineyard districts are planted with several different varieties, to find out which do best.

Having chosen the cultivar, the grape grower then selects a species of rootstock for grafting, prepares the soil, and plans how far apart to space the vines. When choosing a rootstock, the grape grower looks at characteristics such as its susceptibility or tolerance to diseases, its vigour, its resistance to drought, and its adaptability to different soils.

Grafting joins a bud, shoot, or limb of one vine onto another so that their tissues unite and grow.

PRUNING AND TRAINING

Effective pruning and training results in vines that grow in an orderly fashion, in order to:

- exploit light, wind patterns, and rainfall.
- simplify pest and disease control.
- reduce the labour at harvest.
- provide fully-ripened grape clusters.

Pruning is the removal of unwanted canes, shoots, and leaves from the vine stock. Grape growers prune in order to provide the optimum number of fully-ripened, disease-free, grape bunches. There are just two basic techniques:

- spur pruning—cutting away all but the two lowest buds.
- cane pruning—leaving five or more buds on longer canes.

Training involves choosing which canes, shoots, and leaves will remain after pruning, and fastening these down to stakes or wires. By making the vines conform to a desired length, height, and general shape, the grower can optimize the quality of the yield and simplify vineyard management.

There are two basic ways to train vines—low and high. Training low forces the vine to grow close to the ground, where it gets the most warmth from the soil. But if the likelihood of ground frost is high (or if the ground heats up too much during the day, virtually baking the grapes), then the vines must be trained higher, on trellises or wires, to keep them as far away from the ground as possible. Over the centuries and throughout the world's wine regions, grape growers have developed local variations of basic training in order to meet their specialized needs.

THE CYCLE OF GROWTH

Fruit set is the transition from flower to berry.

The vine undergoes a series of changes during the year. During the winter, it is dormant. In the spring, the buds begin to swell, then produce green shoots (budbreak). If the spring is ideal, the vine begins to flower and bloom about eight weeks after budbreak. In the next stage, the fruit sets and the berries slowly develop. At first quite hard and green, as maturity approaches they plump up, soften, change colour, and fill with sugar. The original high acidity rapidly decreases. In the autumn, once the fruit has fully ripened, the leaves change colour and fall. Once again, the vine becomes dormant.

budbreak

flowering

fruit set

ripened

PHOTOSYNTHESIS

Photosynthesis is the process by which the leaves of a plant use light energy to convert carbon dioxide and water into carbohydrates and sugars. The reaction is:

$$6\ CO_2 + 12\ H_2O + 673\ 000 \text{ calories of light energy} = C_6H_{12}O_6 + 6\ O_2 + 6\ H_2O$$

In other words, six parts carbon dioxide plus twelve parts water plus light energy equals one part glucose plus six parts oxygen plus six parts water.

Photosynthesis occurs in the chlorophyl cells of the leaves. The amount of light, and the temperature, affect the rate of photosynthesis. Lack of water decreases the rate of light absorption. Ideal leaf temperature for photosynthesis ranges from 25°C–30°C; above 30°C the rate diminishes, and above 45°C it stops completely.

THREATS TO THE VINE

Vineyards may be attacked by fungal, bacterial, and viral diseases, as well as by insects, mites, rodents, birds, and other local wildlife.

Fungal Diseases

Fungal diseases are related to weather conditions. Common diseases include the following:

- downy mildew (peronospera)—caused by the Plasmopara fungus of American origin, this is the most damaging disease to Vitis vinifera. Before the veraison, the fungus attacks the inside of the green parts of the plant, causing leaves to dry and shoots and berries to be malformed. Downy mildew is prevalent in warm, humid regions as well as cool, wet conditions; it is spread by wind. The fungus can be controlled by pruning and by spraying with bouillie Bordellaise (Bordeaux mixture).

Bouillie Bordellaise is a combination of copper sulphate, slaked lime, and water. It is sprayed often, especially after heavy rains, to prevent fungus.

- powdery mildew (oidium)—caused by the fungus Uncinula necator. It develops between 5°C–28°C in damp conditions, during the night or under prolonged overcast skies. The fungus attacks the surface of the plant, forming a greyish-white powdery growth, which causes withering and staining of leaves and splitting of berries. Prevention and control is managed by spraying with sulphur.
- grey rot (grey mould, slipskin)—caused by Botrytis cinerea. (See the box on the next page.) Grapes such as Cabernet, Merlot, Pinot Noir, and Chardonnay are particularly sensitive, as are Riesling and Sémillon in cold, damp weather conditions. Grey rot can result in the loss of the entire harvest. Preventive measures include careful pruning, giving grapes and leaves good exposure to sun, and spraying the vineyard with fungicides.
- black rot (Guignardia bidwelli)—of North American origin. Under prolonged wet and warm conditions, it attacks the upper parts of the vine, causing black spots on leaves and berries. The decaying fruits wither, harden, and turn black within a week. Black rot can be controlled with applications of sulphur, and by quickly removing and burning all infected debris.
- Eutypa dieback (dying arm disease)—rots the wood by entering the wounds in canes after pruning. Symptoms may not appear until several years after infection. No vine type is immune, and certain varieties (such as Cabernet Sauvignon and Grenache) are particularly susceptible.

Symptoms include cupped leaves, stunted shoots, cankers, and berries that are not of uniform size. To control this rot, the grower must prune below the infected area, or, if the rot has reached the crown area, pull out the entire vine.

ROTTEN OR NOT?

The Botrytis cinerea mould is a rapid-spreading, parasitic fungus with mixed blessings. Temperature, humidity, and the variety of grape will determine whether the appearance of botrytis is a sad or a joyous development. If the varieties grown include Riesling, Chenin Blanc, Gewürztraminer, and Sémillon, there is cause to hope that weather conditions will be warm and humid, with a light wind to disperse the spores. When a drier, sunny period follows, botrytis becomes noble rot, and the affected grapes will be turned into some of the sweetest, most concentrated wines in the world. But if the weather turns cool and damp, grey rot sets in and the outcome will be catastrophic—potentially the loss of the harvest.

As botrytis develops, it penetrates the skin of the grape by entering through its pores. As it grows, it uses up much of the water from inside the grape. This gentle, natural dehydration of the berries results in the concentration of sugars, acids, and glycerol, and gives the wine a very fine, distinctive flavour and aroma. Wines made from botrytized grapes are highly regarded and eagerly consumed. They may be labelled *Botrytis-Affected* or *B.A.*

If, however, the weather turns cool and damp, the skin of the grape breaks, the juice begins to ooze out, and grey rot sets in quickly. Some varieties (including Chardonnay, Cabernet, Merlot, and Pinot Noir) can not tolerate botrytis under any conditions, and will immediately develop grey rot rather than noble rot.

Noble rot is called *pourriture noble* in French, *edelfaule* in German, and *muffa nobile* in Italian.

Bacterial Diseases

Bacterial diseases, which are difficult to control, can cause serious losses in the vineyard. The main one is Pierce's disease (Anaheim disease), most prevalent in California where it was first identified in 1884. This deadly and incurable bacterial disease can destroy entire vineyards. It is transmitted by insects (spittlebugs and leafhoppers), during grafting and from infected pruning shears.

Infected vines show browning, dessication, discoloration, delayed budbreak, uneven cane maturity, and stunted growth. The vine dies within two to five years. To control the disease, the grower must remove all infected vines, spray with insecticide, and use sterilized pruning tools. Ultimately, it is necessary to switch to resistant cultivars.

Viral Diseases

Viral diseases that affect vines include:

- fanleaf degeneration—one of the oldest known infections, it is a combination of related diseases that include yellow mosaic and vein banding. The virus is spread through the soil by nematodes (microscopic groundworms). Fanleaf degeneration stunts the growth of the vine, and shortens its life. Other symptoms include poor fruit set, small, seedless, discoloured berries, and the development of leaves that look like a fan.
- grape leafroll—found around the world, it is most prevalent in California. This is an airborne virus that causes discoloration of the fruit, reduced yields, delayed ripening, increased acidity, and significantly reduced sugar levels. The virus spreads from an infected parent vine during grafting.

Because there is no cure for viral infections, vineyards must be uprooted and replanted with infection-free vines in virus-free soil.

Insect Damage

Numerous insects (cochylis, wasps, mealybugs, thrips, beetles, borers, cicadas, termites, grasshoppers, cutworms, moths, mites, nematodes, etc.) can attack the flowers, fruit, leaves, wood, and roots of the vinestock. Generally, they are controlled with insecticides, which are regulated by law in most countries. Some insecticides are preventative, others kill only on contact and will require several sprayings during the infestation period.

The most common and devastating bug to Vitis vinifera is the North American aphid called Phylloxera vastatrix. This burrowing plant louse feeds on the roots of the vine, injecting a poisonous saliva that eventually kills the plant. Adult, winged phylloxera move from vine to vine, damaging leaves and tendrils. The winged insect can be controlled with insecticide, but the only known prevention is by grafting Vitis vinifera onto resistant American rootstocks.

Grafting does not affect the nature of Vitis vinifera grapes but it does reduce the lifespan of the vine.

Fanleaf virus may have been spread in Europe by the louse, since this disease was unknown before the phylloxera outbreak.

Other Problems

- Birds, deer, gophers, raccoons, rabbits, mice, and bears—difficult to control, and can inflict serious damage upon vines or fruits.
- Poor climatic conditions—seasonal drought, frosts, hail, storms, lightning, and high winds can damage vines and crops.
- Early frost—can kill the young shoots of vines. The area on the stem from which a shoot will develop is called a node. These nodes can produce three shoots; if one freezes, another may form in its place. But secondary shoots will not produce the same number of flower clusters, thereby reducing the crop.
- Coulure—too much cold rain during the flowering prevents many of the blossoms from being pollinated. Coulure results in the dropping of the blossoms and (sometimes) of the young fruits and berries.

CHAPTER 4

Grape Varieties

Climate, soil, and human interference all have an effect on the way a wine will taste, but the grape variety contains the genes that give the wine its basic personality traits. To understand why Chardonnay is different from Sauvignon Blanc is beyond the scope of this book. But to discover what those differences are, the following chart may be helpful. It provides an overview of the world's main grape varieties. For each variety listed, you will find a pronunciation guide, a few words about the grape's general characteristics, other common regional names used, major viticultural areas where it is planted, the type of wines most commonly produced from it, and traditional food matches.

WHITE VARIETIES

Grape	*Synonyms*	*Main Growing Areas*
Aligoté (ah-lee-go-TAY) Relatively hardy, thin-skinned, and high yielding.	Blanc de Troyes, Chaudenet Gras, Giboudot Blanc, Griset Blanc, Plant Gris	France (Burgundy, Savoie), Canada (Ontario)
Auxerrois Blanc (OAX-air-WAH BLAH*N*) High yielding; does well in cool climates.	Still often wrongly referred to as Pinot Auxerrois, it is not of the Pinot family.	France (Alsace), Canada (British Columbia, Ontario)
Bacchus (BAHK-oos) A German cross of Sylvaner-x-Riesling with Müller-Thurgau; high yielding and adaptable to poor soils.	None	Germany, Canada (British Columbia)
Bual (boo-AHL) A high-quality, moderate-yielding grape.	Boal	Portugal (Madeira), Spain
Chardonnay (shar-duh-NAY) Reputed to be named after the village of Chardonnay in the Mâcon area, it is a low-yielding variety, that buds and ripens early and does best in limestone soil. Climatically adaptable but thrives best in cooler regions. It was once incorrectly believed to be of the Pinot family, and still is referred to as Pinot Chardonnay by many producers.	Arnaison, Aubaine, Beaunois, Epinette Blanche, Melon Blanc, Petite Sainte Marie, Weisser Klevner	France (Burgundy, Champagne), Spain, Italy, Australia, Canada, Chile, New Zealand, South Africa, United States
Chasselas (shuss-LAH) With fossilized traces found in Egypt and Lebanon, this may be one of the oldest varieties to be cultivated. Chasselas is happiest in cooler regions.	Dorin, Fendant, Perlan, Golden Chasselas, Gutedel, Moster, Royal Muscadine, Silberling, Valais	Switzerland, France (Loire, Savoie, Alsace) Germany, Canada (British Columbia)
Chenin Blanc (shuh-NE*N* BLAH*N*) A tough-skinned grape susceptible to bunch rot, it grows well in cool and sunny climates.	Blanc d'Anjou, Pineau de la Loire, Steen	France (Loire), South Africa, United States (California)
Ehrenfelser (AIR-en-fell-zer] A cross of Riesling and Sylvaner. An early-ripening variety with slightly higher yields and greater soil adaptability than Riesling. It is excellent for Late Harvest wines as it takes well to noble rot.	None	Germany, Canada (British Columbia)

Wines	*Food Matches*
High in acidity, fast maturing, and undistinguished. Aligoté wines are best consumed within a year or two from the vintage as they have no real ageing potential. In Burgundy, they are traditionally used as the base wine for Kir, mixed four or five to one with the black currant liqueur, Cassis.	Poultry and fish, ham, pork, escargot with garlic butter, andouillette sausage
Dull flavoured, high in acidity and alcohol, Auxerrois should be consumed young. In the French region of Alsace, where the grape is blended in Edelzwicker, it is the fourth most-planted variety.	Cold cuts, pan-fried fish, veal; as an aperitif, with hors d'oeuvres
Full-bodied, low-acid wines, with a fragrant Muscat nose (rich, heavy, dried fruit, flowers, honey). Best drunk young and fresh, they are often used in blending to give thinner wines more body.	Pâtés, terrines, fish in cream sauce
Strongly-scented, golden, sweet, fortified wines. Also used in blends.	Cold consommé, cheese, nuts, sponge cake
Straw-yellow with green reflections (from cooler climates) to deep, brassy gold (from warmer climates or with age). Chardonnay wines can have a ripe apple, green fig, or peach nose. In hotter regions, the aroma is subdued and the wines will be fat and dull. Barrel-fermented or oak-aged examples can develop complex, toasty, and buttery nuances. Many New-World versions are too oaky and lack finesse. Their ageing potential, depending on production methods, is very long.	Depending upon the wine's style, poached fish or seafood with cream sauce, roast poultry, pork, veal
Relatively ordinary but can have fine balance. In Switzerland, the wines are often produced in a slightly spritzy style.	Cheese fondue, raclette, pan-fried lean fish, ham
Easily affected by noble rot, and tough skinned, the grape is excellent for producing luscious Late Harvest wines and icewines, which age very well. The dry wines, which are best consumed young, can be quite pleasant, with a fruity aroma of peaches and bitter almonds. In hot climates, Chenin makes rather dull, candied wine.	With dry wines: cold cuts, fatty fish, suckling pig, goat cheese. With sweet wines: chicken in cream sauce, dessert, rich pâtés
Similar to Riesling, but lighter in the nose and quicker to mature.	As an aperitif, or with fish, seafood, sauerkraut, pork, veal, roast poultry, strong semi-hard cheese, or with dessert

WHITE VARIETIES

Grape	*Synonyms*	*Main Growing Areas*
Emerald Riesling (REEZ-ling) A cross of Muscadelle and Riesling developed by the University of California (Davis) in 1946.	None	United States (California)
Furmint (FUR-mint) The main grape of Hungary's famous Tokaji wine.	Sipon, Moslavac Bijeli	Hungary, Austria
Gewürztraminer (ge-VURTS-tra-mee-ner) Gewürz means "spicy" in German. It was first discovered as a spontaneous mutation of the local grape near the Italian village of Termeno (hence traminer), in Alto Adige. It does best in cool climates, where it is able to maintain some of its acidity. In warmer climates, the wine become too fat, flabby, and dull.	Edeltraube, Formentin, Gentil Aromatique, Red Traminer, Savagnin Rosé, Traminer, Traminer Aromatico, Traminer Musqué	France (Alsace), Germany, Italy (Alto Adige), Canada, South Africa, United States (Oregon)
Grüner Veltliner (GROO-ner FELT-leen-er) This Austrian grape produces well south of Vienna. In California, some wines labelled *Traminer* are made from this grape.	Grünmuskateller, Weissgipfler, Zleni Veltlinac	Austria, Canada (Ontario)
Kerner (CARE-ner) A German hybrid (Trollinger-x-Riesling), it is the fourth most-planted variety in Germany.	None	Germany, Canada (British Columbia)
Malvasia (Mull-vah-ZEE-ah) An ancient variety that has been grown all around the Mediterranean basin, its name originates from the ancient town of Monemvasia in the Peloponnesos.	Malmsey, Malvagia, Malvasier, Malvoisie	Portugal (Madeira), Italy, France, Greece, Australia, United States (California)
Morio-Muscat (MOR-ee-oh MUSS-kat) A high-yielding German hybrid (Sylvaner-x-Pinot Blanc).	None	Germany, Canada (Ontario)
Müller-Thurgau (MEW-ler TOOR-gaow) An early-ripening, low-acid, hybrid variety named after Dr. Müller from the town of Thurgau in Switzerland. It is a widely planted cross of two clones of Riesling.	Riesling Sylvaner, Rivaner	Germany, Austria
Muscadelle (muss-kah-DELL) A heavy-bearing vine, it buds late and ripens early, giving it little time to develop any character.	Colle, Douzanelle, Guillan, Muscadet, Muscat Fou, Musquetter	France (Bordeaux), Australia

Wines	*Food Matches*
With a hint of Muscat in the nose, the wines are pleasant but, compared to Riesling, short in character.	Veal, poultry, cooked cheese
Furmint yields dry and sweet wines with aromas ranging from dried apples and apricots to buckwheat honey. If affected by noble rot, topaz-coloured wines with remarkable acid balance and extremely long ageing capacity are produced.	Chocolate, macadamia nuts, dishes flavoured with paprika
Profound aromas of litchi, lanolin, and rose petal combine with a spicy flavour and a lingering, oily finish. Botrytis-affected and Late Harvest versions from Alsace and Germany can be magnificent, if good acid balance can be preserved.	Foie gras, rich pâtés, crustaceans, fish in spicy sauces, marinated or smoked salmon, Asian foods, all but the hottest curries
The best example is Gumpoldskirchner, a fresh, pleasant, fruity white wine, low in acidity and best consumed young. A related grape, Frühroter Veltliner, which has reddish berries, yields fruity wines with similar characteristics.	Cold cuts, poultry, pan-fried fish, wiener schnitzel
Hearty wines, with good acidity and a slight Muscat nose.	Appetizers, game in rich sauce, chicken dishes, medium cheeses, soft fruits
Sweet and most often fortified (as in Madeira) or partly dried (as in Tuscan Vin Santo).	Cheese, nuts and fruits; as an aperitif, or with dessert
Full flavoured and medium to sweet, with a Muscat nose. The wines often have a short finish.	As an aperitif, or with salads or dessert
The wines have a faint Muscat nose but do not age well. They are used mainly in blends.	As an aperitif, or with cold cuts, pan-fried or baked fish
Used mostly for blending, to stretch Sauvignon Blanc and Sémillon. Australian winemakers use Muscadelle to make rich, dark "Tokay" Liqueur. Rarely available as a varietal.	None

WHITE VARIETIES

Grape	*Synonyms*	*Main Growing Areas*
Muscadet (muss-kah-DAY] Its correct name is Melon de Bourgogne. In France, it is grown exclusively near the mouth of the Loire River in a temperate, wet climate.	Gros Auxerrois, Lyonnaise Blanche	France (Loire), Canada (Ontario)
Muscat (MUSS-kat) One of the few grapes that adapts well to the production of wine, brandy, juice, table grapes, and raisins. It is also the most ancient of white vinifera varieties, with fossils found in Greece, Babylon, and Sumer. It grows well in sunny locations. Many clones and mutations exist, including a few with dark skin (Frontignan, Canelli, Ottonel, Alexandria, Aleatico, Hamburg).	Moscato, Moscatel, Moscatel Branco, Muskuti, Weisse Muskateller, White Frontignan	France, Germany, Greece, Italy, Portugal, Australia, South Africa, United States
Palomino (pal-oh-MEE-noh) The principal grape of Sherry. It has high yields.	Palomino Fino, Jerez Fino, Listán	Spain (Jerez de la Frontera)
Pinot Blanc (pee-NOH BLAH*N*) A mutation from Pinot Gris, it yields well in sunny locations, particularly where the soils are deep and moist but well drained. Its tightly packed bunches are susceptible to rot in overly damp conditions.	Borgogna Bianco, Clevner, Klevner, Pinot Bianco, Weissburgunder	France (Alsace, Burgundy), Germany, Hungary (Mecsek), Italy (Trentino-Alto Adige), Canada (Ontario), Chile
Pinot Gris (pee-NOH GREE] Pinot Gris, with its pinkish skin, is a mutation of Pinot Noir. This vine prefers deep, heavy soil.	Auvernat Gris, Auxerrois Gris, Burot, Grauburgunder, Grauklevner, Pinot Beurot, Pinot Grigio, Ruländer, Malvoisie, Szürkebarát, Tocai, Tokay d'Alsace	Austria, France (Alsace), Germany, Hungary, Italy, Canada (British Columbia, Ontario), United States (Oregon)
Riesling (REEZ-ling] Considered the king of all grapes in Germany (as Chardonnay is in France), it produces best in cool but sunny locations.	Hochheimer, J.R., Jo-berg, Johannisberg Riesling, Laskiriesling, Laski Rizling, Lingelberger, Neiderlander, Olaszriesling, Olasz Rizling, Rajnski, Renski, Rheingauer, Rhine Riesling, Rizling, White Riesling	Austria, France (Alsace), Germany, Hungary, Italy, Canada (British Columbia, Ontario), New Zealand, South Africa, United States (California, Oregon)

Wines	*Food Matches*
Light, dry, and fruity, with high acidity. The wines do not age well.	Crustaceans, molluscs, fried fish; alone as an aperitif
Luscious wines, strongly scented, are produced in Greece (Samos), Portugal (Setúbal) and France (Beaumes de Venise, Rivesaltes, Frontignan, and Lunel). Some excellent dry wines are made in Alsace. Some Muscat wines can be long lived, but dry versions are best young.	Dry versions go well with foie gras, rich pâtés, and as an aperitif; sweet wines are perfect with Oriental food, or with desserts
Except when affected by flor in the Sherry-making process, this grape makes poor, dry white wines for blending.	With dry Sherry: consommé, turtle soup, appetizers, paella
In cooler climates, the wines tend to be soft, fruity, and pleasant with good body and apple flavours. When it is grown in hotter zones, the wines are generally heavy, plain, and uninteresting.	Cold cuts, poached or pan-fried fish, roast poultry, vegetables; as an aperitif
Flavourful and full bodied with good spice, low acidity, and a touch of bitterness in a lingering finish. The wines are most often consumed young, though some examples from Alsace age tremendously well. Italy's Pinot Grigios are lighter than those of Alsace and Germany.	Foie gras, pâté en croûte, poached fish, smoked meat and fish, pasta with cream sauce, pork fried in breadcrumbs, game, cheese
A versatile grape, Riesling produces a spectrum of wines from crisp, dry, austere versions, to rich and sweet Late Harvest wines, botrytis-affected wines, and icewines. Wines from Alsace, the Mosel, and the Rhine tend to be light, fruity, fresh, and delicate, with a nose reminiscent of petrol, lime, apples, and pears. They have a long aftertaste and can age for twenty to fifty years when well made.	Fish, crustaceans and molluscs, sauerkraut, pork, veal, roast poultry, Munster cheese; as an aperitif, or with dessert

WHITE VARIETIES

Grape	*Synonyms*	*Main Growing Areas*
Sauvignon Blanc (SO-veen-YOH*N* BLAH*N*) This variety prefers cool climates. It has small clusters of greenish berries with tender skin and aromatic flavours. It is susceptible to spring frosts and to noble rot.	Blanc Fumé, Fié, Muskat Sylvaner, Sauvignon, Sauvignon Fumé, Surin	Austria, France (Bordeaux, Loire), Hungary, Italy, Australia, New Zealand, South Africa, United States (California)
Scheurebe (SHOY-ray-buh) A German hybrid (Sylvaner-x-Riesling) which has low yields but ripens quite early.	None	Germany, Canada (British Columbia)
Sémillon (say-mee-YO*H*N) Highly susceptible to noble rot, it is the favoured companion to Sauvignon Blanc in Bordeaux.	Blanc Doux, Colomier, Málaga, Riesling	France (Bordeaux), Australia, United States (California)
Silvaner (seel-VAH-ner) A highly productive variety that gives best quality when grown in deep, damp, but well-drained soils.	Bötzinger, Franken Riesling, Gamay Blanc, Gros-Rhin, Grünedel, Grüner Zierfandl, Grünfrankisch, Johannisberger, Österreicher, Sylvaner, Schwabler, Szilvani Zöld	Austria, France (Alsace), Germany (Franken), Hungary, Switzerland, Australia (Victoria), United States (California)
Trebbiano (tre-BYAH-noh) A prolific variety that is widely planted in Italy and is the prime grape for distillation in the Cognac region. It grows well in all but sandy soils.	Albano, Bobiano, Lugana, Greco, Procanico, Rossola, Rossetto, St-Émilion, Ugni Blanc, White Shiraz, White Hermitage	France (Armagnac, Cognac), Italy, Portugal, Argentina, Australia, Brazil, Chile, South Africa, United States (California)
Vidal Blanc (vee-DAHL BLAH*N*) This hybrid (Ugni Blanc-x-Seibel 4986) was developed by J. L. Vidal in Bordeaux. It is well suited to cool climates, a heavy cropper and, with its thick skin, well adapted to late harvesting. In Canada's Niagara Peninsula, it is the primary grape used to produce icewine.	Vidal	Canada (Ontario)

Wines	*Food Matches*
In the Bordeaux region, normally blended with Sémillon to provide acidity. The wines distinguish themselves most in the Loire River valley regions, especially Pouilly-Fumé, Reuilly, Quincy, and Sancerre, where they acquire a flinty, grassy nose, with high acidity and gooseberry nuances. These wines age well. The New World also can produce good Sauvignon Blanc wines.	Chèvre (goat cheese), fatty fish, pork, charcuterie, all sorts of hors d'oeuvres
Wines vinified to complete dryness have high acidity, a pronounced bouquet of black currants, and a fruity flavour. Sweeter versions, produced from overripe, Late Harvest, or botrytis-affected grapes are delicious but short-lived.	Dry wines go best with appetizers, cold cuts, and patés. Sweeter versions do better with game in rich sauce, or Oriental dishes.
Sémillon is used to blend with Sauvignon Blanc in the production of the great Sauternes and Barsacs of France, because it takes so well to botrytis. Dry to semi-dry varietals have a lemony taste.	Poultry, white meat
Sylvaner produces pleasant, ordinary wines with high acidity and a short finish. They are best drunk young and fresh. Many California Rieslings are made from this grape.	Raw oysters, shellfish, fried fish, pork; as an aperitif, or with hors d'oeuvres
Some of the most ordinary of wines, nevertheless they are important for their sheer production volumes. The wines are blended in most countries, but are produced as a single varietal in Italy. Despite high acidity, they have little ageing potential.	None
In the dry style, the wines are generally medium bodied, flavourless, and short-lived. When late harvested, Vidal wines display rich, honeyed apricot flavours that can develop over as much as a dozen years.	Patés, fruits, desserts

RED VARIETIES

Grape	*Synonyms*	*Main Growing Areas*
Aglianico (ahl-YAH-nee-koh) An ancient variety first cultivated by Greek colonists in southern Italy around 700 B.C. It grows best in dry, sunny conditions, as it is late ripening, with deep colour and ample tannin.	Ellenico, Gnanico	Italy
Barbera (bar-BEAR-ah) A prolific high-acid grape requiring a warm and sunny climate. After Sangiovese, it is the most widely planted red variety in Italy.	None	Italy (Piedmont), United States (California)
Cabernet Franc (CAB-air-NAY FRAH*N*) A relative of Cabernet Sauvignon, Cabernet Franc ripens early and grows best in cool climates, where it develops an intense fruity aroma.	Bordo, Bouchet, Breton, Carmenet, Gros Bouchet, Grosse-Vidure, Trouchet Noir	France (Bordeaux, Loire), Italy (Tuscany), Canada (Ontario), United States (California, New York, Washington)
Cabernet Sauvignon (CAB-air-NAY so-vee-NYOH*N*) A dark, thick-skinned, tannic grape with large seeds. It prefers a warm climate and a long growing season.	Bouchet, Petit Cabernet, Sauvignon Rouge, Vidure	The most widely planted red variety in the world. Wherever there are vineyards, someone will plant Cabernet Sauvignon.
Carignan (ca-ree-NYAH*N*) Carignan thrives in hot climates, where it gives high yields that ripen in late mid-season.	Bois Dur, Carignane, Cariñena, Catalan, Mataro, Mazuelo, Monestel, Roussillonen, Spagna, Tinto Mazuelo	Algeria, Morocco, France (Rhône), Spain, United States (California)
Cinsaut (sen-SO) Cinsaut is highly productive, yields high sugar levels, and withstands blistering hot climates.	Cinsault, Black Malvoisie, Cinq-saou, Hermitage, Malaga, Picardan Noir	Algeria, Morocco, France (Rhône), Lebanon, South Africa, Spain
Gamay (ga-MAY) The grape of Beaujolais is productive, yielding berries with tough skin and lots of juice. In California, what is labelled as Gamay Beaujolais is not a Gamay, in fact, but a clone of Pinot Noir.	Bourguignon Noir, Gamay Noir à Jus Blanc, Gamay Rond, Petit Gamai	France (Beaujolais, Loire, Côte Maconnais), Canada (British Columbia, Ontario)

Wines	*Food Matches*
Two of the most famous wines of Italy are Aglianico del Vulture (from Basilicata) and Taurasi (from Campania). Both are hard and closed in their youth, but develop a deep garnet colour, fine bouquet, and balanced flavour with lengthy ageing.	Spicy stews, pasta in tomato sauce, terrine of game, charcuterie, pâtés, strong cheeses
Purple and fruity with good tannin and high acidity. Pleasant and refreshing when young, the wines can develop some charm when aged in oak.	Grilled, broiled, and roast red meats, roast poultry, pork, ham, medium cheese, pasta with tomato sauce
Most often blended with Cabernet Sauvignon or Merlot (or both) to give softness, lightness, elegance, and an enticing fragrance of black raspberries. In the Loire, where the wines are bottled as single varietals, they are lighter and somewhat more herbaceous, with nuances of dill. Cabernet Franc wines do not age as well as Cabernet Sauvignon. They are more pleasant when served slightly chilled to bring out the delicate fruit flavour. The grape is also used to make appealing rosés in the Anjou district of the Loire.	Cold cuts, charcuterie, pâté, light cheese, white meat, fish stew, Arctic char, salmon
Deep purple, high in tannin, and well suited to oak ageing. Aromas of black currant, cedar, printer's ink, eucalyptus, bell pepper, red currant, olive, and elderberry are typical. The wines need long, slow, cellar ageing to soften and develop character.	Roast or grilled red meats (especially lamb), dark poultry, soft cheeses like Brie
Sound, sturdy wines with good colour, an intense aroma, and medium acidity. They lack complexity, though, and are used mainly for bulk wines and blending.	Stewed meats, cold cuts, charcuterie
Excellent wines are made, ranging from red and rosé to Port. The reds have intense colour, meaty flavour, and low tannin. They are often used to blend with other wines because of their consistent results.	Richly sauced red meats and game, spicy pâtés and terrines, medium strong cheese, pasta with meat sauce
Purple, lively, with good acidity, low tannin, and an intense nose of cherry, strawberry, and raspberry. Gamay wines should be drunk young and on the cool side. Some Beaujolais can be aged from three to five years, but those aged longer take on characteristics of old Pinot Noir. When Gamay undergoes carbonic maceration (Beaujolais Nouveau), the fruitiness is intensified, but the ageing potential diminishes to about three months.	Grilled sausage, red meat, charcuterie, hamburgers, sandwiches, pizza, plain pasta with Parmesan cheese, medium cheese

RED VARIETIES

Grape	*Synonyms*	*Main Growing Areas*
Grenache (gre-NASH) A vigorous, high-yielding grape that thrives in hot climates. Its small pinkish to red-purple berries ripen early with high sugar levels and medium-to-low acidity.	Alicante, Cannonau, Carignane Rousse, Granaccia, Granacha, Garnacha Tinto	France (Languedoc-Rousillon, Rhône), Spain, United States (California)
Grignolino (green-yo-LEE-noh) A native of Monferrato in Piedmont, Italy, its yields vary from year to year as the grape clusters ripen at different stages.	Balestra, Barbesino, Verbesino	Italy (Piedmont), United States (California)
Malbec (mahl-BECK) Produces average yields but is quite susceptible to coulure. In the French region of Cahors and the Loire, this grape (known there as Cot) has a deeper colour than Cabernet grapes.	Auxerrois, Cot, Côt, Malbeck, Pressac	France (Bordeaux, Loire, the South West), Argentina (Mendoza)
Merlot [mair-LOH] The most widely planted red variety in Bordeaux. Merlot is an early-ripening, thin-skinned, dark grape that is low in acidity and tannin.	Bignet, Crabutet Noir, Petite Merle, Plant Medoc, Sémillon Rouge, Vitraille	France (Bordeaux, especially Pomerol, Languedoc-Rousillon), Italy, Switzerland, Argentina, Canada (British Columbia, Ontario), South Africa, Chile, United States (California, New York, Washington)
Mourvèdre (moor-VED-ruh) A thick-skinned, deeply coloured grape best suited to warm climates.	Balzac Noir, Beni-Carlo, Damas Noir, Esparte, Estrangle-chien, Maneschaou, Mataro, Negron, Trinchiera	France (Rhône, Provence), Spain, United States (California)
Nebbiolo (ne-BYOH-loh) The great grape of Barolo and Barbaresco. It ripens very late, has a thick skin, and is high in tannin and acidity.	Chiavennasca, Picotener, Pugnet, Spanna	Italy (Piedmont)
Petit Verdot (pe-TEE vair-DOH) High in acidity with low yield, it is quite susceptible to diseases.	Carmelin, Plant des Palus, Verdot Rouge	France (Bordeaux)
Petite Sirah (pe-TEET see-RAH) Not to be confused with the real Syrah of the Rhône region, this California name is given to the French grape, Durif. Its main redeeming feature is that it is resistant to mildew.	Durif, Pinot de l'Ermitage, Sirane Fauchure	United States (California)

Wines	*Food Matches*
Grenache is used mostly for blending, but also produces good, sweet, red wines and excellent, fruity rosés in the Tavel and Lirac districts of France, in Spain, and in California. The wines are generally soft, high in alcohol, and quick to mature.	Poultry, roast white meat, pâté, terrine, cold cuts, charcuterie, fish stew, salmon
Austere, light-bodied wines usually pale red or orange in colour, with a distinct scent of strawberries, flowers, and herbs. In California, the grape is used for rosés.	Veal, poultry, light cheese, salami, air-cured meat, game birds
Produces pleasant, delicate, light-coloured wines, which mature quickly. In Bordeaux, Malbec is used to soften Cabernet Sauvignon in blends.	Veal, pork, poultry, rabbit, pigeon
Merlot wines are straightforward and fruity, with intense colour, a good alcohol level, a plummy aroma, and a silky texture. They mature quickly and should be drunk relatively young. Oak barrels can give the wines a spicy, black-currant-jam nose, with a touch of mint. Merlot is used extensively in the French regions of St-Emilion and Pomerol and is often blended with Cabernet Sauvignon to give roundness.	Roast meats, offal, rich pâtés, beef stew, sausage, game birds, cheese
Yields strong wines with high acidity and high tannin. As the grape lacks individual flavour, it is usually blended with varieties such as Grenache and Cinsaut, to give them structure and body.	Stews, offal, patés
Closed and hard in their youth, Nebbiolo wines are made for ageing, ultimately developing marvellous aromas of violets, rose, tobacco, leather, tar, and truffles.	Game, pâtés, red meat in rich sauces, terrines, tomato sauce, aged Pecorino, Parmesan, or Gruyère
Strongly flavoured, deep in colour, and rich in tannin. Petit Verdot wines age slowly, adding complexity to Bordeaux blends. Rarely available as a varietal.	None
Deep colour, high acid and tannin, a peppery nose, and full body with an uninteresting finish are the reasons this variety is best used for blending.	None

RED VARIETIES

Grape	*Synonyms*	*Main Growing Areas*
Pinot Meunier (pee-NOH mew-NYAY) A mutation of Pinot Noir, this variety is named for the white fuzz under its leaves and buds which makes them look floured, hence meunier ("miller" in French). It flowers late, resists frost, gives high yields, and has the ability to shoot a second time if damaged by frost. It is planted primarily in Vallée de la Marne and Aube for use in making Champagne. It is grown successfully in Germany, where it is made into a pale red Schwarzriesling, and in Canada's Okanagan Valley, where some winemakers produce excellent and complex reds.	Auvernat Gris, Blanche Feuille, Müllerrebe, Plant de Brie, Schwarzriesling	France (Champagne), Germany, Canada (British Columbia), United States (California)
Pinot Noir (pee-NOH NWAHR) The grape of red Burgundy is a finicky grape. It dislikes heat, and cold rain, is quite susceptible to frost damage and grey rot, and (even in the best conditions) yields modestly. Pinot Noir prefers a sunny location in a cool climate, likes a late, warm autumn, and thrives in calcareous soil.	Auvergnat Noir, Blauburgunder, Cortaillod, Klevener, Morillon, Pineau, Pinot Nero, Savagnin, Spätburgunder, Vert Doré	Austria, France (Burgundy, Champagne, Jura, Loire, Alsace), Italy, Hungary, Spain, Switzerland (Valais), Australia, Canada (British Columbia, Ontario), New Zealand, South Africa, United States (California, Oregon)
Pinotage (pee-NOH-TAZH) This hybrid (Pinot Noir-x-Cinsaut) was developed in South Africa.	Hermitage	South Africa
Sangiovese (SUN-joh-VAY-zay) This thick-skinned, light-coloured variety is the classic grape of Tuscany. The name means "blood of Jove." Sangiovese is late ripening, high in tannin, and hardy, with a thick, dark skin.	San Zoveto, Sangiovese Grosso, Sangioveto, Prugnolo Gentile, Brunello	Italy (Tuscany), United States (California)
Syrah (see-RAH) An ancient grape reputed to have originated in Persia. In warm, sunny climates it is reliable and productive.	Shiraz, Syranne, Serine, Hignin Noir	France (Rhône), Australia, Canada, (British Columbia), South Africa, United States (California)
Zinfandel The mystery grape of California was for some time thought to be the Italian grape Primitivo. Modern scholars now believe it is the red Plavac Mali grape.	None	United States (California)

Wines	*Food Matches*
The grape is generally pressed for the agreeable, fruity taste of its white juice. The wines tend to lack finesse, have low acidity, and do not age well. The juice is used primarily for blending into sparkling wines in the Champagne region and in California.	When made as red wines, same matches as for Pinot Noir
Quality can vary considerably depending on where the grapes were grown and what winemaking techniques were employed. The best (as well as the worst) often come from Burgundy. Good Pinot Noir wines are complex, powerful, and elegant, with a profound (almost ethereal) aftertaste. Young wines have a ruby hue, substantial body, and exciting red-berry aromas, often hinting of cherry, strawberry, and wild rose. Mature wines develop an orange hue and a bouquet of barnyard (the French call it *merde du lapin* which means "rabbit droppings"), dry figs, buckskin, tar, humus, and tea leaves. Pinot Noir grapes are also used in the production of Champagne.	Beef, caribou, venison, game birds (partridge, pheasant, turkey), Brie, Camembert and Livarot cheeses
Pinotage can yield intricate, high quality wines. When young, they are often quite plummy, implicitly sweet, full, and fruity, with hints of acetone in the nose. Once mature, they develop more delicate flavours.	Stewed beef or pork, kidneys, roast poultry, medium hard cheese
Chianti Classico, Brunello di Montalcino, and Vino Nobile di Montepulciano are three of the best-known wines of Italy. Blood orange in colour, with good tannin, these wines mature exceptionally well.	Grilled red meats, steak tartare, liver, game, grilled sausage, spicy stews, pizza, pasta, medium cheese such as Taleggio
Some of the great reds of the northern Rhône region, such as Cornas, Côte Rôtie, Crozes-Hermitage, and Hermitage, are made solely from Syrah. These wines need plenty of time to develop. They are heady, and rich in colour, tannin, and fruit extract. Syrah wines mature to a bouquet of violets, chocolate, coffee, creosote, licorice, and black peppercorns.	Spicy stew, strongly flavoured furred game (bear, moose, wild boar), hard cheese
Full bodied, inky in colour, and intensely flavoured, with substantial tannin and fairly low acidity. Alcoholic reds demand long ageing. The aroma hints at cherry, blackberry, and black raspberry, cowhide, cedar, and peppercorns. Late-picked versions also give hints of raisins, prunes, and cloves. Most Zinfandel production is turned into characterless blush wine for quick mass-market sales.	Full-flavoured dishes, spicy stews; chocolate for the adventurous

CHAPTER 5

WINEMAKING

There was a time when wine was made solely for immediate consumption as a wholesome and restorative drink, made from vines planted nearby. In those days, there was a complete lack of understanding of the process of fermentation. Winemakers relied on the traditions of their forebears, on ambient yeasts (most did not even know these yeasts existed), on good weather, and on luck.

Today, this perspective has changed dramatically. Most wine is made to be sold to many and various markets; winemakers have access to a multitude of grape varieties, clones, and hybrids; fermentation is clearly understood; and a variety of yeast strains is available. The chemistry of wine, with all its corrective practices, is applied carefully to produce better wines. Today's reality is that winemakers make wine for profit, not simply to drink with their families. Consequently, market demand often dictates the style and type of wine that is produced. The trick is to make the best wine at the lowest price, while making a profit with the least amount of labour.

English	French	Italian	German	Portuguese	Spanish
red	rouge	rosso	rot	tinto	tinto
white	blanc	bianco	weiss	branco	blanco
blush	rosé	rosato	bleichert weissherbst rotgold	rosado	rosado
dry	sec	secco	trocken	seco	seco
sweet	doux	dolce	süss	doce	dulce
sparkling	pétillant	spumante	spritzig	espumante	espumoso

Wine is the alcoholic drink produced by the complete or partial alcoholic fermentation of fresh grapes. It contains many interdependent constituents which affect its quality. Of these, acids, alcohol, sugars, and phenolic compounds are the most important.

Acids

There are two basic acids—tartaric and malic—and some of less significance. They are present in ripe grapes and are "passed on" to the wine. Acidity, which must be balanced by sugar and alcohol, has a profound effect on wine.

It provides the pleasant yet refreshingly lively bite, gives structure, brings out the fruit flavours, prolongs the aftertaste, helps wine maintain its colour, and preserves wine's living components, giving it a longer lifespan. Too much acidity makes wine sour and unpleasant. A lack of acidity leaves it tasting lifeless, flat, fat, and flabby.

Some acids are fixed, others are volatile. Fixed acids contribute to the tartness and overall mouth-feel of wine, while volatile acids are instantly perceived in its aroma.

- tartaric acid—a fixed acid. In wine produced from grapes grown in warm regions, tartaric acid accounts for more than half the total acidity. Some of it precipitates as an acid salt called potassium bitartrate (or cream of tartar) during the winemaking process. Chilling the wine (cold stabilization) forces more potassium bitartrate to precipitate.
- malic acid—a fixed acid. It is the sourest of all acids and, when present, gives wine a distinct green-apple taste. Grapes grown in cooler regions, either farther from the equator or at higher altitudes, will have a higher concentration of malic acid. Winemakers often reduce it by innoculating the finished wine with a microbe that kick-starts a malolactic fermentation. Once complete, malolactic fermentation leaves wine with a lower level of malic acid and a higher proportion of lactic acid. This gives an overall softer taste and thicker texture.

There are details on malolactic fermentation later, in the section on Fermenting.

- lactic acid—a fixed acid. It is not found in grapes, but is created during alcoholic fermentation and by bacterial action on malic acid. Lactic acid is a weak acid (it is found in milk) with a slight odour.

Other fixed acids are citric, succinic, and sulfurous, but these play a less significant role in winemaking.

- acetic acid and ethyl acetate—highly volatile acids. They are the by-products of alcoholic fermentation, and they exist in all wines to some degree. Below a certain level, volatile acidity is imperceptible on the palate but will nevertheless lift the aromas and flavours and add an element of complexity to wine. When the level of volatile acidity gets too high, it turns wine vinegary. Volatile acid is more easily detected in dry, light-bodied white wines than in strong, aromatic reds. Winemakers monitor volatile-acid levels very carefully.

Other volatile acids include formic, butyric, and propionic. When present in higher than trace concentrations, they are clear indicators of spoilage.

Alcohol

Wine has several types of alcohol. The most important of these is ethanol (ethyl alcohol). It is most often simply referred to as alcohol. Alcohol affects body, aroma, and flavour. It heightens the perception of sweetness. Alcohol is usually indicated as a percentage of the total volume, e.g., 10%, 10% by volume, 10% alc./vol., etc. In table wines, it reaches levels ranging from 7%–14%. The amount of ethanol produced depends on the amount of sugar and on the nutrient level in the grapes. To a lesser degree, alcohol is affected by the yeast strain used, and by the temperature during fermentation.

Methanol is another form of alcohol present in wine, produced when juice is fermented with grape skins.

Fusel alcohol describes a group of alcohols (propanol, butanol, and phenylethanol) present in tiny amounts. They are responsible for some of the more complex aromas in wine.

Sugars

Sugar comes in many forms. The primary and most important ones are glucose and fructose. These are fermentable sugars. If these sugars are not converted completely to alcohol during the fermentation, the wine will contain residual sugar and will taste sweet. Dry wines should not contain more than 1% residual sugar.

Sweet flavours in wine are derived from other elements besides sugar: alcohol, glycerin, and some minerals that combine with organic acids. These by-products of fermentation are occasionally mistaken for sugar—they have a smoothness that adds to the perception of sweetness.

Phenolic Compounds

Phenols play a role in wine's colour, its taste (bitterness and astringency), and its body.

- tannin—a name given to phenolic compounds found in stems, skins, and seeds, as well as in oak barrels. Tannin levels are determined by the grape variety, the duration of skin/stem contact before and/or during fermentation, the severity of the pressing, and the time spent in barrel.
- anthocyanin—a pigment found in red wine. As wine ages, anthocyanins become incorporated into tannins. The wine begins to change colour and lose its astringency.

Wine contains a multitude of other trace minerals and inorganic compounds which contribute to character, taste, and colour. These include: boric acid, bromide, chloride, various phosphates, silicate, sulfate, hydrogen sulphide, mercaptan, aluminum, calcium, copper, iron, magnesium, manganese, potassium, rubidium, sodium, and zinc. Although all are found in minute amounts, they are of considerable importance and generate much debate over their therapeutic value within a healthy diet.

TURNING GRAPES INTO WINE

The process of transforming grapes into wine involves six major steps: harvesting, crushing, pressing, fermenting, clarifying, and maturing.

HARVESTING

Harvesting normally takes place through September and October in the northern hemisphere, and in March or April in the southern hemisphere.

Grapes can be mechanically harvested, but are often picked by hand. Hand picking is far more costly—it is time consuming and labour intensive. It demands, on average, from one to ten days for one person to harvest a single hectare. That same area can be picked in less than five hours with the aid of machines. Mechanical harvesters use flexible rods that slap or shake the vines to loosen the berries, which then fall into catching frames. But hand picking is preferable as it allows for better quality control—unripe, damaged, diseased, or decayed berries can be removed, leaving only the sound fruit.

In some cases, such as the steep slopes of the Rhine and Mosel rivers in Germany and the Douro in Portugal, mechanical harvesting is either dangerous or impossible.

Precisely when to harvest is one of the most critical decisions that grape growers and winemakers have to make. The grapes have to reach a desired level of maturity. Picked too soon they yield a tart green wine; left too long on the vine and they give a wine with high alcohol and insufficient acidity. The most sought-after element is balance.

CRUSHING

As soon as the grapes arrive at the winery, they are dumped into a crusher/de-stemmer. White grapes may be given a measured amount of dissolved sulphur dioxide. This retards enzyme activity, which would oxidize the wine and cause browning; it does not alter aromas and flavours if used sparingly. With red grapes, browning is not a problem.

As the grapes pass through the crusher/de-stemmer, the skins are gently broken while the stalks are removed. Stalks contain bitter-tasting phenols, which can adversely affect the taste of the finished wine. Some winemakers purposely leave some (or all) of the stalks in to soak for a period of time and so increase the proportion of phenols and tannins. This is intended to give the wine more complexity and longevity. Most winemakers, however, avoid the stalks, since bitterness takes away much of the appeal of a wine in its extreme youth, which is when most of it is drunk.

The Must

The mixture of grape pulp, juice, skins, seeds, and stem fragments that makes it through the crusher/de-stemmer is called the must.

THE COMPOSITION OF MUST

For every 1000 g of must the following proportions are average:

water	700 g–780 g
glucose and fructose	100 g–250 g
organic acids (tartaric acid and malic acid)	2 g–5 g
potassium bitartrate (cream of tartar)	3 g–10 g
other minerals	2 g–3 g
pectin and nitrogen	0.5 g–1 g

This may be sent to a holding tank to allow for a short period of skin contact—to extract additional flavour. Otherwise, it will go directly to the press.

As the level of alcohol in wine is directly proportional to the amount of sugar in the must, winemakers regularly measure sugar levels. Grape sugars are calculated in the vineyard. However, the most precise measurements are taken in the lab once the grapes have been crushed and the must is ready to be fermented. A hydrometer (a sealed tube marked with a scale and weighted at one end) is used to determine the specific gravity (relative density) of the must. Pure water has a specific gravity of 1.000. By comparing the specific gravity of the must to the known density of water, winemakers can precisely measure the amount of alcohol that will be produced during fermentation, and can make adjustments as necessary.

Wine-producing countries have developed their own systems for measuring sugar levels: the French use Baumé degrees, German winemakers calculate in Oechsle, while their Canadian and American counterparts use Brix. Yet another scale, the Balling scale, is similar to the Brix scale.

One degree Brix corresponds roughly to 18 g/L sugar.

Specific Gravity	Baumé	Oechsle	Brix	% Alcohol
1.065	8.8	65	15.8	8.1
1.080	10.7	80	19.3	10.0
1.090	11.9	90	21.5	12.1
1.105	13.7	105	24.8	14.3
1.120	15.5	120	28.0	16.4

PRESSING

Pressing can be done before, during, or after fermentation. Grapes for red wines are usually pressed during or after fermentation, when there has been enough extraction of colour and tannins from the skins.

The degree to which grapes are pressed ultimately affects the quality and the taste of the wine. The gentler the pressing, the better the wine will be.

Even before pressing, a substantial amount of the juice—more than half—can pass through tiny perforations in the crusher/de-stemmer, or in a stainless steel cage inside the press. This naturally flowing juice is called the free-run juice. It is the sweetest, least astringent juice, considered to be of the highest quality.

Several types of press are used in modern winemaking. Modern state-of-the-art wineries are equipped with computer-controlled presses which regulate the whole cycle. Some common presses include the basket press, the continuous press, and (the most gentle) the horizontal pneumatic membrane press.

This last type contains an airbag that runs the entire length of the press. While the must is pumped in, the airbag is collapsed. Once the press is loaded to capacity and the cage is closed, the airbag is inflated and the must is gently pressed through the perforations, and funnelled into fermentation vats. Skins, stems, and seeds are retained inside the cage.

The airbag will be inflated and collapsed several times, until all that is left inside the cage is dry pomace. This is discarded, or sent for distillation into marc, or used as organic mulch for vineyard fertilization. In some regions, it has water and sugar added, and is fermented into a coarse wine given to harvesters and winery employees as a daily drink.

Brandy made from pomace is called marc; brandy made from wine is called fine (pronounced feen).

Pomace is the debris (skins, stems, seeds, pulp) that is left after pressing.

Settling

Settling removes particles (skin fragments, bacteria, yeast cells) that contain bitter phenolic compounds which could cloud the juice. It is sent to a settling vat, and may be cold settled, filtered, or centrifuged.

- cold settling—inhibits oxidation, stops yeast and enzyme activity, and helps the unwanted particles settle to the bottom of the vat. It involves cooling the juice to around 8°C–12°C, and takes twenty-four to forty-eight hours.
- filtering—a simple procedure. The juice is pumped through filters to remove solids.
- centrifuging—spins the juice rapidly inside a chamber. The solid matter is separated by the centrifugal force (rather than by gravity, as in settling).

Filtering and centrifuging are fast, aggressive methods. They can over-aerate the juice and remove many of the colloids, which are aromatic components; stripping the wine of colloids reduces the character of the wine.

Once clear, the juice is transferred to fermentation vats.

FERMENTING

Fermentation begins as soon as yeast comes into direct contact with sugar. There are thousands of strains of wild and cultured yeasts: different yeasts give different characteristics to the wine. The most appropriate yeasts for each variety find their way into the vineyard and attach themselves to the bloom on the skins of the berries. Unfortunately, other natural airborne micro-organisms (such as acetobacter, which longs to turn wine into vinegar), and wild yeasts will also stick to the bloom. So, most winemakers use cultured yeasts to ensure that their results will be more predictable.

To start the process of fermentation, the juice is transferred to a fermentation vat. This can be a stainless steel tank (perhaps insulated and wrapped with a temperature-control jacket), a glass- or epoxy-lined cement vat, an oak cask, or even a plastic container.

The juice is innoculated with the chosen yeast culture and the rest is up to nature.

Under warm conditions (20°C–25°C), fermentation will be completed in under two weeks. Under cool conditions (below 20°C), the process takes much longer.

Yeast

Thousands of strains of wild and cultured yeasts exist.
Winemakers require a strain that:

- quickly starts the fermentation process.
  This is of particular importance when sweet wines are being made, since high sugar levels make the initial fermentation difficult.
- has the ability to ferment all the sugar to alcohol without the risk of a stuck fermentation.
- is cold tolerant, if the wine is to be cool-fermented in temperature-controlled tanks.
- produces few undesirable metabolites (i.e., acetic acid, sulphur dioxide).
- can produce the right type of desirable secondary metabolites (i.e., oily alcohols, esters, glycerol).
- has the ability to leave the acids alone, or to convert them to bring the wine into balance.
- precipitates dead yeast cells and other undesirable solid matter, thus clearing the wine.
- kills undesirable wild yeasts.

Alcoholic Fermentation

Fermentation is a combination of aerobic (oxygen present) and anaerobic (oxygen absent) reactions.

After yeast is added, the juice lies quiet for twenty-four to thirty-six hours.

Then it begins to ferment, slowly at first, but soon quite vigorously—the juice looks as if it were rapidly boiling. During this period (which can last from three days to as long as a month), grape sugars are transformed into ethyl alcohol and carbon dioxide. As more sugar is converted and less remains, the fermention slows down.

At the beginning, some oxygen is required to get the yeasts to start multiplying. Afterwards, anaerobic fermentation takes over. A biochemical reaction begins, creating enzymes which transform the grape sugars into ethyl alcohol and carbon dioxide.

$$\underset{\text{sugar}}{C_6H_{12}0_6} \xrightarrow{\text{yeast}} \underset{\text{alcohol}}{2C_2H_5OH} + \underset{\text{carbon dioxide}}{2CO_2}$$

Some of the components of the juice (e.g., certain essences, oils, pigments, proteins, and tannins) stay unchanged and remain in the wine.

The mechanics, the extent, and the result of the fermentation are all determined by the type of wine to be made. Here are three examples:

- complete fermentation produces a dry wine, while partial or arrested fermentation results in a wine with residual sugar.
- fermentation in an open vat yields a still wine, while fermentation in an airtight container will produce a sparkling wine.
- if white wine is to be made, white grapes may be used; red grapes may also be used if the skins are removed immediately, before any of the pigment is released into the juice. To produce red wine, red grapes must be used, and the skins must remain in contact with the juice long enough for the pigments to be released.

The type of container used, the temperature, the speed, and the duration of the fermentation will determine the eventual style of the wine. A cool fermentation (10°C–13°C) produces wines with more fruit character, while a slightly warmer fermentation yields less fruity but more complex and full-bodied wines.

Fermentation Containers

- Stainless steel vats—come in a variety of sizes, are easy to clean, and have optional cooling systems that help winemakers control the fermentation temperature to preserve natural fruit flavours.
- Oak barrels—small, can be used only for a limited number of years (and are therefore more expensive in the long run), and more difficult to sanitize. Barrel fermentation, however, yields wine with more body and complexity.

Malolactic Fermentation

After the alcoholic fermentation, certain wines—especially those grown in cool climates—may undergo a secondary fermentation, called malolactic fermentation.

This is a bacterial fermentation caused by a range of bacteria and is (usually) intentionally induced by winemakers.

It works by converting some of the wine's more aggressive malic acid into the much softer lactic acid, thereby reducing the sour taste and increasing the suppleness. Grape varieties with a low natural acidity are generally not treated with this process for fear of lowering the acidity too much.

Malolactic fermentation can occur spontaneously. Too much oxygen, sulphur dioxide, low temperatures, or a low pH will inhibit it. In some cool-climate regions (such as Burgundy), the wine may be left in the barrel on its lees to undergo a spontaneous malolactic fermentation. In this case, the bacteria grow slowly throughout the autumn—until temperatures drop to winter levels—at which point the fermentation stops. It resumes again only when temperatures rise in the spring. Throughout history this process was referred to as the "flowering" of the wine.

The sediment that settles at the bottom of a container is called the lees. Wine lees can consist of bits of seeds, pulp, stems, skin, dead yeast cells, and insoluble tartrates.

Malolactic fermentation can also be induced by inoculating the wine with a cultured strain of bacteria. This provides somewhat more predictable results. In the right environment, cultured strains dominate any unwanted bacteria in the wine, and cause the right biochemical reactions required to improve the wine.

The wine can be treated with sulphur dioxide or sorbic acid to control unwanted bacteria that may still be present.

CLARIFYING

Even after the fermentation is complete, particles of skin, pulp, and yeast may remain in suspension. Living yeast cells or other bacteria may spoil wine, or cause it to re-ferment after bottling. The result would be a rapid increase of carbon dioxide, which could push out the cork (or even explode the bottle) with its pressure.

Clarification is the process by which wine is cleared of these particles. This can involve several steps:

- racking—transferring wine from one container to another by carefully siphoning off (or pumping out) the clear wine, leaving all the sediment in the bottom of the barrel.
- fining—the addition of a coagulant, to attract particles and bind them together into much larger particles, which can then be removed by filtering. Fining agents used in the wine industry include egg albumen, gelatin (boiled-down bone glue), bentonite (an absorbent powdered clay), isinglass (animal tissue taken from fish bladders), casein (milk protein), silica, and powdered ox blood.
- filtering—the removal of the suspended and/or settled particles, usually using cellulose-mesh filter pads.
- centrifuging—using a machine to spin the wine, forcing the particles to move to the outside edge of the spinning chamber, where they are trapped behind a screen while the wine is pumped out from the middle.

MATURING

Maturing refers to the storing and ageing of wine prior to its sale. Wine can be stored in stainless steel, in oak barrels, or in glass bottles. Most wines, no matter how simple, benefit from a short period of maturing in the winery cellar. Six months is adequate for most whites and straightforward reds. The more complex the wine, the more it will benefit from further ageing. Tannic red wines need more time to achieve additional complexity, or to soften up.

The maturing wine is racked off, leaving the sediment in the bottom of the barrel. The wine will be racked several times to avoid the development of hydrogen sulphide, which would give it a rotten egg smell.

Oak Ageing

Wine is aged in wooden barrels for two reasons. Firstly, barrels are porous: the grain in the wood allows for slow and gentle exposure of the wine to air, which softens the tannins, tones down the acids, and allows the development of complexity in the bouquet. Secondly, wooden barrels impart additional extract to wine: vanillin, lactones, and tannins that differ from those of the grapes.

The extract is the sum of all the non-volatile solids that are in wine.

Oaks Used in Barrel Making

Oak is preferred over other types of wood because of its tight grain and its complex extracts. Only white varieties of oak are used in winemaking; the red varieties are too porous.

Much like the vine itself, the oak tree is affected by the climate and by its general environment. The rate of growth determines grain width, porousness, and concentration of extract. French oak displays the greatest range of characteristics.

Some of the mostly widely used oaks are:

- Limousin oak—from France, noted for its tight grain. It is used primarily in the ageing of cognac, armagnac, brandy, rum, and other distilled spirits, as well as Sherry and Port. It can be used for reds with strong personality (Cabernet Sauvignon, Syrah, Zinfandel, and Carignan) and for some whites, most notably Chardonnay.
- Nevers and Allier—light-coloured, tight-grained woods from central France. Barrels made from these oaks contribute delicate vanilla flavours and an element of elegance. They are used for Chardonnay and Sauvignon Blanc, as well as Cabernet blends, Syrah, Pinot Noir, Gamay, and many, many others.
- Bourgogne wood—also from central France, this offers characteristics similar to those of the Nevers and Allier forests, perhaps with a hint less elegance.
- Tronçais—from France, tightly grained. It releases its clean nutty-vanilla perfume and bitter tannins slowly. It is preferred for varieties like Pinot Noir and Chardonnay.
- Vosges oak—from Alsace in the northeastern corner of France, wood of a much tighter grain. Vosges barrels tend to develop a tougher structure in the wine, along with a roasted nuttiness in the bouquet.
- American white oak—particularly from Minnesota, Oregon, and Wisconsin, it is distinguished for its rich, aromatic extract and strong, distinct flavour. In North America, it still is used more for whiskey than for wine. In Spain, American oak is preferred for the way is enriches Tempranillo and the sweeter Sherries. Australians often use it for Shiraz and Cabernet Sauvignon.
- Slovenian oak—somewhat open grain, which tends to release more tannin than many delicate grape varieties are able to subdue. It is often used in Italy for Sangiovese and Nebbiolo, two varieties with overpowering personalities of their own.

The type and the age of the barrel, and the length of time that the wine is kept in barrel, varies according to the wishes of the individual winemaker. Two years in oak is not uncommon for some varieties, while others are ruined by long wood ageing. Winemakers use their artistic judgment as well as economic factors to determine how long a wine will remain in cask.

Barrel size and construction are significant considerations in winemaking. Large casks are used both for the fermenting and ageing of wine; small barrels (with their lower rate of evaporation and oxidation) are used mainly for the final maturing. The thickness of the barrel staves and the degree of charring of the inside of the barrel have a significant effect on the wine's style and complexity.

New barrels impart a large amount of extract, but each year the amount of extract decreases. After the third year, it decreases rapidly. Barrels from four to six years old are used primarily to oxidate the wine slowly, adding minimal extract. Winemakers may put some portion of a wine into new barrels and the rest into older barrels, to minimize the effect of the extract on the wine's flavour.

Since wooden barrels are porous, some wine evaporates with time. This creates a space inside, called ullage. This space must be topped up to prevent overexposure to air, i.e., oxygen, which could cause the wine to spoil.

Cold Stabilization

Prior to bottling, many white wines (and some reds) undergo cold stabilization. The wine is chilled to just below 0°C, causing potassium bitartrate (cream of tartar) to crystallize. Filtering removes the crystals. Cold stabilization done at a winery prevents these tartrates from crystallizing during winter transport, or in a cold cellar. Tartrates found in a bottle are harmless, but try to convince a restaurant patron that these are not glass shards, or sugar, in the wine!

Once the wine is ready to be bottled, it may receive one final filtering. It is then bottled, labelled, and prepared for shipment.

MAKING WHITE WINE

White wines are made, for the most part, from white (green) grapes. They can also be made from red (black) grapes. Anthocyanins (the red pigments found in the skins of red grapes) are not particularly soluble in water and at cold temperatures. With gentle and quick pressing, negligible amounts of anthocyanin are released into the juice.

The procedure, outlined below, minimizes contact with the grape skins.

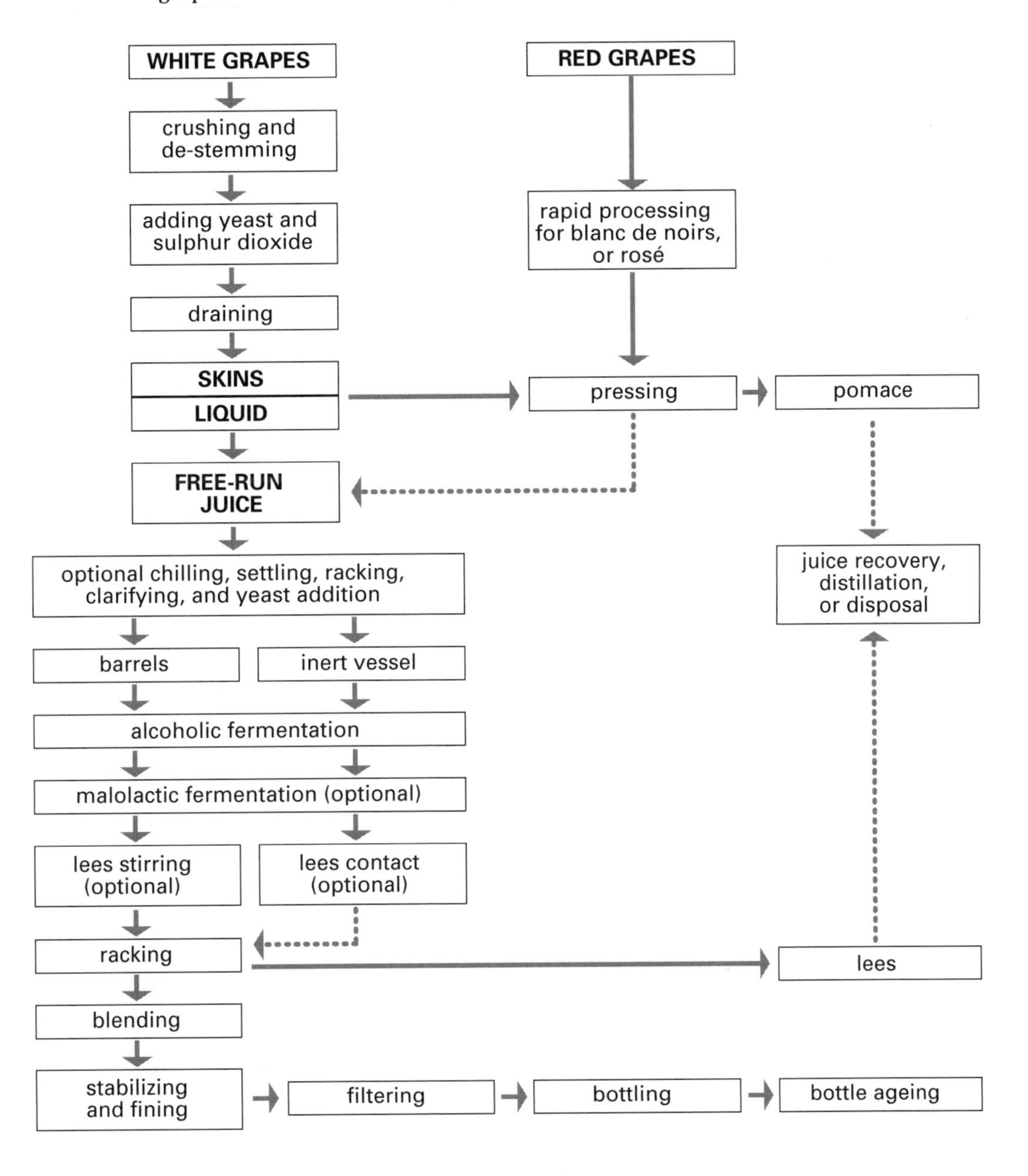

MAKING RED WINE

Red wines can only be made from red (black) grapes, since the procedure depends on contact with the grape skins.

Removal of the stalks is not compulsory—some winemakers like to increase tannin levels by leaving bunches intact.

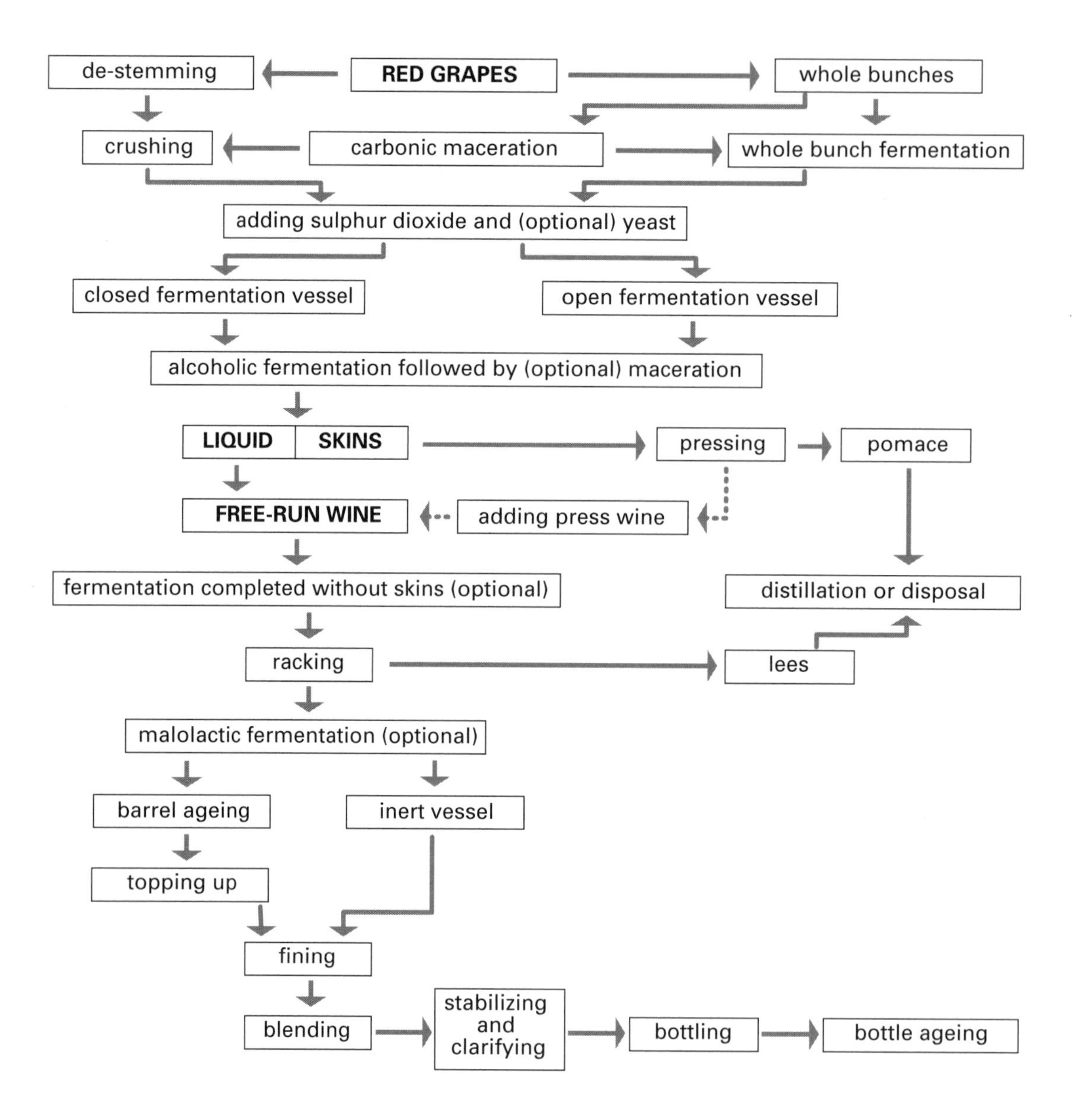

Fermenting Red Wine

After the grapes are crushed, the must is transferred to a fermentation tank where it immediately undergoes its first fermentation. During fermentation, solid matter (skins, seeds, stalks) tends to float to the surface of the vat, and forms a cap. To ensure adequate extraction of colour and flavour from the skins, the must is pumped over the cap. This circulation of the must over the cap is called remontage. Winemakers may also use large paddles to punch down the cap, several times a day, in order to mix it with the must. This practice is called pigeage.

The must may have a short period or a long period of skin contact, depending on the ripeness of the harvest, the style of the wine desired, and the grape variety. Pinot Noir, for example, having a large berry with more pulp than skin, requires longer skin contact to extract maximum colour, tannin, and flavour. Smaller berries with thick skins, such as Cabernet Sauvignon, need less skin contact to gain sufficient colour and tannin.

Pressing after Fermentation

Following the primary fermentation, the free-run wine is racked off the pomace. The pomace will be pressed several times to extract all the remaining wine, which is called the press wine. Press wine has more colour and tannin than free-run wine, but it is quite harsh and cannot be drunk on its own. Some or all of the press wine may be added to the free-run wine to give more body, flavour, and complexity.

Carbonic Maceration

To macerate is to soften by soaking.

Carbonic maceration is done to create soft, fruity, easy-drinking reds with reduced acidity, meant to be consumed while young and fresh. Although a range of grape varieties undergo carbonic maceration, the most famous is Gamay, used in making Beaujolais and Beaujolais Nouveau.

Carbonic maceration takes place in an anaerobic environment. In this process, carbon dioxide (hence the word carbonic) is used to exclude oxygen. Entire grape bunches are placed into a stainless steel vat filled with carbon dioxide, and sealed. The weight of the grapes presses on the berries at the bottom of the vat, causing them to split and expel juice. Whole berries in the centre of the vat become immersed in this juice and are macerated.

Yeast (on the skins of the split berries, or added) begins an alcoholic fermentation with the juice. This releases carbon dioxide, creating heat and additional pressure. The temperature in the vat gets higher; this extracts colour and tannin from the grape skins. As fermentation progresses, juice from the bottom is mechanically pumped to the top of the vat, helping the berries in the middle to soften further. As they collapse, their juice and skins begin to ferment. When all the grape sugars have been converted to alcohol, the alcoholic fermentation stops.

Sugar may be added prior to the completion of the alcoholic fermentation.

Meanwhile, berries at the top of the vat undergo an independent, intracellular fermentation. Unlike the alcoholic fermentation taking place below, this intracellular one is not activated by yeast, but by the surrounding carbon dioxide. The berries slowly absorb the carbon dioxide, and undergo cellular changes. The level of the malic acid is reduced by 15% –35%, without the production of lactic acid. When the whole berries absorb enough carbon dioxide, the intracellular fermentation stops.

The whole process takes three to seven days. The free-run wine is then pumped from the vat, leaving the pomace and some whole berries behind. Winemakers will press the pomace and remaining berries, and add the pressed wine to the free-run wine.

MAKING ROSÉ WINE

A rosé (pink, blush) wine can be made in any of three ways.

Direct Pressing

Red grapes are crushed and gently pressed, releasing a minimal amount of pigment into the juice. The resulting wine is a pale rosé or vin gris ("grey wine").

Bleeding

After several hours of soaking with red grape skins (but before these skins are completely macerated), a portion of the juice is drained off, to make a wine with a deep pink (almost pale red) colour. The primary purpose of this method is to increase the proportion of red skins to juice, which concentrates the colour and tannins in making red wine. The rosé is more of a by-product.

Blending

A small amount of red wine is blended with white wine.

MAKING SWEET WINE

The obvious difference between dry wines and sweet wines is that the sweet wines have residual sugar. Left to its own devices, yeast will convert all the natural or added sugars in wine into alcohol, until the alcohol level gets so high it stuns the yeast—around 14% to 18% alc./vol. However, some sweet wines have alcohol levels as low as 5% alc./vol. (Moscato d'Asti) while others hit highs of around 20% (Sherry and Port).

There are two separate and distinct aspects to consider with respect to the making of sweet wines: where the extra sugar comes from, and what processing technique was used to produce the wine. The following considers both aspects.

Botrytis-affected Wine

Under certain conditions, some grape varieties—the most famous are those used to produce the Sauternes of France, Trockenbeerenauslese from Germany, and Tokaji Aszu produced in Hungary—will be affected by the fungus Botrytis cinerea, known as noble rot. The botrytis fungus absorbs water from inside the grape, which causes the grape to shrivel. Sugar and acid are concentrated but balanced. Late in the season, the shrivelled grapes are picked, crushed, and fermented like white wines.

Fermentation may stop naturally, or with intervention, leaving a high percentage of unfermented residual sugar.

Botrytis-affected wine has a unique bouquet and flavour reminiscent of dried fruit, honey, custard, and truffles. Production is usually quite low and is always labour intensive.

Late Harvest Wine

Grapes that are not affected by botrytis, but are left on the vine until later in the harvest time, will begin to shrivel and "raisin" anyhow, creating intensely high sugar levels. Austria, Canada, France, and Germany regularly produce some of the world's finest examples of Late Harvest wines.

Icewine

Grapes left on the vine into the winter will freeze on the vine. If hand picked while naturally frozen (at a minimum of –8°C) and pressed while still frozen, the volume of juice will be extremely small. (Water content is greatly reduced when the shards of ice are pressed out.)

Grape yields for icewine are less than 5% of that which can be gathered at normal harvest time. But the sugar level will be higher than achievable by any other method. Icewines exhibit rich aromas reminiscent of litchi, buckwheat honey, orange peel, apricot, and dried fig.

Raisining

In this process, the grapes are harvested as soon as they are ripe, and are then laid out in the sun on straw mats, or on special trays stacked in well-ventilated buildings. They shrivel into raisins, losing up to half their water content. These dried grapes are pressed or crushed, and fermented. The resulting wine is then aged, becoming strong, rich, and high in alcohol.

Adding Sugar

Sweet wines can be made by simply adding refined sugar or sugar syrup to a finished wine. The technique leaves much to be desired, though, particularly in the area of taste. Wine produced in this way may be cloying—the sugar is often out of balance with the acidity. Except for the practice of chaptalization, most respectable wine regions forbid the addition of sugar that is not natural grape sugar.

For details on chaptalization, see below, under Corrective Practices.

Back Blending

This widely-used technique was perfected by German winemakers in the last century. At harvest, a small portion (around 10%) of the juice from crushed grapes is sulphured to prevent fermention, and set aside. The remainder is then fermented fully. The unfermented juice (sweet reserve or süssreserve) is then blended back into the fermented wine; the blend is filtered to remove any remaining yeast, preventing further fermentation.

Some features of this method are that it is inexpensive, the grape sugars are balanced by the fruit's natural acidity (so no additional acidification is necessary), and the fruit flavours of the grapes complement the fruit flavours of the wine. However, back blending lowers the alcohol content of the wine.

This method is permitted for use with Late Harvest wines of the highest level of quality, provided that the süssreserve comes from the same vineyard as the fermented wine to which it is added. This technique is also used to make the great sweet Sherries of Spain.

Arresting Fermentation

The must is permitted to ferment until it has reached the desired alcohol level. Fermentation can be stopped by sulphuring (treating with sulphur dioxide), or by rapid chilling, to stop the yeast from working. Either process is followed by thorough filtering to remove all remaining yeasts. This technique is employed in the production of relatively inexpensive sweet wines.

Fermentation can also be arrested by fortification: a distilled spirit, usually grape brandy, is added to the fermenting must in sufficient quantities to shock the yeast into inactivity. There are more than a hundred different wines produced using this method, including Port, Madeira, Marsala, Malaga, Commandaria, Pineau des Charentes, Floc de Gascogne, and Ratafia.

MAKING SPARKLING WINE

The two by-products of fermentation are alcohol and carbon dioxide. If the must is fermented in a sealed tank, the gas will not be able to escape into the atmosphere and will be dissolved back into the wine. Sparkling wine is simply a wine which contains bubbles of carbon dioxide.

There are four ways to produce sparkling wine. The best and most complex method is the méthode champenoise, developed in the region of Champagne. Other, cheaper, bulk processes have been developed: the transfer method, the charmat method, and carbonation.

Méthode Champenoise

With the méthode champenoise (a.k.a. classic or traditional method), still wine undergoes a second fermentation in the bottle. This method produces sparkling wines with tiny, long-lasting bubbles.

The process is explained in detail in Chapter 10, Champagne.

Transfer Method

Here, too, a second fermentation takes place in the bottle. Afterward, the now sparkling wine will be transferred into a large tank, chilled, filtered, sometimes treated with a dosage of sugar, and finally re-bottled for sale.

Charmat Method

Also known as cuve close, this process is fast and cheap. The wine is produced in a sealed, pressurized tank, and bottled under pressure. The wine's bubbles do not last long in the glass.

Carbonation

The quickest and cheapest way of producing sparkling wine is to simply pump carbon dioxide gas into a tank of wine, then bottle the wine under pressure.

CORRECTIVE PRACTICES

In a perfect world, every vintage would be great and all winemakers would have to do is pick the grapes and bottle the wine. But it is a rare vintage that presents ideal conditions, so winemakers must carefully monitor every procedure and be ready to deal with minor or major problems.

Corrective practices deal with problems that relate to imbalances in sugar, alcohol, acid, or colour:

- alcohol adjustment.
- acidification.
- de-acidification.
- decolourizing.

Not all of these practices are used worldwide. Some are prohibited in certain wine regions while others, though allowed, are shunned by the wine producers themselves. And some of these techniques are used more to create special styles of wine rather than deal with shortcomings.

Alcohol Adjustment

Alcohol adjustment is a common winemaking practice used in many regions. If poor weather produces grapes with low sugar levels, winemakers will often raise the sugar level so that the final product will have the desired alcohol level. They can do this by increasing the sugar concentration in the must, or in the wine.

Chaptalization

Refined white cane sugar (table sugar) or liquid invert sugar from sugar beets is added to grape juice, or to must prior to or during fermentation. The purpose is to boost the alcoholic content of the finished wine.

Chaptalization has its uses and its misuses. When carried out judiciously, it can add some body and complexity to wine, augment its bouquet, and possibly increase its ageing potential. But if too much sugar is added, the alcohol level will be raised beyond that which can be hidden by the natural fruit flavours. The wine will taste candied, spirity, and heavy, distorting the true fruit character of the grapes.

Chaptalization is used predominantly in cooler regions where a high level of grape ripeness is not always achieved. In regions where chaptalization is forbidden by law, winemakers creatively use grape concentrates.

Addition of Grape Concentrates

Where chaptalization is prohibited, grape concentrates are usually approved for use. A portion of the juice is boiled to eliminate some of the water, and so concentrate the sugar. This is then added to the must, and fermentation proceeds as usual. Traditionally, this technique is limited to use in red wines, since it darkens the colour.

Rectified grape concentrate is a sweet concentrate made neutral in colour and flavour. It can be used to boost either red or white wines as it has little effect on colour or taste. It is costly, though, so its use is somewhat limited.

Partial Freezing

The must or the grape juice is chilled to the point where it just begins to freeze. It is the water which freezes, not the sugars, acids, or fruit extract. As ice crystals begin to form, they are removed, leaving a more highly concentrated must or juice.

Cryo-extraction

The grapes are harvested during the normal harvest period. A portion is taken immediately into an industrial freezer, frozen solid, and then pressed. The concentrated juice is added to the remainder of the harvest, to increase sugar and concentrate flavour. This practice is used extensively in France, in the Sauternes region, and in New Zealand. In the United States, it is used to make fake icewines.

Bleeding the Wine

In the case where winemakers think their red wine may lack colour and concentration (usually after a wet vintage), they will drain away some of the juice prior to, or at an early stage of, the fermentation. The smaller proportion of remaining juice is then macerated with all the skins, resulting in more colour and tannin in the wine. The juice bled away is often made into a rosé wine.

Centuries ago, the practice was widespread in Bordeaux. The run-off was sold as vin clairet, hence the name claret.

Reverse Osmosis

Where other methods tend to remove water from must or from wine, reverse osmosis extracts the concentrate by passing the must or wine through a semi-permeable membrane filter made of cellulose acetate, leaving behind the water.

Acidification

Acidification is used extensively in hot climates or during hot vintages, when the harvested grapes have excessively high sugar levels and very low acidity. To avoid flat and flabby wines, tartaric or malic acid is added to the wine, before fermentation. Alternatively, green, under-ripe berries can be added to the must. A small amount can make a big difference.

De-acidification

A must or a wine with too much acidity can be de-acidified (completely or to a small degree) in different ways: by adding salts (such as calcium carbonate or potassium bicarbonate), by inducing malolactic fermentation, or by blending with a low-acid wine.

Decolourizing

In white wine production, minimal air contact and the addition of sulphur dioxide can prevent the skins and juice from browning. When browning occurs despite preventative actions, the wine can be treated with charcoal—but charcoal treatment will strip the wine of any character. Decolourizing is not used in the production of quality wines.

PART II

The Business and Service of Wine

The chapters contained in Part II introduce and examine a variety of topics related to the business and service of wine. A wine professional needs to be expert in several areas.

Wine Tasting

Drinking wine is simple. Recognizing the distinctive nature of wine takes more time, effort, and concentration. The difference between the two processes is what separates the casual drinker from the serious taster.

Few people spend time looking at wine, consciously smelling it, or thoughtfully savouring it. Most consumers just drink it. But that is their choice. They are your customers and they want to enjoy themselves.

Professional tasters are technicians. They examine wine for a variety of reasons:

- to judge its overall quality and determine if it has any serious faults.
- to identify what grape varieties have been used.
- to recognize its origin.
- to describe its state of maturity and predict its ageing potential.
- to evaluate the vintage.
- to differentiate between the effects of grape-growing practices and winemaking techniques.
- to assess the wine's commercial value.
- to ascertain its most appropriate food partners.
- to enjoy the nuances and complexity as fully as possible.

Consumers are subjective—they like a wine or they don't. But professional tasters must always try to remain objective, separating their personal preferences from confirmable qualitative judgments, based on certain measurable standards.

TASTING CONDITIONS

It is helpful to taste wine in ideal (or as near as possible) conditions. These include having good lighting in the room, keeping the wine at a preferred temperature, and absolute freedom from distractions such as odours or chatter.

Lighting

Bright, natural light, such as on a sunny day in a white or light-coloured room on the north side of a building is ideal. Rooms in direct sunlight are not ideal—any momentary cloud cover can dramatically change the lighting conditions.

In the southern hemisphere the room should be on the south side.

In a wine cellar, the light is less than ideal. So, many winemakers or cellar masters use the tastevin, a small, shallow, bumpy, silver cup. This reflects candlelight through the wine, giving a good indication of its colour and clarity.

Temperature

Most wines are best tasted at a temperature within the standard cellar temperature range of 8°C–14°C. Much colder and the wine "closes up," in the sense that it stops giving up its flavours and smells. If the room is too warm, the wine gives up subtle nuances too quickly. At the very least, wines being compared should be tasted at the same temperature.

Distractions

- The worst distraction is any smell that pervades the tasting room. Perfume, smoke, cooking aromas, flowers, air fresheners, fresh paint, strong cleaning fluids, etc., all interfere with good judgment.
- Some tasters enjoy listening to soft music while analysing wines. It is necessary, however, to focus on the task at hand. The most annoying form of noise is the chatter of fellow tasters. Keep quiet. Insist that others do, too.
- Even the most experienced professionals can not taste properly when they have a cold or flu, are tired, or are under pressure.
- A taster must have a fresh palate. Any strongly flavoured food or drink interferes with the ability to recognize critical nuances. It is always preferable to taste before eating, rather than just after.

HOW TO TASTE WINE

Wine tasting is detective work. The evidence is in the glass—you must determine the facts and present a logical analysis of the situation.

- Like a detective, you must consider every sample meticulously, identically, and conscientiously. Keep a notebook with you in case tasting sheets are not available.
- Arrange the wines in some order. In general, taste sparkling wines before still wines; light wines before heavy wines; and ordinary wines before fine wines. Taste dry whites before dry reds; dry before sweet; and young wines before older wines.
- When tasting several wines in the same category, group them together according to some similarity—such as all those from the same grape variety, or growing region, or vintage.
- All the glasses used by any individual taster must be identical, i.e., of the same size and shape. Use wine-tasting glasses approved by the International Standards Organization (ISO), if available. If not, any glass with a round-bottomed bowl and tallish sides will do. Ceramic, plastic, metal, or styrofoam simply will not do.

There is a sample of a layout for tasting notes on page 73.

A wineglass has three parts:

- the bowl, which contains the liquid.
- the base, which prevents the bowl from falling over.
- the stem, which is the only part of the glass you should hold.

Handling a glass by the bowl results in fingermarks, which take away much of the pleasure of looking at a wine.

More important, though, holding a glass by the bowl warms the wine. This can change the delicate flavours and speed up their deterioration.

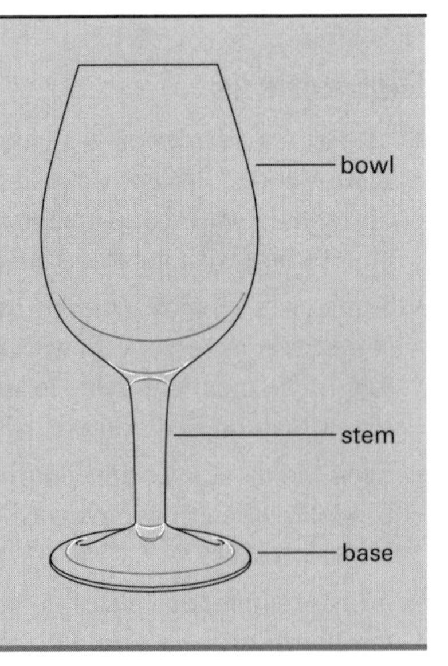

Appearance

The colour and clarity of a wine help indicate its quality, nature, and maturity.

Colour

Hold the glass upright and look at the wine. Its colour is a function of grape variety, method of winemaking, and age.

- Colour in a white wine can range from almost colourless and silvery with a greenish tinge, to golden, to dark brown. Red wine can range from an opaque blue-black purple, through shades of red, to mahogany, to dark brown.
- Depth of colour gives an indication of the concentration of fruit extract in the wine; the more concentrated the fruit, the deeper the colour of the wine.
- White wine tends to darken with age, while red loses redness and gains brownness as it gets older.
- Wine that is young tends to be uniform in colour from the rim to the centre of the bowl. Older wine appears lighter in colour at the rim, and darker at the centre.

Clarity

Tilt the glass inward toward yourself so that you can look into it rather than through it. Note the wine's clarity.

- A healthy wine should have a bright, reflective surface. Any dullness in the appearance suggests that the wine may have some flaw which, inevitably, will affect its taste.
- There should be no sediment in the glass.

Nose

A wine may have a simple and straightforward aroma or a very complex bouquet. The terms *aroma* and *bouquet* are often used interchangeably, but traditionally there is a distinction between the two. The more general terms *nose*, *smell*, or *scent* may be used.

AROMA AND BOUQUET

Aroma is the scent that distinguishes each grape variety. Just as a McIntosh apple smells different from a Spartan, so a Chardonnay grape smells different from a Riesling, or a Cabernet Sauvignon from a Pinot Noir. A grape's unique aroma is not lost during fermentation. Very young wines can be identified by characteristics associated solely with the grape variety. The wine's aroma is straightforward and one dimensional.

Bouquet, on the other hand, combines several smells, just as a bouquet of flowers includes several varieties, and a bouquet garni employs a number of different herbs. Bouquet does not exist until after the fermentation. It comes from the interaction between the fermenting juice and the yeast, and from oak barrels if they are used to ferment or to age the wine.

- By increasing the wine's surface area, you can increase its smell. So, swirl the wine around to coat as much of the inside of the glass as possible. Try snapping your wrist with a gentle twist, no more than once or twice—more is unnecessary.
- Put your nose into the glass as deeply as possible. The glass should touch your face both above and below your nose. Inhale gently through your nose—aggressively sniffing the wine does nothing for you, and quickly tires out your olfactory apparatus.

  You may also exhale through your nose (retro-olfaction) to assess the aroma and bouquet of the wine after it has been warmed by your body.
- Decide whether the nose (the smell) is pleasant or unpleasant. Its intensity can be mild, moderate, assertive, or aggressive. It may be fruity, vegetative, floral, nutty, woody, earthy, herbal, spicy, perfumed, or caramelized. If there are any flaws, the wine may smell sour, mouldy, oxidized, chemical, or simply artificial.

The human nose can recognize thousands of various smells, unlike the human palate which recognizes only four tastes (sweet, sour, salty, and bitter).

Describing Smells

Even experienced tasters have great difficulty putting what they smell into words. Often, the problem is not recognition, it is articulation. The aroma wheel was developed by Ann Noble at the University of California (Davis) to assist in standardizing the language of the nose. It has become the wine taster's official list of smell descriptions.

Look at the diagram on the facing page.

- Start with the inner ring. Determine if the primary smell is fruity, vegetative, floral, chemical, nutty, etc.
- Move to the middle ring. If the smell is fruity, for example, decide if it is citrus, berrylike, tropical, or other.
- Finally, look at the outer ring. If the fruity smell is berrylike, for example, decide if it is more like blackberry, raspberry, strawberry, or black currant.

Do not limit your thinking to what is on the list. It provides a starting point. "Berry fruit" could also include blueberry, red currant, bramble, mulberry, gooseberry, loganberry, or any other berry you have ever come across.

Of course, if you have never smelled a gooseberry, there is no way you will recognize its aroma in a wine glass. It is not part of your "smell vocabulary." The aroma wheel can not help you to recognize smells you have never experienced. As a professional taster, you must develop a solid vocabulary of smells by becoming alert to your surroundings, and by making an effort to find "new" smells.

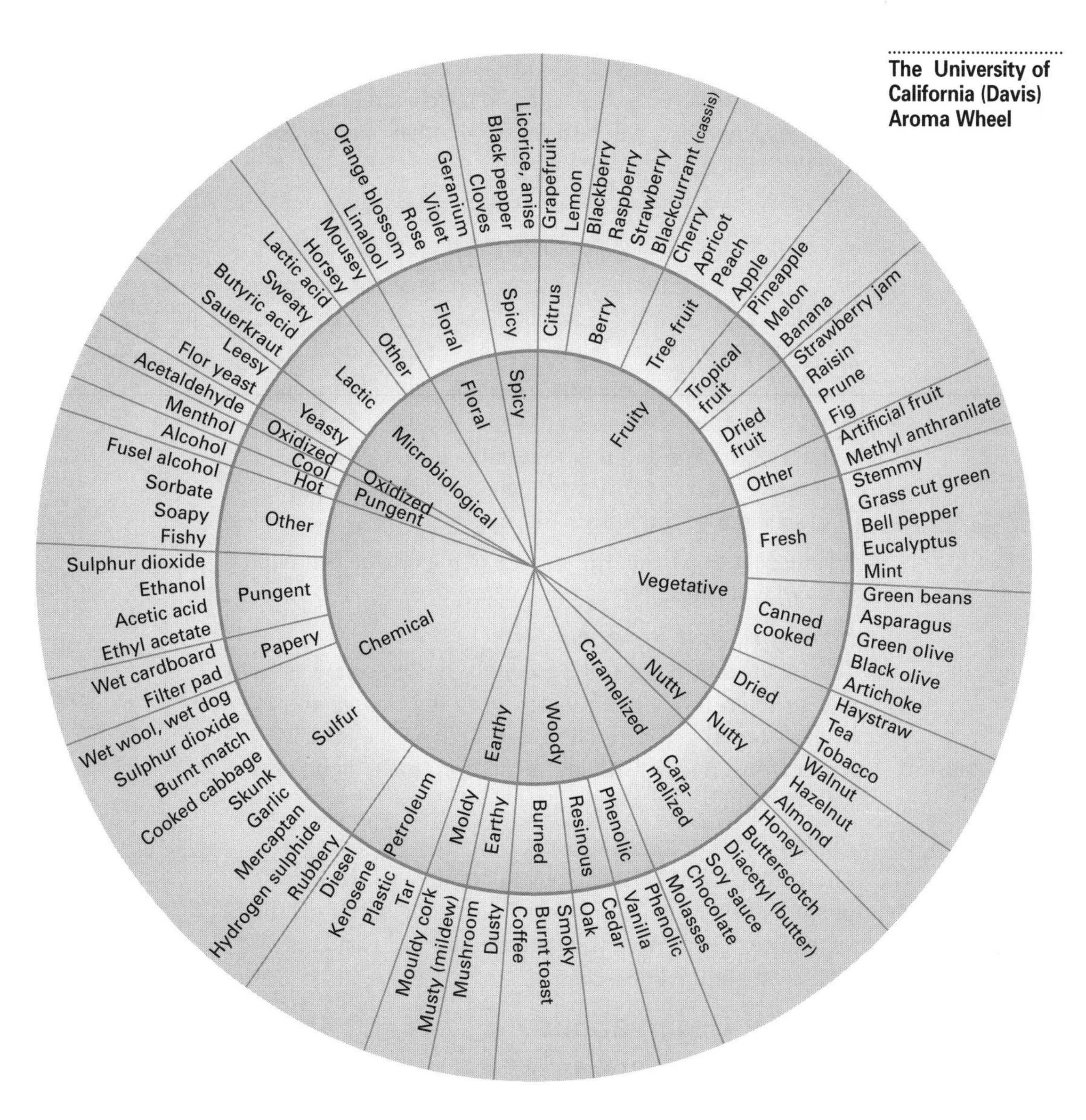

The University of California (Davis) Aroma Wheel

Taste

Virtually all the flavour of a wine is observed by smelling. Flavour is simply a confirmation of what has already been noted in the smell. But a wine's structure, balance, and finish can only be judged by tasting.

- To get the most out of the taste, take a tiny sip and swish it around thoroughly, ensuring that some enters every crevice of your mouth. Imagine you are chewing the wine as you might a bite of food.
- Lower your chin to suck in a bit of air through the wine, as though you were slurping hot soup off a spoon. (Loud slurping sounds are unnecessary.) By "aerating" the wine, you are able to re-experience the smell, which enters your nasal chamber via the back of your throat.

Recognizing Tastes

Asian cultures also accept spiciness—the burn of ginger or chili pepper—as a taste.

The palate registers only four basic tastes: sweetness, saltiness, sourness, and bitterness. All other "tastes" are a function of smell.

Alcohol has a barely noticeable sweetish taste. High alcohol levels can bite or burn at the very back of your throat.

- Sweetness is perceived at the tip of the tongue.
- Salt reacts with the upper sides of the tongue.
- A prickly sensation at the sides and beneath the tongue indicates sourness, i.e., acidity. The more prickly the sensation, the more acidity there is in the wine.
- Bitterness can be identified at the back of the tongue, and by a furry dryness in the area between the inside back of the lips and the front of the gums. The most frequent cause of this astringence is tannin; the taste associated with tannin is woody.

A large central part on the surface of the tongue is numb to the various tastes.

Structure and Balance

Tannin, in combination with alcohol, gives the wine its body and structure. Wine has balance when the alcohol and the primary tastes are in equilibrium—such a wine is called harmonious. A wine with low alcohol will seem light, one with too much will seem heavy.

Finish

Aftertaste (finish) is the taste you get after you swallow or spit out the wine. The longer it lasts, the better—if the wine tastes good. Wine will reveal additional elements of complexity in the finish.

General Observations

Regarding the overall quality of the wine, subjectivity must enter the picture.

- How much do you like the wine? Do you consider it to be poor, fair, good, very good, or excellent?
- Note the maturity of the wine. Decide whether it is over the hill, past its peak, at its best, still needing time to develop, or has potential with significant ageing.
- What foods would you like to eat with the wine?
- How much would you pay for the wine?

Tasting Notes

These can be kept in a notebook, on tasting sheets provided at a wine tasting, or on forms that you can keep in a binder. There are many formats for tasting sheets: some allow you to assign a numerical value for features of the wine, or to make written notes. On the next page is an example of a general form, which can be photocopied.

Tasting of ______________________________

At ______________________________ **Date** ______________

	WINE	*TOTAL* (20)	*APPEARANCE* (4)	*NOSE* (6)	*TASTE* (8)	*OVERALL QUALITY* (2)
1.						
2.						
3.						
4.						
5.						
6.						

WINE FAULTS

Of the hundreds of identifiable components in wine, some can lead to the development of minor flaws or serious faults.

Haze

When unstable proteins are not removed, they can form a haze. This may indicate that the wine has an undesirable decayed flavour.

Cloudiness

Cloudiness is distinct from haze, and need not always be considered a fault. Some winemakers believe filtering strips wine of its character, and avoid the process. Unfiltered wine, particularly old red wine, will throw a sediment. Improperly handled (or clumsily decanted), the sediment will cause a cloudiness which, though perfectly harmless, is unpleasant to look at, or to taste.

Off Colour

Darkening or fading colour can indicate laccase—an enzyme associated with botrytis. It produces undesirable changes in both white and red wine. Whites darken, while reds fade and turn brown. Laccase can be prevented by allowing the wine to rest in a sealed vat saturated with carbon dioxide after fermentation. Deprived of air, laccase cannot activate.

Bubbling Wine

Wine improperly fined and filtered (or bottled with too little sulphur dioxide) permits bacteria to work uncontrolled. The result is that small bubbles form. (There may also be a sauerkraut smell.) In certain cases, bubbles in wine are not a fault—the vinhos verdes of Portugal and the frizzantes of Italy traditionally are produced by a delayed malolactic fermentation.

Burning Match

This pungent smell, along with an aggressive tingling sensation in the nose, indicates excessive use of sulphur dioxide. Sweet wine is more prone to have this odour—sulphur dioxide is often used to stabilize wine that has natural residual sugar. The smell does diminish with time. Swirling the wine in the glass speeds up the evaporation.

Vinegar or Acetone

Acetic acid (vinegar) and ethyl acetate (nail polish remover) are two highly volatile compounds that are easily detected at room temperature.

They are created during primary fermentation, during malolactic fermentation, and when wine is exposed too long to the air. In minute amounts, they enhance wine's aromas and flavours, and add a degree of complexity. Too much, however, destroys the flavour and finish of the wine.

Sauerkraut

Biochemical processes are never perfect or completely predictable. Occasionally malolactic fermentation goes wrong, resulting in the creation of minute amounts of diacetyl. This bacteria, which is desirable in producing the distinctive aroma of sauerkraut, is unacceptable in wine.

Geranium

If sorbic acid has been used to stabilize the wine after primary fermentation, and lactic bacteria start to grow, the bacteria will change the sorbic acid into a compound that smells, unfortunately, like geraniums.

Rotten Eggs

The presence of hydrogen sulphide gas in wine produces an odour of rotten eggs. The element sulphur (which is approved in organic viticulture) is widely used to control mildew. Although grape growers generally stop spraying long before the harvest, the fruit can still trap trace amounts of sulphur. Certain strains of wild and cultured yeast will metabolize the sulphur to form hydrogen sulphide gas. As the gas is quite volatile, it can be eliminated with an aerated racking. If left untreated, however, it will combine with alcohol to form ethyl (or methyl) mercaptan. This is less volatile than hydrogen sulphide, and lingers in the wine.

Skunk, Manure, Burnt Rubber, or Rancid Garlic

Ethyl (or methyl) mercaptan gives wine a variety of bad odours, the most common ones being skunk, manure, burnt rubber, and rancid garlic. Different odours can develop depending on the type of wine.

Mouldy/Corky

Prior to their sale to wineries, corks are treated with chlorinated compounds in order to bleach them and to remove a variety of natural microflora, especially fungi. If any chlorine remains on the corks after their final wash and some of the fungi survives the treatment, a compound called trichloroanisole is produced. Even the smallest traces (measured in parts per billion) of this will spoil the bottle. The wine takes on a musty nose and unacceptable taste.

Madeirization

Exposure to warm temperatures is harmful to wine. The taste becomes dull, with caramelized sugars. Whites deepen in colour, reds fade and turn brown.

Oxidation

Exposure to air—either through a bad cork or as a result of leaving a bottle standing upright too long—results in a loss of intensity of character, and ultimately an unpleasant stale odour.

Oxidation can also take place when wine is fermented or aged in wood barrels. If the amount of oxygen in contact with the wine over an extended period is minute, then the oxidation will be beneficial. But too much exposure to air is harmful.

All wines oxidize eventually.

CHAPTER 7

WINE MANAGEMENT

Wine management is an essential element of the hospitality industry, regardless of whether the restaurant is a fine dining room, a small family-run café, or part of a large hotel. In these different venues, wine may be in the charge of a sommelier, a food and beverage manager, a server, or an owner. Because effective wine management can raise or maintain a restaurant's revenues and standards, whoever takes charge of wine must be competent, efficient, and thorough—that is, both a committed student of wine and a wine professional.

As the wine professional, you will do more than just pull corks and fill glasses. You may be required to do any combination of the following tasks:

- stock and maintain a wine cellar.
- design a wine list.
- set wine prices.
- market wine in the restaurant.
- manage wine service staff.
- sell and serve wine.

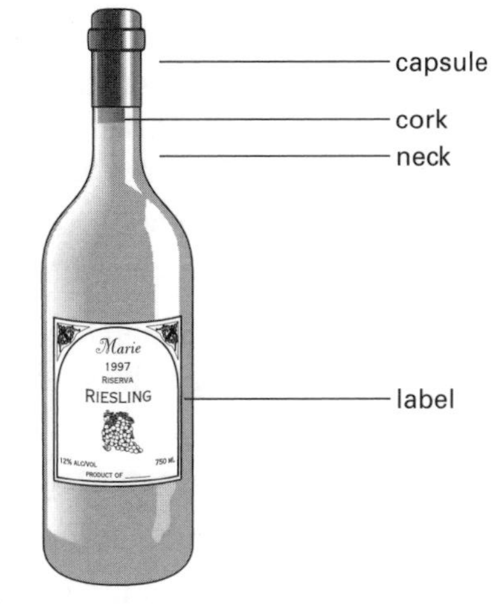

This chapter focusses on the first five tasks. Chapter 8 shows you how to sell and serve wine.

As the wine professional, you must clearly understand the restaurant's theme, and the chef's style. Consider, for example, the following questions:

- Who is your typical customer?
- What type of food is featured on the menu?
- Is the menu always the same, or does it change with any regularity?

Once the restaurant's needs have been determined, the wine professional can proceed to buy wisely, and create a wine list that enhances the menu.

Purchasing Wine

When buying wine, the following general guidelines apply.

- Sample the wine before making a purchase. It is the best way to ensure satisfaction.
- Ask to see the wine merchant's tasting notes to help you in your selection.
- Become familiar with the wines available. Visit wine regions; meet with wine distributors, producers, shippers, and wholesalers; attend as many trade tastings as you can.
- If buying wine by the case (or mixed lots in boxes), open the cases to check you are getting what you paid for.
- Avoid bottles stored upright: the cork dries out, allowing air to enter and oxidize the wine. Purchase only those bottles that have been stored on their sides.
- Avoid bottles used in window displays, or those stored under bright lights. Heat and light cause a wine to age prematurely.
- Avoid bottles with soiled labels, leaking capsules, and protruding corks.

Make an exception for aperitifs. Because of their high alcohol levels, they should be stored upright to keep the liquid from touching the cork or screw cap.

Once the purchases have been made, you will be expected to keep an inventory of the wine cellar.

Cellar Arrangement

Here are some guidelines for cellar arrangement.

- Wine bottles should be placed horizontally (to prevent the cork from drying out). Labels should be upright to make sure they do not become damaged during handling, and to allow personnel to remove bottles gently, checking for sediment without turning or disturbing them too much.

- Aperitifs, liquors, and spirits should be stored upright to keep the liquid from touching the cork or screw cap.
- Whites and sparkling wines should be stored in the lowest part of the cellar, where it is coolest and has the least temperature fluctuation.
- Reds should be stored above the whites, in progression according to the wine's sturdiness. Light, fruity reds can be laid just above the whites, while full-bodied wines like Recioto and vintage Port can be placed on higher shelves.
- Top-selling wines should be easily accessible.
- Wine should be stored logically by country and region.

Ideally, the wine professional should be able to enter the cellar and remove any bottle without turning on the light.

Some form of bin card system is essential in maintaining an efficient inventory to identify the wine, vintage, date of purchase, cost price, and number of bottles remaining in stock. Columns are provided to register withdrawals, with the name or signature of the person removing the wine. A bin card system can be replaced by a computer program, with daily printouts.

Entering details regarding the movement of wines is not difficult, and is invaluable for security.

The cellar door should be fastened with secure locks, and keys should be issued only to bonded personnel.

Storing Wine

Proper storage is essential if wine is to be kept for any length of time. First, it allows wine to mature properly, gaining in quality and value. Second, it protects the capital invested in a cellar. Astute purchasing of wine for a cellar (though not without risk), is the best investment you can make in the hospitality business. The cost of a good wine purchased young can return many times over when it is mature and rare. (A bottle of 1964 Petrus purchased for $30 in 1967 was worth $2000 in 1996.)

There are two basic types of storage: short-term storage for consumption immediately and in the near future, and long-term storage for ageing and maturing.

Short-term Storage

Short-term storage is for establishments that have a substantial wine inventory, and a long-term storage cellar far from the point of sale. The short-term cellar holds all the wines needed to fill present orders from the restaurant.

In it, wines should be kept close to their ideal temperature for consumption. Whites and sparkling wines should be kept in a room or refrigerator, at about 4°C–5°C. (Refrigerated units for display and storage in the restaurant are available and they are a good investment.) Reds should be kept at a temperature closer to that of the restaurant's dining room.

Long-term Storage

Long-term cellaring need not be in a vaulted underground cave. Any place will do, as long as it provides the proper environment for wine to undergo its chemical and physical changes slowly and peacefully. The basic requirements for proper storage are as follows:

- Darkness—lighting should be sufficient to allow movement around the cellar without disturbing the wine. Bright light and sunlight must be avoided. Low voltage incandescent bulbs are preferable to fluorescent lighting.
- Temperature—the ideal is between 8°C–14°C. White wines stored below 5°C for too long risk a precipitation of crystals of tartrate. Above 16°C, some white wines will start to madeirize and age prematurely. Red wines above 19°C will do the same. The warmer the temperature, the faster the wine will age. Fluctuations of temperatures should be avoided at all cost.
- Humidity—this should fall between 60% and 80%. Below 50%, corks begin to dry out. Above 80%, corks go mouldy and labels rot.
- Tranquility—quietness and freedom from vibration is necessary for the wine to mature properly. Sudden loud noises will disturb it.
- Cleanliness—a wine cellar should be well ventilated, without any strong odours. No items other than wine, spirits, and liquors should be stored in the cellar.

Ageing Wine

Good cellar management also means knowing when certain wines will need to be consumed and which ones are likely to improve with age.

When cellaring wine, keep in mind that older is not necessarily better. Some wines do not benefit from long ageing. Some qualities, such as fruitiness and fresh aroma, are better when the wine is young. (The varietals Gamay, Muscadet, and Sylvaner taste better in their youth.) An integral part of the wine professional's job is to know which wines can age well and, more importantly, when a wine has reached its peak and should be featured on the wine list.

A varietal wine is a wine produced from a single grape variety.

Mechanics of Ageing

The mechanics of ageing are baffling. Wine is a living organism and its molecules are constantly morphing—some disappear while new ones are created. Changes continue until the wine is dead.

Tannins are the primary reason that red wine has the potential to age. During the ageing process, tannins precipitate, leaving a better-balanced wine with greater complexity. Like all elements in wine, tannins must be present in the right amount—in balance.

The majority of whites are meant to be consumed young. Few improve with ageing. Examples of white wines that do age well are prime white Burgundies or Chardonnays, vintage Champagne, and some Rieslings. It is also well known that Sauternes, Tokaji Aszú, icewines, and botrytized wines can keep from ten to twenty years. These have a perfect balance of sugar and acid. In contrast, a young, dry, Pinot Blanc with low acidity has no ageing capacity.

If a wine begins to pass its prime, you must do the utmost to sell the wine quickly. This can involve aggressive marketing, e.g., featuring it in a promotion and/or lowering its price. The idea is to minimize losses.

DESIGNING A WINE LIST

The ultimate goal of the wine professional is to have an assortment of wines that:

- offer a range of options to the customers.
- complement the menu.
- generate revenue for the restaurant.

The wine list should have rhyme and reason. It must fit the menu, and it must offer value and quality. Successful wine professionals aim for a variety that can create excitement and a cutting-edge image for the restaurant.

A key factor in increasing wine sales is creating a wine list appropriate to the restaurant's size, budget, and typical customer. The wine list should accomodate the level of interest and knowledge of the customers. They must be able to make good and effortless decisions, whether or not capital is available for a long wine list.

Presentation

The wine list is your calling card. It should be attractive, organized, and easy to read. Every wine list should contain the following information (at least) about each wine.

A good wine list shows:

- the vintage.
- the full name of the wine.
- the name of the producer or shipper.
- the country and region of origin.
- the bottle size, if not standard.
- the price.

Check the spelling in the wine list, carefully.

Categories in the Wine List

The list can be categorized according to whites, reds, rosés, and sparklers, with additional pages for fortifieds, spirits, liqueurs, beers, ciders, and non-alcoholic beverages.

Alternatively, it can present the wines by country and region, grape variety, style, or recommended food matches. The list may even display the wine labels.

Do not attempt wine descriptions in the list. These are inadequate, and can appear condescending to your customers. Information about the wines should be given to the service personnel so that they may provide descriptions to customers if requested.

HOUSE WINES

House wines reflect the wine professional's ingenuity in ensuring quick sales and turnover. They should meet the following guidelines.

- Red, white, and (if possible) rosé should be offered.
- They should be sold by the glass and (if possible) also in carafes of various sizes—litre, half, and quarter.
- They must complement a wide range of menu items.
- In most cases, the house wines are the lowest-priced (not necessarily lowest-profit) wines on the list, so they should offer good value.

SETTING WINE PRICES

Wine is the only item in a restaurant for which there is a comparable retail price. Therefore, particular care needs to be taken in setting wine prices. Many restaurants have poor wine sales because their wines are overpriced.

Several methods can be used to set prices. Each method has its merits. The wine professional's choice of pricing sytem will depend on the restaurant's style.

Markup on Wine

In the examples that follow, loss from overpouring, spillage, and breakage is predicted and added to the pricing calculation, at a rate of 3% of the cost price.

Straight Markup

A simple markup calculated on a percentage of the price is a common system, especially for those wines with a lower price. Some restaurants will mark wines up as much as 200% (i.e., three times the cost price). Markups over 200% can be justified only for limited-release, rare, and (in some instances) single-vineyard wines. Lower markups, say below 100% (two times the cost price), help to increase sales.

Example of a Straight Markup

Cost (C)	Loss (L)	Markup (%)	Markup (M) (C + L) × %	List Price (C + L) + M
$10.00	$0.30	100%	$10.30	$20.60
$15.00	$0.45	100%	$15.45	$30.90
$50.00	$1.50	100%	$51.50	$103.00

Sliding-scale Markup

A sliding-scale markup means the percentage is reduced as the cost price of the wine increases. The markup is lower for the more expensive wines, but the dollar profit remains higher. This method appeals to the discerning customer.

Example of a Sliding-scale Markup

Cost (C)	Loss (L)	Markup (%)	Markup (M) (C + L) × %	List Price (C + L) + M
$10.00	$0.30	125%	$12.88	$23.18
$15.00	$0.45	100%	$15.45	$30.90
$50.00	$1.50	80%	$41.20	$92.70

Cost-plus Markup

In the cost-plus markup method, each wine is marked up by a certain percentage, then a fixed dollar amount is added. This allows for a higher-priced wine to be sold at a more reasonable price.

Example of a Cost-plus Markup, $5 fixed

Cost (C)	Loss (L)	Markup (%)	Markup (M)	List Price
			(C + L) × % + (F)	(C + L) + M
$10.00	$0.30	50% + $5	$10.15	$20.45
$15.00	$0.45	50% + $5	$12.73	$28.18
$50.00	$1.50	50% + $5	$30.75	$82.25

Flat Markup

With the flat markup method, a fixed dollar amount is added to the cost of the wine. The relatively low markup on higher-priced wines will attract the discerning customer and encourage repeat business.

Example of a Flat Markup

Cost (C)	Loss (L)	Markup	List Price
			(C + L) + M
$10.00	$0.30	$10.00	$20.30
$15.00	$0.45	$10.00	$25.45
$50.00	$1.50	$10.00	$61.50

Wines by the Glass

To determine the selling price for a glass of wine (e.g., house wine), consider the size of the average portion, and calculate its price. Then add on enough to cover a 2%–3% loss for overpouring, spillage, and waste. Finally, establish a price using one of the various markup methods.

Example of Pricing Wine by the Glass

For a single portion (180 mL) from a bottle (1000 mL) costing $12.00, with 3% loss:
(1000 ÷ 180 × $12.00) = $2.16
$2.16 + $0.06 = $2.22
Using a straight markup method,

Cost (C)	Markup		List Price
$2.22	200%	$4.44	$6.66

Although most restaurants serve 180 mL (six ounces) per glass, many are moving to smaller portions. This can create more sales and encourage moderation. The size of the glass should be clearly indicated on the wine list.

The recommended portion for aperitifs and fortified wines by the glass is 60 mL–90 mL (two to three ounces).

MARKETING IN THE RESTAURANT

Promoting or merchandising wine in the restaurant heightens the customers' awareness of wines in general and (possibly) of certain wines in particular. It can be done in many ways, including the following:

- Display wines near the main entrance and throughout the restaurant. Wine racks, wine bottles, wine paraphernalia, and promotional posters are effective in communicating the restaurant's focus on wine.
- Use a specialized theme to direct the merchandising. For example, the restaurant could promote new wines; wines from a specific country, region, or winery; wines from a particular winemaker.
- Offer inclusive dinner packages that pair specific wines to certain menu items. These tend to generate an interest in new wines and increase overall sales.
- Include wine agents and distributors as resources to be used in planning special promotions. Reps are always willing to educate restaurant staff about their particular products, if it means greater sales.

Other ideas that help promote wine are the following:

- tent cards—to promote theme dinners or dinner packages, or highlight wine reviews from newspapers.
- wine book—though once considered old fashioned, a book of wine labels (for the wines in the wine list) does help customers in their choices. Present it with the wine list.
- menu inserts—can feature specific wines. Provide all important information (i.e., vintage, full name of the wine, name of the producer or shipper, country and region, bottle size, and price). Promoting particular wines at their peak will not only increase sales but also turn over inventory, ensuring that wines do not languish in the cellar.

In addition to being the main source of up-to-date wine information for the service staff, you also will be expected to provide the best example of how to sell and serve wines.

Staff Training

General wine training should deal with:

- sales techniques (e.g., how to sell up, and how to sell second bottles).
- presenting, opening, and serving wines.
- the correct glasses to use (if there is a variety).
- proper serving temperatures.
- correct pronunciation.
- recent price changes.

Educating staff about wine involves:

- conducting tastings on a regular basis. Each tasting should consist of no fewer than three wines.
- conducting specialized seminars. These could focus on wine categories, grape varieties, regional differences, characteristics of recent vintages, and traditional (or innovative) food and wine combinations.
- making sure that the staff keep notes. They should learn to describe the colour, nose, body, taste, character, and food matches of the wines on your list.

As their understanding of food and wine grows, your staff will become adept at selling wine with confidence and flair. As a wine professional, you must encourage your staff by providing useful information, corrective criticism, and incentives. The most obvious incentive to increasing wine sales comes to them in the form of tips.

Example of Tips From Increased Wine Sales

This example is based on a 15% tip on a \$20 wine sale (\$3), assuming that the server works five shifts per week for fifty weeks in the year. Obviously, as the price of a wine sale goes up, so does the tip.

Extra Bottles Sold Per Shift	Extra Annual Tip Income
1 × (\$3 × 5 × 50)	\$750
2 × (\$3 × 5 × 50)	\$1500
5 × (\$3 × 5 × 50)	\$3750

Attitude

All staff selling wine must maintain a professional attitude that allows the wine to take centre stage. Of course, the wine professional must set the example. Basic guidelines are as follows.

- Speak clearly and directly, so customers can understand you, even when the restaurant is busy and noisy.
- Remain confident, friendly, agreeable, and diplomatic at all times. Address customers by their formal titles and refrain from initiating meaningless conversation and familiarities. Even regular customers must be formally addressed, unless they suggest otherwise. Gossip, coarse language, and the discussion of personal problems is inappropriate.
- Think before speaking, and choose words carefully.
- Be reliable and punctual.
- Maintain good health. If suffering from a virus or cold, do not serve food or beverages. Muffle coughing or sneezing with a handkerchief, turn away from the customers' view, and wash your hands immediately.
- Do not smoke while at work. This interferes with any wine tasting required, and the smell will offend customers and colleagues alike.
- Do not drink, except to taste wine as required.
- Maintain good posture and alertness. Do not lean against walls or furniture, and do not sit with customers.
- Dress for the job. In a traditional dining room, this usually means a white shirt or blouse, black pants or skirt, black socks or stockings, and polished black shoes. A black bow tie is optional.
- Maintain cleanliness and grooming. Hands must be kept clean and manicured at all times; long hair should be tied back, and short hair should be trimmed regularly. Heavy colognes and perfumes are unacceptable. The job of serving usually demands long, hard hours and constant rushing around, so you are likely to perspire. For this reason, spare clothing that is fresh and pressed as well as an extra pair of polished shoes should be available.

Tools of the Trade

Serving food and wine requires that certain tools be kept on hand at all times. Avoid sharing them with others.

Basic tools include:

- clean, white napkins to wipe the bottle lip after pouring
- candles (for decanting wines with sediment) and matches
- two reliable pens and a writing pad to take food orders
- a professional-quality corkscrew

Traditionally, a tastevin was used to taste wine, but today this is more a symbol of the position rather than a tool of the trade. A wine glass is more efficient.

Corkscrews

Wine professionals generally carry two corkscrews for everyday use: a single-lever corkscrew (waiter's friend), or an Ah-So corkscrew. These are suitable for most corks.

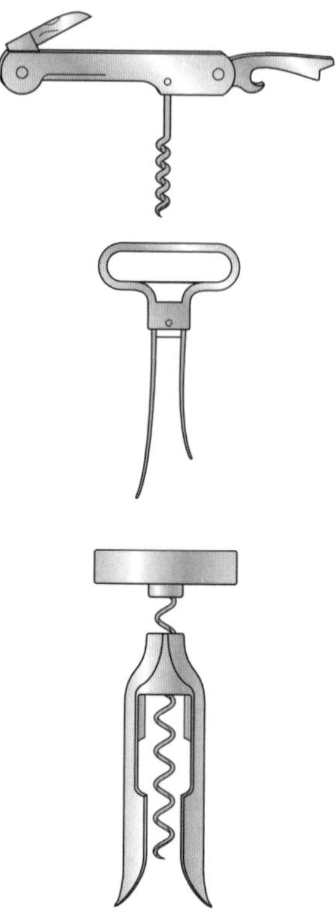

- The single-lever style is widely used. It can be carried in a pocket, is relatively easy to master, and has a knife for cutting capsules. When purchasing a single-lever corkscrew, make sure it has a sharp tip. Its helix (screw) should be at least 5 cm long (five twists) to completely penetrate long corks without crumbling them.
- Some wine professionals prefer the Ah-So, which has two flat metal prongs to insert along each side of the cork. Once it grabs the cork you simply twist upward.
- a screwpull corkscrew, useful for poor quality or deteriorated corks. The screwpull has two plastic pegs that sit on the bottle's neck and direct the long, teflon-coated helix down the centre of the cork. It is quiet, easy to use, and an excellent device for stubborn corks. It is, however, fragile and bulky.

Other types of corkscrew (T-bar, butterfly, reverse action) are not useful to the wine professional—they are clumsy, detracting from an atmosphere of relaxation and good taste.

Other Professional Tools

- The slipped-cork extractor helps remove a broken cork that has sunk into the bottle. It has a long wire with a hooked end. To use it, insert it into the bottle neck, catch the cork with the hook and gently pull the wire to bring the cork out of the bottle.
- A can of non-stick cooking spray may be kept on hand to coat the helix of a corkscrew. This keeps it from squeaking when pulling waxed corks.

WINE SERVICE

Selling and serving wine is the focus of the wine professional's role. In this area, education and management of staff can increase standards and revenues. And, as the wine professional, you are expected to provide the best example of how to sell and serve wine.

ARRIVING EARLY

The job extends beyond regular serving hours. You should arrive early to ensure that:

- all wine glasses are spotless. Smudges and dirt must be rinsed from each glass in tepid water and without detergent. A clean, pressed linen cloth should be used to wipe them.
- menus and wine lists are clean and up-to-date. Find out what is temporarily unavailable, and pass this information along to staff and customers.

FROM RESERVATION TO MEAL'S END

The following procedure outlines the professional method of selling and serving wine—from the first phone reservation to the end of the customer's stay. Individual restaurants will designate responsibilities in different ways, but it is the job of the wine professional to ensure that every task is carried out professionally, regardless of who performs it.

Before the Meal

- Selling wine begins with the phone reservation. If a reservation is made for a party of six to eight, ask whether it is a special occasion, and if the host would like a special wine, or bottle of Champagne, ready at the table.
- When customers arrive, greet them, confirm their reservation, guide them to a table, and present them with a wine list and menus.
- Give customers a few moments to settle themselves. Then proceed to the table, and stand to the right of the host to offer aperitifs. An aperitif stimulates the appetite and helps the customers to relax.
- When conversing with customers, use affirmative language and assume that everyone wants wine. For example, "Good evening, ladies and gentlemen. May I take your aperitif order?" If a customer refuses an aperitif, reply, "Perhaps you would like some wine with your meal. I'll leave you to consult the wine list. I am at your service if you need any assistance." Passive language such as "Would you like something from the bar?" increases the likelihood of refusal.
- Aperitifs, like wine, should never be filled to the brim of the glass. If a customer requests a specific brand name, take the bottle to the table and stand to the right of the customer to display the label and then pour. Correct procedure impresses those who know their wines and spirits.
- While serving the aperitifs, use the time to familiarize yourself with the customers. An understanding of your customers enables you to serve them better and to increase wine sales. Because appearances can be deceiving, do not make assumptions based on their attire. Dialogue is a much better indicator. For example, customers talking formally among themselves may be having a business meal, while those who are relaxed are likely close friends or family members. By understanding the situation, you will be better able to suggest an appropriate wine—Champagne for special occasions, excellent vintages for important business meetings, casual wines for casual dining.
- Once the food order is received, take the wine order. Direct your attention to the host and be attentive to her or his wishes.

Recommend wines if guidance is requested. Point out any new wines recently added to the wine cellar, or wines currently being promoted. Always make sure that the price of any recommended wine reflects the price range of the meal ordered; never suggest the most expensive wine. Consider all the dishes chosen, and provide suggestions that will create harmony on the palate throughout the entire meal. As well, explain the qualities of the wines you recommend, mentioning the nose, body, and flavour.

- When the host has chosen, confirm the selection.

  You may praise the choice, for example, "Goat cheese and Sauvignon Blanc make a wonderful match," or "This Cabernet is perfect with the rack of lamb."

  If the host makes an unconventional selection (e.g., a Sauvignon Blanc with roast beef) take the order without comment. The sale of the wine has been accomplished and the customer may prefer white wine.

  If the selection is unfortunate (e.g., a dessert wine with the main course), it is your job to intervene, explaining that there may be a more suitable wine. This must be done in a positive and respectful manner—both in words and actions. For example, "Yes, that icewine is excellent. But it is probably too sweet for a cheese soufflé. May I make a suggestion?" If the host agrees, make the suggestion and explain how it will complement the food choices—for example, "We have a wonderful Chardonnay from Niagara. It is so delicate it will not overpower the souffle, yet its fresh aroma and flavour have a natural affinity to Cheddar." If the host insists on the original choice, simply fetch the wine and serve it with grace.

Selling Up

Once the wine order has been placed, the wine professional may then try to "sell up," i.e., encourage customers to do any of the following.

- *Buy two types of wine if a variety of dishes are being ordered.*

  For example, if the host's choice complements the main course, but the first course is unaccompanied, or if some customers have chosen a fish course while others have selected red meat.

- *Buy a more expensive wine, that may be a few dollars more but offers better value.*

  You might say, "Instead of the house wine, you may be interested in trying the Riesling Spätlese. Although it's a bit more expensive, it is wonderful with the pork."

- *Order two bottles of a wine if the party consists of more than four guests.*

  Ask the host if a second unopened bottle of the same should be held in reserve.

During the Meal

The serving of the wine is the restaurant's opportunity to demonstrate its knowledge and skills. It is also the stage at which customers will quickly lose confidence if the wine is served in a sloppy manner.

Wine Glasses

Wine glasses should be clear glass (without any engraving or decoration), the proper size, and have a stem long enough so that fingers need not touch the bowl. They should narrow slightly at the rim, as this shape helps retain the bouquet to be "nosed" or sniffed.

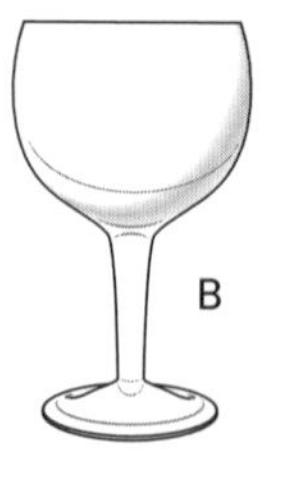

- The best all-round glass (A) has a tulip shape. It can be used for still wine—red, white, and rosé.
- The classic red wine glass (B) has a larger bowl than the tulip glass. More surface area allows more of the wine's bouquet to quickly reach your nose.
- The classic white wine glass (C) is slimmer and smaller in size. This minimizes the likelihood of the wine warming up too quickly.

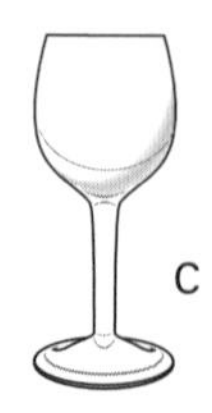

- The traditional glass used for fortified wine (D) is smaller than the others, and quite narrow from bowl to rim.
- Sparkling wine should be served in tall fluted glass (E), not the saucer-shaped coupe glass. The coupe allows the wine to warm up and the bubbles to disappear too quickly.

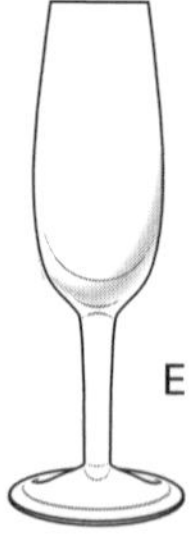

SERVING TEMPERATURES

For all the flavours of a wine to be enjoyed, it must be served at a certain temperature. The ideal serving temperatures are as follows:

4°C–7°C	inexpensive sparkling wines
6°C–8°C	vintage and finer Champagnes; rosés; fino Sherries; light, fruity whites
9°C–11°C	complex, dry whites; white Ports; dry Madeiras; naturally sweet wines
10°C–12°C	complex, dry whites; medium-sweet wines; young, fruity reds
11°C–13°C	top-growth, sweet whites
14°C–16°C	simple reds
16°C–18°C	mature reds; tawny and wood Ports; oloroso Sherries; medium-sweet Madeiras.
17°C–19°C	vintage Ports; old, sweet Madeiras

Opening Wine

For specific advice on serving white wines, decanting red wines, and serving sparkling wines, see the following pages.

- Present the unopened bottle to the customer, at a 45° angle to display the label. When serving and holding a wine bottle, your hand should never be placed over the label. If the label is extremely dusty, use a napkin to clean it off. (The bottle can be left dusty, as this adds to its aged appearance.)
- Read the label aloud to confirm the name of the wine and the vintage. Only after the customer has approved the wine, can you proceed to open the bottle.
- If you are right-handed, place a service napkin over your left arm. Place the bottle on the table or service trolley. The bottle should always be opened in full view of the customer.
- To open the wine using the lever corkscrew, hold the bottle by its neck. With the blade, cut the capsule along the underside of the bottle's lip. While cutting, avoid moving the bottle as this may disturb any sediment. Remove and discard the top of the capsule.

The capsule may be made of metal, wax, or plastic.

- Wipe the bottle top and the cork with a clean napkin. Pull out the helix of the corkscrew. Wrap the folded napkin around the neck of the bottle and hold it with your left hand.
- Insert the point of the helix into the centre of the cork and give the corkscrew a clockwise turn. Make sure that you turn (rather than push) the corkscrew throughout the four or five twists required. Stop once the helix reaches the bottom of the cork—not below it.
- Hook the lever onto the bottle's lip and hold it there with your left forefinger over the service napkin. Lift the cork out in a firm, slow, and silent motion until three-quarters of the cork is removed. Grasp the cork with your right hand and gently, evenly, quietly, wiggle it from the bottle.

  If some or all of the cork is pushed into the wine, take the bottle to the kitchen and remove the cork with a cork extractor. If the cork cannot be removed from a bottle of white wine, the bottle must be replaced. Reds, on the other hand, can be decanted (see next page).
- Untwist the cork and discreetly sniff it, to make sure the wine is healthy. Set the cork on a side plate next to the host for inspection.

  Corked wine will smell musty or vinegary, and must not be served. In this situation, excuse yourself, take the corked wine to the kitchen, and replace the wine.

Pouring Wine

- Use a napkin to wipe the inside and outside of the bottle's mouth, to remove any bits of cork or dried wine. Again, fold the napkin over your arm.
- Pour a tiny amount of wine into the host's glass. Wait for the host to taste and approve the wine.
- If all is well, serve the wine. Pour first for the customer to the right of the host, till the glass is half full. After pouring, twist your wrist slightly to prevent drips.
- Continue in a counterclockwise direction. Wipe the bottle's lip with the napkin as you pass behind each customer before pouring again.
- Once everyone is served, place white wine bottles into an ice bucket and reds on a coaster to the right of the host, or on the service trolley. Do not wrap a napkin around the bottle, as this deprives the customers of its appearance.
- Before the first bottle is empty, ask the host if another is needed. Remove empty bottles immediately. If the host requests a different wine, bring new glasses.

In some establishments, the sommelier will taste the wine first, to make sure that it is sound.

Placing the wine to the the host's right lets the host refill the glasses. It is, however, an integral part of the wine professional's job to remain alert when customers need their glasses refilled, or wish to place additional wine orders.

Serving White Wine

- Fill an ice bucket with a combination of ice and water. Place the bottle in the bucket, and the bucket to the right of the host.
- When the wine has reached the desired temperature, open the bottle as explained previously, but do so with the bottle in the bucket.
- When serving, make sure water does not drip from the wet bottle—wipe it with a napkin.
- When you are finished serving, put the bottle into the bucket with a clean, dry napkin across. Make sure the bucket is in a safe place away from passersby.
- When the bottle is empty, and no other white wine is to be served, remove the wine bucket.

Decanting Red Wine

Decanting (pouring wine from its bottle into a decanter) is necessary when a young wine needs to be aerated, an old vintage has deposited some sediment, or cork parts have slipped into the bottle. Decanting is usually done over a candle, to illuminate the sediment. Some people use a decanting funnel and a piece of muslin cloth, but many experts feel that muslin strips the wine of its character.

- When using a candle, hold the wine bottle in one hand and the decanter in the other. Position the bottle well above and just in front of the flame: do not let the candle heat the wine!
- Tilt the decanter slightly. Steadily pour the wine into the decanter until the sediment reaches the neck of the bottle.
- Discard the bottle, and serve from the decanter.

Serving Sparkling Wines

Sparkling wines do not need to be tasted to assess their quality. Nosing the wine is sufficient.

- Set the sparkling wine bottle in an ice bucket filled with ice and water, to the right of the host. Drape a clean napkin over the bottle.
- Pull the bottle from the ice bucket and dry it with the napkin. Wrap the napkin around the neck to absorb any possible overflow of foam.
- Cut the foil below the wire. Untwist the wire and remove the wire cage and the foil. Hold the bottle at an angle, pointing the cork away from yourself, and away from the customers.
- Firmly hold the cork with your thumb on top, and twist the bottle with your other hand. The pressure inside the bottle pushes the cork out. Hold the cork firmly to ease it out slowly. Avoid making a popping sound.
- Fill the flutes slowly, or wine will overflow. Once everyone is served, place the bottle back in the ice bucket and drape it with a napkin.

THE ORDER OF SERVING WINES

There is a natural order in which wines should be served. When choosing wine to taste on its own, or to complement a meal, follow these guidelines.

Serve:

- chilled wine before room-temperature wine.
- dry wine before sweet wine.
- light-bodied wine before medium- or full-bodied wine.
- simple wine before complex wine.
- young wine before an older vintage.

HOW TO READ A WINE LABEL

BOTTLE SIZES

Miniature (20 mL–50 mL) A sampler size. Used mainly for spirits in hotel minibars. One Canadian winery bottles icewine in the 50 mL size.

Nip or **Quarter** (187 mL) One-quarter of the standard 750 mL bottle. Often seen as the single-serving portion on airlines and in hotel minibars.

Split or **Half-Bottle** (375 mL) One-half of the standard bottle.

Half-Litre or **50** (500 mL) Most often used for ordinary wine of no special provenance. Some producers have promoted its use with quality wines with little success. The term 50 refers to centilitres, the common European measurement.

Standard Bottle (750 mL) The generally accepted format for most wines.

Litre (1000 mL) The usual format for vermouths and ordinary wines.

Magnum (1500 mL) A double bottle.

Marie-Jeanne (2000 mL) Almost equal to three bottles.

Tappit Hen (2000 mL) Almost equal to three bottles. This name is used by Port producers. Tappit is an old Scots term, meaning "crested," and refers to a drinking vessel with a knob on the lid.

Double Magnum (3000 mL) A double double, i.e., equals four bottles.

Jeroboam (3000 mL) The double magnum of the Burgundy and Champagne regions has this name. It is equal to four standard bottles. See also below.

Dame-Jeanne (About 4000 mL) Equals just over five bottles.

Jeroboam (4500 mL) The common name for the six-bottle size, as it applies to the Bordeaux-shaped bottle.

Rehoboam (5000 mL–6000 mL) Not seen very often.

Imperial (6000 mL) The equivalent of eight bottles in the Bordeaux shape. Used for Cabernet Sauvignon and Merlot throughout the world.

Methuselah/Mathusalem (6000 mL) Same size as Imperial, it has the Burgundian sloped-shoulders format. Used in Burgundy and Champagne, and for Chardonnays and Pinot Noirs from countries other than France.

Salmanazar (9000 mL) The equivalent of a whole case of twelve bottles.

Balthazar (12 000 mL) Equals sixteen bottles.

Nebuchadnezar (15 000 mL) Equals twenty bottles.

Sovereign (About 26 000 mL) Equals about thirty-five bottles.

*W*INE AND FOOD

*T*he ideal combination of wine and food occurs when each enhances the other. The wine may be good, the food fresh and well prepared, but if the two complement each other, the combined effect becomes splendid and far more interesting. To disregard the dish when choosing the wine, or vice versa, is to do a disservice to the food, to the wine, and especially to the consumer. A poor wine and food combination is detrimental to both. For example, a fruity, young Beaujolais, if paired with delicious, fresh oysters will undoubtedly result in a disagreeable metallic taste on the palate, ruining the wine-tasting experience as well as any possible enjoyment of the food. High levels of iodine in the oysters clash with even minor amounts of tannin in the red wine. On the other hand, a tart Muscadet or Chablis, without any tannin, pairs remarkably well—the wine's acids perfectly offset the sea-saltiness in the oysters.

A deeper understanding of what works and what does not can be gained only with experience and a thorough study of basic wine and food compatibilities. The research has already been done over the centuries. It is up to us not to ignore it.

Of course, combining wine and food is as subjective as enjoying art, music, and literature. But trendy twinnings and avante-garde alternatives can lead to culinary clashes and digestive disasters.

The duty of the wine professional is to guide and to advise customers to make informed and wise choices relative to income and taste preferences. Naturally, if customers choose to disregard the advice, that is their choice. After all, it is their money and their palate. The customer is always right.

BASIC RULES

Here are some basic rules that will help to make a safe and compatible choice.

- Red wine with red meat, white wine with white meat or lean fish. No matter what anyone tells you, this is still the "Golden Rule." There are exceptions, but they simply underscore the importance of this first and foremost rule.
- Strong sauces and strongly flavoured dishes demand full-flavoured, full-bodied wines.
- Delicate foods and light dishes are easily overwhelmed by strong tastes. Suggest light and fruity wines.
- If a dish is a regional speciality, a wine from the region is likely the most compatible choice.
- Sweet wines go especially well with rich patés and certain fresh fruits.
- Sparkling wines are the perfect aperitif. They also go well at the end of a meal, if they are sweet.
- When serving more than one wine at a meal, proceed in order: white before red, dry before sweet, light before heavy, simple before complex, younger before older.
- Avoid heavy, alcoholic wines at lunch or during hot summer days.
- Bananas, citrus fruits, pickles, and artichokes clash with most wines.
- Curry powder and very spicy food annihilate the flavour of wine. Suggest beer.
- Coffee and chocolate tend to distort the taste of most wines. The exception is a sweet, fruit wine like Cassis or Framboise.
- Salads and any dishes containing vinegar dull the delicate balance of most wines. Suggest mineral water with salads.
- Smoked fish or meat, asparagus, and eggs are difficult to match with wine.

WINE COMPATIBILITIES

Certain wines always go better with some particular dishes. To list the best wine to match with each dish would take more space than is available here. Some classic pairings are given on the following page. See also Chapter 4, Grape Varieties.

Dry White Wine
As an aperitif, or with shellfish, seafood, charcuterie, white meats, soft cheeses.

Sweet White Wine
As an aperitif, or with foie gras, rich patés, most desserts, rich fish dishes, strong cheeses.

Dry Rosé Wine
Shellfish, seafood, fried fish, fish in sauce, charcuterie, white meats, soft cheeses.

Sweet Rosé Wine
Desserts, soft cheeses, charcuteries.

Light Red Wine
Fatty fish, white and red meats, game birds, soft cheeses.

Full-bodied Red Wine
Red meat, furred and feathered game, medium-strong cheeses.

Sparkling Wine
As an aperitif, or with white meats. If it is sweet, with desserts. Sparkling wine tends to go well throughout a meal.

Dry Fortified Wine
As an aperitif, or with consommés, smoked meats.

Sweet Fortified Wine
As an aperitif, or with rich patés, strong cheeses, desserts, nuts.

FOOD COOKED WITH WINE

Just as garlic, salt, and pepper add flavours and aromas to most dishes, wine can do similar magic by adding an extra dimension to the culinary art. In the kitchen, wine can be used as a tenderizing ingredient, for colouring certain foods, to add flavour, and as a natural preserving agent. It is part of the composition of a number of classic sauces and dishes. It can be used for basting, glazing, and deglazing. Chefs use wine as they would use herbs and spices, i.e., according to the taste and aromatic characteristics of the intended food creation.

Wine used in cooking is heated to a temperature that will cause the alcohol to evaporate, leaving only its flavours in the dish. "Raw" wine is rarely added to a dish, except in the preparation of certain cold sauces, coulis, or fruit syrups.

PART III

WINE REGIONS AND WINES

The chapters contained in Part III examine key wine-producing regions of the world. With the exception of chapters treating the distinct regions of Champagne, Port, and Sherry, each chapter deals with one country. Of course, the number of countries that produce wine is far too great to be covered in this manual. The chapters in Part III, therefore, focus on the wine-producing countries that are most significant—in terms of both quality and quantity.

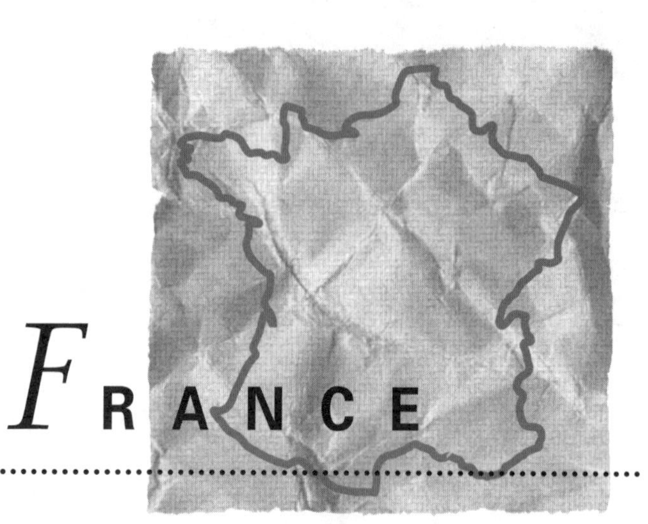

FRANCE

For centuries, France has been considered the heart of the wine world. The dedication of the French people to food and wine, as well as the situation and history of the country, has created many of the wines we prize today. France is second to Italy in annual production (54 600 000 hL) and consumption (65 L per capita) of wine. Approximately two-thirds of France's production is red wine. The country's one million hectares under vine include some of the most influential regions of wine production: Bordeaux, Burgundy, Rhône, Loire. (The unique region of Champagne will be treated in the next chapter.) French wines continue to be the benchmarks for other countries.

blanc (blah*N*)—white
château (shah-TOE)—wine estate, usually controlled by a house or castle located on the estate
clos (cloh)—vineyard enclosed (or at one time enclosed) by a wall
commune (kuh-MEWN)—parish or township
côte (coat)—slope or hillside
cru (crew)—growth of a vineyard; also a classification of a particular terroir
département (day-par-te-MAH*N*)—one of the main administrative divisions
domaine (doe-MEN)—estate
grains nobles (gre*N* NUB-luh)—botrytis-affected grapes
pays (pay-EE)—country
perlant (pair-LAH*N*)—very lightly sparkling
pétillant (pay-tee-AH*N*)—slightly sparkling
rouge (roozh)—red
rosé (ro-ZAY)—pink
tonneau (tuh-NO)—barrel
vin mousseux (vahn moo-SOOH)—sparkling wines other than Champagne

GEOGRAPHY

France covers 543 970 km^2, located mostly between 51°N and 42°N. The country has a diversity of terrain and climate very suitable for grape growing. Most areas of France make wine: three-quarters of the country's ninety-five administrative départements grow grapes.

The land is irrigated by a multitude of rivers. About a third of the country is mountainous, with the Massif Central (a great mountainous plateau) in the south central area.

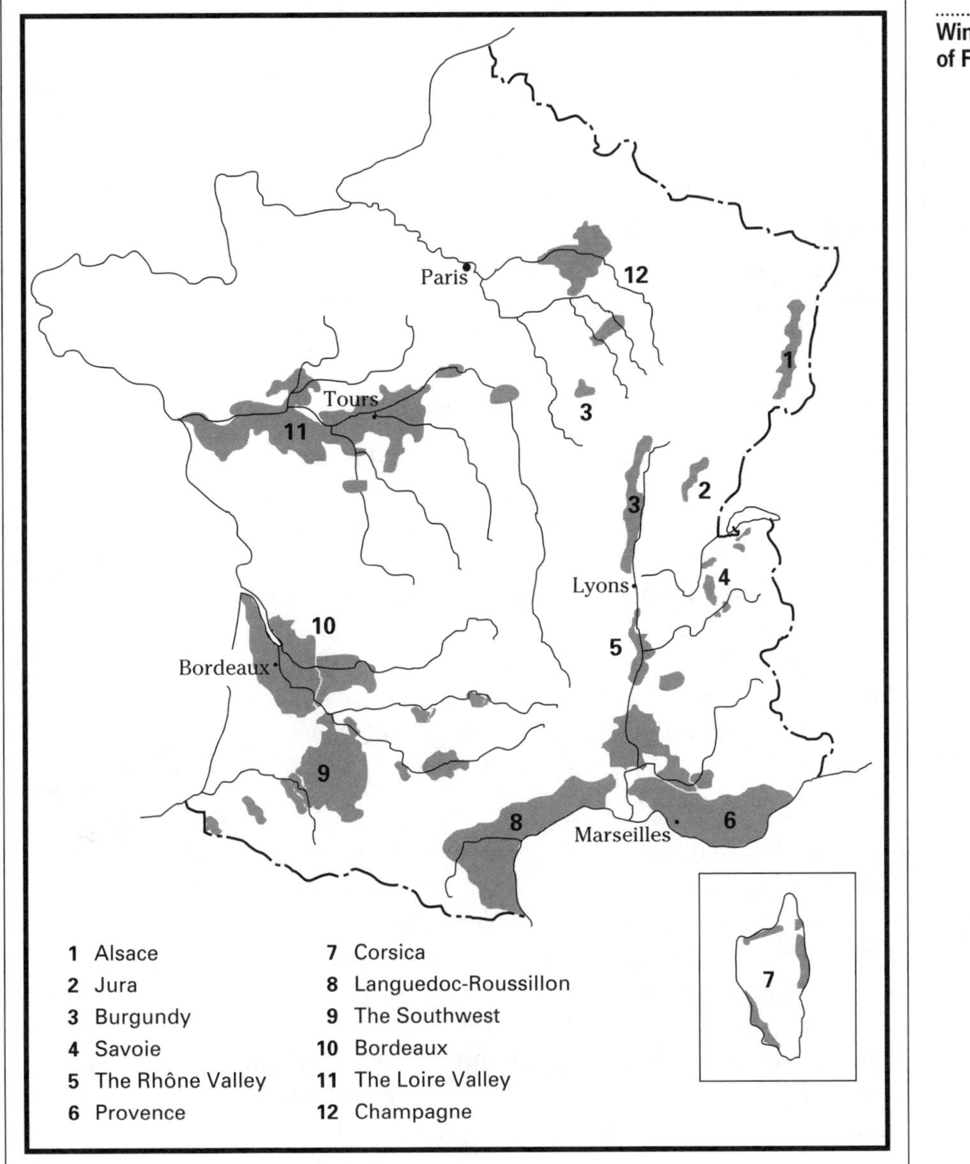

Wine Regions of France

Macroclimate refers to the regional climate; mesoclimate refers to a much smaller area, like a particular vineyard; microclimate refers to very restricted areas under and around the vine canopy.

France has three major macroclimates:

- the western, Atlantic coastal climate, with its mild winters and warm summers.
- the southern climate along the Mediterranean Sea, with its mild winters and hot summers.
- the eastern, continental climate, with its cold winters and warm summers.

Within each macroclimate are numerous mesoclimates that, along with soil conditions, determine what grapes are grown and with what results. From their experience of grape growing and winemaking over the centuries, the French have developed a sophisticated understanding of terroir.

Terroir is the total environment of a vineyard, including climate, soil, and surface features of the terrain.

HISTORY

Viticulture in what is now France began long ago. In the seventh century B.C., when the Phocaeans (Greeks from Asia Minor) laid the foundations of Massalia (Marseilles) on the Mediterranean coast, they found native vines growing among olive groves—the local Ligurian tribes already knew how to make wine. The Greeks (and with them, their wine) made incursions as far north as Vienne on the Rhône River.

Centuries later, the conquering Romans found that the Gauls also made wine. However, the local tribes favoured ale and mead, because these were easily made. Wine was probably reserved for the few, particularly chieftains. During the Gallo-Roman era (50 B.C. to A.D. 400) the Gauls became competent grape growers, and some of their wine was praised in Rome. The Gauls also invented the barrel, a more practical container than the cumbersome amphora. Gallo-Roman boats travelled the Garonne, Loire, Rhône, Saône, and Moselle rivers, loaded with wine barrels for the thirsty Roman soldiers.

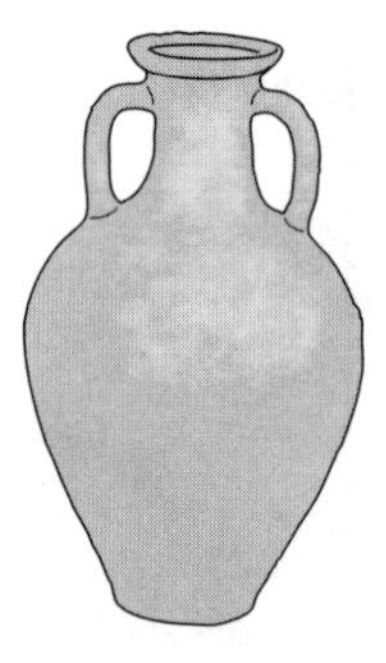

amphora

During the fifth century, successive waves of barbarians (Visigoths, Franks, Burgundii, etc.) from the east assailed the Roman Empire, destroying the Gallo-Roman culture and most of the vineyards. The Christian church became the guardian of grape growing and winemaking, when Clovis I became king (481–511 A.D.) of the Franks—monasteries and some cathedrals had their own vineyards. During the eighth century, Charlemagne (c. 742–814) encouraged the Church to plant new vineyards in Burgundy and along the Rhine.

Royalty continued to foster the development of the wine trade. King Louis IX (1226–1270), was a cooper in his spare time, and kept a record of the wines he tasted. (If he liked the wine, he was given a barrel of it.) His favourable assessment substantially increased the price of the wine—and the worth of the vineyard.

A cooper makes and repairs wooden barrels.

During the Middle Ages (476–1500), vines could be found in most regions of what is now France, and French wines were exported to England, the Baltic, and areas in what is now Belgium and Germany.

The eighteenth century brought great developments in the wine trade: glass bottles and cork became more available. This made it easier to age wines for longer periods, and store and label them with their source. Some vineyards became famous for wines that aged well and fetched high prices. Bordeaux's wines were classified into six categories, and two-thirds (200 000 tonneaux) of its wines were exported to England, Holland, the Scandinavian countries, and the French colonies. Winemaking became a lucrative business and vineyards expanded. In 1778, approximately 1 600 000 ha were under vine; by 1829, this had grown to over 2 000 000 ha.

In the nineteenth century, however, the industry suffered from downy mildew, powdery mildew, and the arrival of phylloxera. Eventually, most vineyards were replanted with hybrids, or with French vines grafted onto resistant American rootstocks. Meanwhile, artificial wines were made to fill the void. (Artificial wines were made without fresh grapes, and contained artificial flavouring and colouring.) In attempts to stop such fraudulent practices, the first definition of wine was written into French law, in 1889. According to the legal definition, wine is "the result of full or partial fermentation of fresh grapes or the juice of fresh grapes." In the same decade (in 1885), Bordeaux developed a growth classification (cru classé) system to protect its wine trade.

The twentieth century brought more upheavals: grape growers revolted in 1907 and 1911 when sales were low and prices collapsed; production fell during World War I (1914–1918) when labour was unavailable; production in the 1930s was so high it created surpluses. In 1935, the government stepped in to stabilize the market, and began the regulation of yields, plantings, and other aspects of wine production and sale that it still controls today.

LAW AND REGULATIONS

Since 1935, French wines have been controlled by a code of law known as Code de Legislation des Appellation d'Origine Contrôlée (AOC), complemented by various other laws, such as Vins Délimités de Qualité Supérieure, Vin de Pays, and more recently, Vin de Table.

The objective of the AOC laws is twofold:

- to geographically define wine-producing areas, so that the origin of a wine may be clearly and accurately stated.
- to establish production standards for each defined wine area, thus guaranteeing the origin and quality of a wine.

These laws and regulations conform to the requirements of the European Union (EU).

Regulatory Bodies

From the planting of the grapes to the sale of the wine, the French wine trade is subject to a series of controls carried out by four organizations:

- INAO (Institut National des Appellations d'Origine)—the National Institute of Appellation of Origin.
- ONIVINS (Office National Interprofessionel des Vins)—the National Interprofessional Wine Office.
- DGCCRF (Direction Generale de la Concurrence, de la Consommation et de la Repression des Fraudes)—the Department of Competition, Consumption, and Unfair Trade.
- DGI (Direction Generale des Impots)—the Tax Department.

Both the INAO and ONIVINS are part of the Ministry of Agriculture. The DGCCRF and DGI are part of the Ministry of Economic Affairs and Finance.

Production

All permits relating to vine growing and wine production are issued by the Ministry of Agriculture, with the assistance of the INAO and ONIVINS. The DGI checks the planting, grape varieties, harvest, wine stocks, and vinification. ONIVINS controls the movement of plants and propagation material. DGCCRF conducts analysis at the production and marketing stages to determine that the wine is genuine. Prosecutions are stiff and inflexible.

Movement of Wines

The French government controls the transportation of wines, both to collect taxes and to certify that the wines are authentic. Tax must be paid before the wine is sold. Bottled wines must be labelled and sealed for transportation with a revenue stamp printed on the seal issued by the DGI. Wines transported in bulk must have accompanying DGI documents stating the identity of the wine, consignor, consignee, carrier, date and time of departure, and latest date and time for delivery. Each merchant is required to keep a complete register of inward and outward movements, a processing register, and a bottling register. Registers are compared with stocks by officers of the DGI and the DGCCRF.

A consignor delivers goods to a consignee, who accepts the goods.

Labelling

As a member of the EU, France abides by the EU regulations regarding labelling, and goes further. The anti-fraud service (DGCCRF) monitors the legality of wine labels.

Wine Classification

In its regulation of wines and their production, France uses the two categories of wine designated by the EU:

- table wine (vin de table).
- quality wine (Vin de Qualité Produits dans une Région Déterminée or VQPRD, that is, quality wine from a specific region or area).

Within each of these two EU categories, France recognizes two subcategories:

Table Wine

- Vin de table—table wine.
- Vin de pays—country wine.

Quality Wine

- Vin Délimités de Qualité Supérieure (VDQS) —demarcated wine of superior quality.
- Appellation d'Origine Controlée (AOC) —wine from demarcated and regulated places or areas.

FRENCH WINE PRODUCTION	
vin de table	40%
vin de pays	14.5%
VDQS	1%
AOC	30%
for distillation	14.5%

Vins de table are considered the lowest rank, vins de pays of a higher rank, VDQS of an even higher rank, and AOC wines of the highest rank. These subcategories and their rankings are the key to understanding how French wines are produced and presented.

History of French Wine Law

- The general legislation of 1905 (codified into law in 1908, revised in 1911, 1919, 1927, and 1935) regulated the approximately 400 AOCs in existence.
- In 1936, Châteauneuf-du-Pape (in the Rhône), Cassis (in Provence), and Arbois (in Jura) were the first to receive the AOC.
- French law has recognized the term *vin de pays* since February 8, 1930.
- The term *Vin Délimités de Qualité Supérieure* appeared for the first time in a Tax Bill in 1945 and was recognized in a law on December 18, 1949.
- Since 1979, ONIVINS has been entirely responsible for regulating vin de pays and vin de table.

Vin de Table

Vin de table must meet the following requirements and must stipulate on the label:

- the volume the container is to hold.
- the alcohol content, which must fall within the range 8.5%–15% by volume.
- the company name of the bottler, and the address.
- for wine produced and made in France, the words *vin de table de France* or *vin de table Français* if the wine is to be sold in France, and *Product of France* in all EU languages if the wine is to be exported.
- for wine made from a blend of EU wines, the words *mélange de vins de different pays de l'Union Européenne.*
- for wine made in France from imported grapes, the words *vin obtenu en France à partir de raisins récoltés en...*

Blending with wines imported from countries outside the EU is not allowed.

A wine label may also give a brand name, a description of the character of the wine, and recommendations for drinking. *Vin de table* labels show neither a vintage nor a region of origin.

By blending from a range of sources, makers of vin de table can produce wines of consistent quality year after year, and in a style that suits their markets.

Vin de Pays

Vin de pays must meet the following requirements:

- the natural alcohol content must be a minimum of 10% by volume in the Mediterranean regions, and 9%–9.5% in the other regions.
- under analysis, the wine must have appropriate levels of volatile acidity, sulphur dioxide, and certain additives.

- vin de pays must be made from approved grapes that have not exceeded the maximum yield; using approved cultivation and vinification methods; and from grapes originating from a delimited area.
- in taste tests by an organization appointed by the French Ministry of Agriculture, vin de pays must show typical character and satisfactory colour, aroma, and taste. Also, the wine must satisfy ONIVINS regional representatives.

On the label, vin de pays must show:

- the words *vin de pays* followed by the name of the place the wine is produced.
- the container's volume, the wine's alcohol content by volume, and the bottler's name and address.

The label may also give the name of the producer, name and address of the owner (excluding the words *château* or *clos*), the types of grapes, brand name, vintage, and so on.

Vin de pays is further categorized as:

- regional—vin de pays bearing the name of an area covering several départements. There are three areas that may label their vins de pays in this way.
- départemental—vin de pays bearing the name of the one département in which it is produced (e.g., Savoie). Around eighty-eight départements may label their vins de pays in this way.
- zonal—vin de pays bearing the name of a distinct production area within the département. There are ninety-eight production areas that may label their vins de pays in this way, of which most are smaller than the département.

VDQS Wine

These wines can be of very good quality and value; they aspire to the higher AOC level. The INAO strictly regulates and monitors VDQS wine production. The conditions under which VDQS wine must be produced are based on local customs and defined in law.

VDQS laws stipulate the area of production, grape varieties, minimum alcohol content, maximum yield per hectare, methods of cultivation, vinification methods, analysis, and tasting by an official committee of experts. If a wine meets the standards of VDQS wines, the committee grants a stamp which shows the letters *VDQS*, an inspection number, and (in the centre) a hand holding a glass.

AOC Wine

These are the aristocracy of French wines. Designation as an AOC wine does not simply indicate that a wine comes from one of the 400 AOCs in France. AOC wine is guaranteed to be of high quality, and to be produced in a way that preserves local traditions and emphasizes the uniqueness of the region or terroir. For a wine to be considered an AOC wine, it must meet not only the requirements of the VDQS designation, but also abide by the even stricter rules that govern such issues as pruning, chaptalization, the use of fertilizers, irrigation, stocking, ageing, etc. Each aspect of the wine's production must be carefully documented in the approved manner.
The INAO co-operates with local grape growers and wine producers to constantly revise and improve AOC standards.

WINE REGIONS

ALSACE

Located in the northeast of France along the German border, Alsace is one of the most beautiful wine regions of the country. The 12 900 ha of vineyards are situated on the sheltered east-facing slopes of the Vosges Mountains.

GEOGRAPHY

Alsace's vineyards are found in a narrow band (2 km–5 km wide) that runs 120 km from Wissembourg, past Strasbourg, and south to Thann. The continental climate gives the region mild springs, hot dry summers, long warm autumns, and cold winters. Usually the winter snow is sufficient to protect the vines, and frost damage is unusual. With the protection of the Vosges Mountains to the west and the Black Forest to the east, the vineyards benefit from warmer temperatures than those in surrounding areas. Alsace is sunny and dry, with an annual precipitation of only 500 mm.

Touring Alsace

Blessed with a wonderful climate, this picturesque area offers pine-clad mountains crested with fairy-tale châteaux and steep-roofed villages. The gastronomy of Alsace is renowned: pâté de foie gras and choucroute garnie are its two admitted masterpieces. Two- and three-star Michelin Guide restaurants string the roads that wind through the vineyards. The area produces some of the best French white wines, well known for their dry, fruity character.

The soil is a very complex mixture of granite, gneiss, sand, limestone, marl, chalk, and sandstone. The best growths are situated on limestone slopes between altitudes of 200 m–400 m.

HISTORY

Although there were no extensive vineyards, patches of vines were cultivated around local tribal villages early in the region's history. Other vines were brought by Norse traders via the Danube and the Rhine.

With the Roman conquest, vineyards expanded, viticultural methods improved, and commerce developed. During the Gallo-Roman period, the Ill River was an important route for the wine trade. Wines were highly valued—the law of the time stipulated a double fine for killing a grape grower, and cutting off one hand for stealing grapes.

When the Roman Empire fell in 476 A.D., the Franks annexed the area, converted it to Christianity, and built churches and monasteries. This was an important factor in the proliferation of vineyards: 108 villages produced wine in 800 A.D., 160 in 900 A.D., and 430 in 1400 A.D. During the Middle Ages, the local wines (known as vins d'Aussay) were famous. Considered among the best of Europe, and sold at high prices, they were exported as far as the Elbe, Switzerland, and the Adriatic.

Over the centuries, the territory of Alsace was often disputed. During some of these periods of strife, Alsace produced very ordinary wines to blend with German wines. In the midst of all this, Alsace suffered the pests (eudemis, cochylis, phylloxera) and diseases (downy and powdery mildews) that the rest of Europe did. Viticulture in Alsace only started its recovery in 1946, when vineyards began systematically to replant premium grapes. In 1962, Alsace was granted AOC status.

LAW AND REGULATIONS

According to regulations that outline the requirements for Alsace AOC wines, all such wines must be bottled in Alsace in a tall narrow flute bottle. They are labelled *Alsace* or *Vin d'Alsace* and known by their grape variety (except in the case of Edelzwicker, which is a blend of white grapes).

The following are additional appellations:

- Grand Crus—Riesling, Muscat, Gewürztraminer, and Pinot Gris wines from certain vineyards with high reputations. This appellation was created in 1975.
- Vendange Tardive (VT)—late-harvest wines made from Gewürztraminer, Riesling, Muscat, or Pinot Gris grapes with naturally high sugar levels. Depending on vinification, these wines can be dry, but most are sweet.
- Sélection de Grains Nobles (SGN)—wines made from individually selected grapes affected by noble rot. These are rich, luscious sweet wines.
- Crémant d'Alsace—a white or rosé sparkling wine made by the méthode Champenoise.

GRAPE VARIETIES

Many of the grape varieties grown in Alsace are those grown in Germany, thanks not only to a shared history but also to a similar climate in some areas.

VINEYARD PLANTINGS	
Riesling	23%
Gewürztraminer	19%
Tokay Pinot Gris	6%
Pinot Blanc	20%
Sylvaner	18%
Pinot Noir	7%
Chasselas	3.5%

- Riesling—this variety can produce some of the best dry wines of the region. Its crisp acidity helps the wine age. Riesling wines have 150 or so aromatic substances. Young wines may have aromas of lemon and flowers. Aromas of gasoline can develop from older vines grown on limestone soils; lime, pineapple, and quince develop in older wines; and great vintages have a hint of honey.
- Gewürztraminer—this thrives in Alsace; nowhere else can this grape flaunt most of its 500 aromatic substances. When the vine is planted in limestone soil it has aromas of flowers, and can have heavy aromas of pineapple, faded roses, litchi, frankincense, and geranium. With relatively low acidity, Gewürztraminer wines have little ageing potential unless made into Vendange Tardive or Sélection de Grains Nobles.
- Tokay Pinot Gris—this grape produces wines with aromas of violets and lemons in their youth, and buckwheat honey when aged. A great Tokay Pinot Gris can be mistaken for a light Gewürztraminer, but a slight bitterness on the finish helps the good taster discern the difference. This variety is related to Gewürztraminer—and has similar low acidity—but is not related to Furmint, which makes Hungary's famous Tokaji.

- Muscat—this late-maturing and finicky grape is said to be the mother of all pure viniferas, but is being planted less in Alsace than in previous years. Of the two most-planted Muscats, the white Muscat d'Alsace (or Muscat Blanc à Petits Grains, as it is known) produces the better wine, and the Muscat Ottonel (a clone obtained from Moreau-Robert at Angers in 1852) is easier to grow. The wines are mostly dry, semisweet, or sweet, and have a rich nose of rosewood and flowers, as well as apricot jam and orange-blossom honey.
- Pinot Blanc—also known as Klevner, this grape makes supple wines that are soft, well-balanced, and easy to drink. Regulations allow wine made from Auxerrois (a Pinot Blanc clone), and blends of Auxerrois and Pinot Blanc wines, to be labelled as Pinot Blanc. Both vines are not very demanding but capable of producing well under mediocre conditions. Pinot Blanc is used extensively to make the sparkling wine Crémant d'Alsace.
- Sylvaner—this grape is most popular in the northern département of Bas-Rhin. When planted on a good site, it can develop wonderful qualities. The wine is refreshing, and at its best when it is young. Crisp acidity and fruity character make it a favourite aperitif and lunch wine.
- Pinot Noir—this is the only red grape authorized by the Alsace AOC. When the vines are severely pruned and the yields low, it makes a wine with a raspberry nose, intense colour, and character. It is also vinified as a rosé, such as in Crémant d'Alsace rosé.
- Chasselas—also known as Flambeau d'Alsace, this grape is not classified by the AOC. It is rarely bottled as a varietal but is exported for blends with Müller-Thurgau and another variety called Knipperlé.

WINE PRODUCTION

Approximately 8000 growers cultivate vines in the region. The vineyards are not large: 62% of them cover less than one hectare. Nearly a third of the growers sell their grapes, and another third belong to co-operatives that make wine. The average annual wine production in Alsace is approximately 1 600 000 hL.

Because of Alsace's constant good climate, vintages are predictable and often excellent. The wines are good value and among the most reliable of French dry white wines.

URA

Located in the northeast of France, about 60 km from the Swiss border, the wine region of Jura is found in the province of Franche-Comté ("Free County"). It has 1500 ha of vineyards. Jura is the birthplace of Louis Pasteur. Here he spent most of his childhood and later conducted his experiments on yeast and bacteria.

GEOGRAPHY

On a plateau near the Jura Mountains and at altitudes of 250 m–500 m, the vineyards of Jura form a north-south strip (80 km long, 6 km wide) that includes the towns of Arbois and Lons-le-Saunier. The climate is continental, with very cold, snowy winters; warm, sunny, wet summers; and long, warm, dry autumns. This mountainous area has many forests, rivers, and streams. Its soil is primarily limestone and clay, with white and yellow marl, as well as chalk, for the topsoil.

HISTORY

As part of the province of Franche-Comté, Jura shares much of its history with Burgundy, which lies to the east. Passed from hand to hand like merchandise through marriage and inheritance, this rich and strategic region has had many nationalities: Germanic, Burgundian, Austrian, Spanish, and French. The Spanish lost the region to France in 1678 under the Treaty of Nimégue.

GRAPE VARIETIES

In addition to some Pinot Noir and Chardonnay brought from the region of Burgundy, there are varieties unique to Jura: Trousseau and Poulsard for red and rosé wines, and Savagnin (related to Traminer) for white wines.

The following are appellations of Jura:

- Arbois—the best-known of Jura's AOC wines. Most of these wines are red, but whites and deep rosés (called locally *vin gris*) are also made. There are five types of wine authorized as AOC Arbois: red and rosé wines from Poulsard and Trousseau grapes; white wines from Chardonnay and Savagnin grapes; vin jaune ("yellow wine); vin de paille ("straw wine"); and sparkling wines made using the méthode Champenoise.

MAKING VIN JAUNE

The origin of vin jaune lies in 120 years of Spanish rule. Vin jaune has a nose similar to that of fino Sherry, with a touch of oxidation, a tangy taste, and great length. After the wine has finished fermenting, it is placed into 228-L barrels and kept for a minimum of six years without racking or topping up. During this time, a film of yeast develops on the wine's surface. Because the temperature in the cellars rises and falls between 8°C and 18°C, this yeast sinks to the bottom of the barrel then rises again (in summer) to grow. About 40% of the wine is lost through evaporation. Each barrel is tested twice a year to protect the wine from developing excess volatile acidity or sickness.

- Arbois Pupillin—from the commune of Pupillin, these wines are richer and have more character than Arbois.
- Côtes du Jura—the general AOC for red, rosé, white, yellow, straw, and sparkling wines from this region.
- Macvin—a mistelle (mixture of grape juice and alcohol) rather than a wine. A mixture of one-third marc and two-thirds grape juice is aged. The AOC was granted in 1991. Macvin is drunk as an aperitif, like Pineau des Charentes of the Cognac region.
- Vin de paille—made from late-harvest grapes that are either placed on straw mats or hung to dry for three months after the harvest. The grapes shrivel, losing about two-thirds of their volume, and then are crushed and fermented. This rare, expensive wine is extremely sweet with a nutty taste.
- L'Étoile—a small appellation that produces white, yellow, and straw wines, but also a well-known sparkling wine made from Chardonnay. The table wines have a slightly oxidized nose.

Mistelle and Vin de Liqueur (VDL) are fortified wines made by mutage, i.e., by fortifying the grape juice during or before fermentation. Mutage arrests the fermentation and retains the natural sugars. Mistelle and VDL have alcohol levels between 15% and 22%, and a minimum sugar level of 170 g/L.

Burgundy

Known as Bourgogne in French, Burgundy is a province in the east of France and includes the départements of Côte d'Or, Saône-et-Loire, Yonne, and Nièvre. It annually produces 3 000 000 hL of wine, and includes some of the best wine regions of France. Because the area includes a wide range of mesoclimates and soils, there is no one characteristic wine. There are hundreds of AOCs in Burgundy, and many of its wines are unsurpassed globally.

GEOGRAPHY

To the southeast of Paris, Burgundy includes the main centres of Dijon, Beaune, Chalon-sur-Saône, and Mâcon. With the exception of Chablis (located in the Yonne département), the vineyards of Burgundy cover 19 600 ha and stretch for 180 km from Dijon in the north to Highway 47 southeast of Lyons.

The climate is mostly cool and damp, with occasional hot summers. Winter frost can be severe, and spring hailstorms make news around the world. When the weather is co-operative (a warm June and September and a hot summer sprinkled with light rains), the vintage can be magnificent. A cold September can stop the area's famous grape, Pinot Noir, from reaching maturity. Of course, weather factors vary from place to place; for this reason, Burgundians refer to each vineyard as a *climat*.

As with climate, the soil of Burgundy varies tremendously. Thus it will be discussed separately for each subregion.

HISTORY

After the Roman conquest in 51 B.C., this rich agricultural region prospered. When the Roman Empire collapsed, the area was overtaken by Germanic tribes known as the Burgundii. During the fourth century, the Burgundians converted to Christianity. Various churches and abbeys (Cluny, Cîteau, Clairvaux) of the region were centres of learning in the Middle Ages.

Burgundy's golden age began in the fourteenth century under Philip the Bold (1342–1404), Duke of Burgundy and fourth son of the king of France. Life was magnificent at the court of Burgundy, with lavish banquets and sumptuous wine. The duchy was one of the richest and most powerful in all France, encompassing the Netherlands, Artois, Picardie, Franche-Comté, and Luxembourg. The nobility bequeathed lands and vineyards to the Church, some of which (Clos de Vougeot, Clos de Tart) are still famous today. Cistercian monks became renowned grape growers who turned poor land into remarkable vineyards. In the fifteenth century, the duchy became part of France.

The French Revolution (1789–1802) brought an end to the high life. Properties were destroyed. Then the Franco-Prussian War (1870–1871) played havoc with the vineyards, especially in the battles of Nuits and Dijon. World War I (1914–1918) did not damage the region's vineyards, but did drastically reduce the labour available to tend them. The vineyards suffered more damage during World War II (1939–1945). Today, Burgundy is prospering anew, spoiled by a demand that exceeds its production levels.

GRAPE VARIETIES

The grapes used in Burgundy are Chardonnay, Aligoté, Pinot Blanc, Melon de Bourgogne (Muscadet), and Sacy for white wines. Pinot Noir, Gamay, Tressot, and César are used for the reds.

WINE PRODUCTION

Most of Burgundy's wines are dry red or dry white, but the region does produce some rosés and sparkling wines. Burgundy has a hierarchy of appellations, ranging from the wide and lowest-ranking regional appellations, to very specific source and highest-ranking appellations.

Regional and Generic Appellations

- Bourgogne Grand Ordinaire—the lowest AOC. The wines are red, white, and rosé made anywhere in the region. All the grapes mentioned above are permitted.
- Bourgogne Passe-tout-grains—an AOC applying to wines made from two-thirds Gamay and one-third Pinot Noir.
- Bourgogne Aligoté—AOC wine made from Aligoté grapes.

- Bourgogne—AOC red or rosé wine made from Pinot Noir, (or Gamay if they are produced in Beaujolais). White Bourgogne is made from Chardonnay or Pinot Blanc.

Subregional Appellations

- Red and white wines from specific areas of Burgundy (e.g., Bourgogne Irancy, Bourgogne Haute Côtes de Beaune, Bourgogne Côtes de Nuits).

Village Appellations

- Red or white wines from specific designated communes or villages (e.g., Volnay, Chassagne-Montrachet, and Gevrey-Chambertin).

Premier Crus

- Wines from specific vineyards with quality soil and sun exposure. These wines are identified by the commune, then the vineyard (e.g., Chambolle-Musigny, Ile des Vergelesses; or Savigny-lès-Beaune, Bas Marconnets).

Grands Crus

- Wines from specific vineyards with excellent soil and high reputations. These are identified by the name of the vineyard alone and each constitutes a separate AOC (e.g., Chambertin, Clos de Vougeot, Corton-Charlemagne).

If an AOC and village do not have the same boundaries, the wine is declassified to the subregional or regional appellation.

Chablis

Including the most northerly of Burgundy's vineyards, this area is located near the town of Auxerre in the Yonne département. The 4047 ha of vineyards centre on the small town of Chablis, which is only 24 km from the fringe of the Champagne area, and 104 km from the Côtes d'Or. The continental climate (with cold winters and warm summers) gives Chablis its particular crisp style. This area is on the edge of the Paris basin, with its characteristic soil conditions: the soil is poor and chalky with a mixture of clay and fossilized small oyster shells.

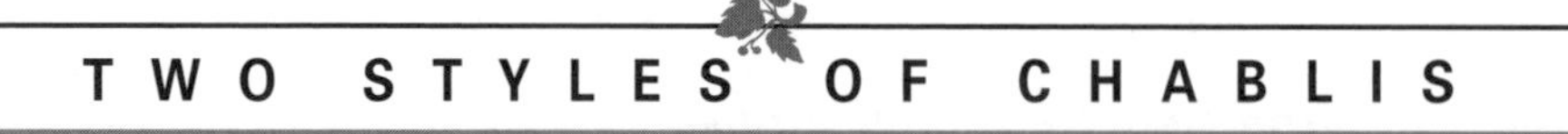

There are two styles of Chablis. The first is traditionally vinified and matured in small oak casks (feuillettes). The winemaker believes the oak adds complexity and character to the wine. The other style is made in stainless steel or concrete vats. In this case, the winemaker believes oak distorts the authentic Chablis character.

The area's wines that receive the Chablis AOC must be made from Chardonnay grapes. There are four classifications of AOC Chablis wines:

- Petit Chablis—lively, affordable quaffing wines that tend to be acidic and should be drunk young.
- Chablis—wines from the majority of Chablis' vineyards. They are generally thin, crisp, and refreshing, with an elegant, fruity nose.
- Chablis Premier Cru—wines from specific vineyards that, in some cases, produce wine as fine as Grand Cru but with less concentration in character (e.g., Mont de Milieu, Côte de Léchet, Vauligneau, Berdiot).
- Chablis Grand Cru—wines from seven specific vineyards (les Vaudésir, Preuses, Clos, Grenouilles, Bougros, Valmur, Blanchots) totalling only 100 ha, and located on the chalky southwestern slopes of a hill just outside of the town of Chablis. These wines are first class, with a deep golden colour and a greenish tinge.

In addition to the Chardonnay wines, the area of Chablis also produces white wines made from Sauvignon and Aligoté. Sauvignon de St-Bris is an example with a VDQS appellation.

Sparkling wines made by the méthode Champenoise are entitled to the AOC Crémant de Bourgogne.

Côte d'Or

The patchwork of soils produces some of the best Pinot Noir and Chardonnay. This area (11 000 ha) of vineyards produces 400 000 hL of red and white wines.

The basis for Côte d'Or is an escarpment: most of the vineyards are found on its slope, facing southeast. Here, with a soil of mainly limestone with a thin layer of marl, are the grand cru vineyards. Lower parts of the area have limestone-clay soils, poorer drainage, and greater vulnerability to frost.

Côte d'Or is subdivided into two areas: Côte de Nuits (from south of Dijon to Corgoloin), and Côte de Beaune (from Ladoix to Cheilly-lès-Maranges).

Côte de Nuits

The narrow strip of vineyards (with several hundred growers) runs roughly north to south for 19 km and covers 1600 ha. The main villages are Marsannay-la-Côte, Fixin, Gevrey-Chambertin, Morey-St-Denis, Chambolle-Musigny, Vougeot, Vosne-Romanée, and Nuits-St-Georges.

The area produces (almost exclusively) deep red wines from the Pinot Noir grape. The wines can vary considerably in quality from grower to grower, but the great Pinot Noirs have an expansive bouquet, depth, and a wonderful strawberry-jam nose. The wines tend to be bigger, more austere, and slower to develop than those from the Côte de Beaune.

The AOC wines include:

- Côte de Nuit-Villages—wines from the vineyards of Fixin, Brochon, Premeaux-Prissey, Comblanchien, and Corgoloin.
- Hautes Côte de Nuits—wines made from vineyards at the top of the hills and on the west-facing side. The reds are more rustic and the whites are mostly vinified by co-operatives.

Côte de Beaune

The position of the vineyards and the composition of the soils are slightly different than for the Côte de Nuits. The climate is slightly warmer. The slopes are not as steep, and are composed of pebbly clay with chalky, red soil that is rich in iron oxide, marl, and chalk. The main villages are Ladoix-Serrigny, Pernand-Vergelesses, Aloxe-Corton, Chorey-les-Beaune, Savigny-lès-Beaune, Beaune, Pommard, Volnay, Monthélie, St-Romain, Auxey-Duresses, Meursault, Puligny-Montrachet, St-Aubin, Chassagne-Montrachet, Santenay, and Maranges.

Red wines (especially Pinot Noirs) dominate the north part of Côte de Beaune—with the exception of the great white Corton-Charlemagne. Further south (in the area of Meursault, Auxey-Duresses, Puligny-Montrachet, and Chassagne-Montrachet), great Chardonnays are produced, each with its own character. Meursault is rich and buttery; the Montrachets are leaner but racy, with a touch of smokiness. Corton-Charlemagne is the elegant balance between the two. Pinot Noirs from the Côte de Beaune tend to be lighter in colour and have a more exuberant aroma than those of Côte de Nuits. Also, they mature faster and do not age as well.

The AOC wines include:

- Côte de Beaune-Villages—red wines, either separately or in a blend, from lesser-known villages.
- Hautes Côtes de Beaune—lesser red and white wines produced by regional co-operatives from the top of the hills and the west-facing slopes of the Côte de Beaune.

Côte Chalonnaise

This subregion is a southern extension of the Côte d'Or and includes approximately 1000 ha of vineyards. It takes its name from the nearby town of Chalon-sur-Saône. The soils are similar to those of Côte d'Or, with outcrops of limestone and marl. Some hillsides with sun exposure and favourable soil have their own village and premier cru designations.

The entire AOC produces an annual average production of 50 000 hL. Three-quarters of this subregion produces reds, with a majority of Pinot Noir, plus some Gamay used in the making of the Passe-tout-grains. The whites are made from Chardonnay and Aligoté.

Passe-tout-grains are made with one-third Pinot Noir (minimum) and two-thirds Gamay.

The wines from this region are an affordable alternative to the more expensive wines of the Côte d'Or. The five appellations for this area are Bourgogne Aligoté-Bouzeron, Rully, Mercurey, Givry, and Montagny. The Mercurey in particular can show the true strawberry-jam character of the Pinot Noir. The Aligoté from Bouzeron is a good wine and has the rare ability to moderately improve in the bottle. A significant amount of Crémant de Bourgogne is also made.

The Mâconnais

South of the Côte Chalonnaise is the Mâconnais wine area. It covers an area of 4500 ha and produces an annual average of 250 000 hL of wine, mostly white. The warmer, sunny climate soaks the undulating vineyards. Summers can be hot, but winters are still cold. In the south, the soil is richer in marl, and has a mixture of clay and chalk. This is where the Chardonnay grape distinguishes itself.

The most famous white wine of this subregion is the AOC Pouilly-Fuissé, made in a district around the villages of Pouilly and Fuissé. This dry, fragrant wine is golden in colour, with greenish tints. It can age for ten years—although most is drunk while young.

Villages have their own AOC appellations (e.g., St-Vérand AOC, Mâcon-Lugny AOC, Mâcon-Viré AOC). The nearby villages of Loché and Vinzelles have a small but reputable production of pleasant white wines named Pouilly-Loché and Pouilly-Vinzelles, respectively. The village of St-Vérand is also commendable for its wines, as are Lugny and Viré. (These last wines are very good value even if they do not have the Pouilly-Fuissé reputation.) Mâcon-Village AOC wines are always white and should be drunk young. Red and rosé Mâcon wines are made from Gamay and Pinot Noir grapes. In addition, Crémant de Bourgogne is made in the area.

Beaujolais

Located in the département of the Rhône, south of the Mâconnais, the vineyards of Beaujolais cover an area 60 km long and 12 km wide. The 22 000 ha produce an average of 1 300 000 hL of wine, mainly red from the Gamay Noir à Jus Blanc. There are over sixty communes making wines. The climate is warm and fairly dry; the soil is granite with outcrops of clay, volcanic rocks, and limestone in the southern part. The hills reach elevations of 1000 m and shelter the vineyards, which lie at about 500 m.

Most of Beaujolais' wines undergo a form of carbonic maceration, which is called maceration Beaujolaise traditionnelle. This technique has the drawback of not allowing the wine to age gracefully.

Some Beaujolais is vinified using the conventional method and aged in vats before bottling. Of the crus, Fleurie, Juliénas, Morgon, and Moulin-à-Vent may be aged for five years. In great vintages they may resemble Pinot Noir.

The area's AOCs include the following:

- Beaujolais—the basic appellation, for wines which are best drunk when young.
- Beaujolais-Villages—from the northern vineyards, the better wines produced on granitic soil.
- Beaujolais Supérieur—wine with alcohol reaching naturally 10% for reds and 10.5% for whites.
- Beaujolais Crus—the wines of ten specific villages in the northern portion of Beaujolais. The vineyards are St-Amour, Juliénas, Chénas, Moulin-à-Vent, Fleurie, Chiroubles, Morgon, Régnié, Brouilly, and Côte de Brouilly. The soil of these vineyards (granite mixed with schist and porphyry) yields richer, full-bodied wines, each with its own character and the potential to age well.

BEAUJOLAIS NOUVEAU

This is wine that is best when very fresh and young, hence the custom of marketing it immediately after its vinification. Beaujolais Nouveau cannot be released for sale until the third Thursday in November after a harvest. It is a rather smart marketing scheme.

- Coteaux du Lyonnais—south of Beaujolais and to the west of the town of Lyon, this large area produces about 13 000 hL of AOC fruity, red wines, from Gamay grapes.

SAVOIE

Sometimes known as Savoy, this region has 1500 ha of vineyards in the French Alpine region near Switzerland.

Savoie became French (by a popular free vote) in 1860.

GEOGRAPHY

The area lies east of the Rhône River, south of Lake Geneva, and north of the Isère River. Nearby is Chambéry, famous for its Vermouth. As one would expect of an alpine area, the climate is generally cold. The vineyards lie in the valleys at altitudes between 200 m–500 m, on rocky soil with outcrops of limestone and clay. The soil of the vineyards near Lake Geneva is thick clay on chalk.

GRAPE VARIETIES

Most of the wines are white—from Altesse (Roussette), Aligoté, Chasselas (Swiss Fendant), Jacquere, Petite Sainte Marie (Chardonnay), Molette, Gringet, and Bergeron grapes. The rosé and red wines are made from Mondeuse, Gamay, Persan, Joubertin, and Pinot Noir.

WINE PRODUCTION

The wines are crisp and fruity and should be drunk young. The appellations for Savoie include:

- Vin de Savoie—the broad AOC for red, rosé, white, and sparkling wines from the region.
- Vin de Savoie Cru—awarded in 1973 to eighteen communes (Marin, Marignan, Ripaille, Ayze, Frangy, Charpignat, Marestel, Chautagne, Jongieux, Abymes, Apremont, Arbin, Chignin/Chignin-Bergeron, Cruet, Montmélian, St Jean-de-la-Porte, Ste Jeoire-Prieuré, and Ste Marie d'Alloix).
- Roussette de Savoie—white wine made from Roussette blended with Chardonnay and Mondeuse, from vineyards around Frangy and Cruet. The four communes of Frangy, Marestel, Monterminod, and Monthoux are classified as *Roussette de Savoie crus*.
- Seyssel—exclusively white wines from the area of Seyssel, of which the majority are sparkling.
- Crépy—wines (table or sparkling) made in the area of Crépy from the Chasselas grape.

RHÔNE VALLEY

Flowing from Lake Geneva, joining with the Saône at Lyons, and running south to the Mediterranean Sea, the Rhône is one of France's most important wine rivers.

For 200 km (from Vienne in the north to Avignon in the south), the Rhône valley vineyards cover 58 000 ha. On average, they produce 3 000 000 hL of wine annually, most of which is red.

GEOGRAPHY

The Rhône Valley region is divided into two subregions: the southern Rhône from Avignon to Montélimar, and the northern part from Valence to Vienne.

The climate and soil both vary significantly along the river, and so affect where vines are planted and the character of the resulting wines. Most of the vineyards are found along the Rhône; a few outlying vineyards (including some on the banks of the Drôme River, a tributary of the Rhône) are considered part of the region.

Northern Rhône

In the northern Rhône region, the winters are mild, the summers hot, and the autumns warm and long. There is considerable rain in the spring. The vines are pruned and staked to resist the mistral, a cold, strong, northerly wind.

Most vineyards are planted on the west side of the river. These steep east-facing slopes are composed of granite and schist, with topsoil that varies from clay to sand. The base of the east bank is granite, with a thin topsoil of loose layers of flint and chalk, which are sometimes dislodged by heavy rain. Many of the vineyards are planted on terraces, called *chalais*.

Southern Rhône

The southern Rhône has a flatter landscape. The climate is Mediterranean with very hot long summers, long sunny autumns, and dry warm winters. The springs are short with torrential rains. On the west bank of the river, the soil is composed of sand, chalky sandstone, and pebbles. The wider east bank, around Châteauneuf-du-Pape, has varied soil of chalk, sand, and sandstone, covered by a thick layer of large round pebbles and small rocks, locally called *galets*.

Châteauneuf-du-Pape

Built in the twelfth century, the fortress that gives the appellation its name was presented by Frederick Barbarossa to the bishops of Avignon. One of these bishops, Jacques Duèse, became Pope John XXII and established a papal vineyard in Châteauneuf-du-Pape ("Pope's new castle"). Years later, the château suffered devastation, and was bombed in World War II.

The baron Le Roy de Boiseaumarié, owner of the Châteauneuf-du-Pape Château Fortia, was instrumental in formulating regulations that became the system of Appellation Controlées. He proposed the delimitation of good vineyards, selection of grapes, method of pruning, as well as the quantity and quality of the wines from Châteauneuf-du-Pape. Today, the AOC of Châteauneuf-du-Pape includes the vineyards of the village of the same name and some neighbouring vineyards.

HISTORY

Grape vines were first brought to the area by the Greeks. The vines thrived under the local Allobroges tribe, who had the wisdom to plant them on the land sloping to the river. The grape which the writer Pliny (A.D. c. 23–79) called *Allobrogica* produced a powerful, scented wine that sold for high prices and rivalled the best of Rome. Pitch or burnt resin was sometimes added to the wine; the result was named *vinum picatum*. Scholars have variously proposed that Pliny's Allobrogica is the grape known today as Petite Syrah; or the ancestor of the Pinot Noir; or the Mondeuse, a fruity, red grape known today as Grosse Syrah.

Roman veterans settled on the hills around Narbonne where they planted the south Rhône valley with grapes and olive trees. It is believed that the Syrah grape was imported from Shiraz in Persia by Roman soldiers.

In the wine history of the region, the next important era was 1305–1377, when the papal court moved to Avignon. Vineyards expanded to serve the clergy with wine for Mass and festivities.

GRAPE VARIETIES

A multitude of grapes are used in the Rhône valley. They contribute to the unique complexity of the wines and permit the winemaker to exercise some creativity. Consequently, a Châteauneuf-du-Pape with thirteen authorized grapes will produce a wine that varies with each winemaker. Of the varieties planted in the region, Syrah, Cinsault (or Cinsaut), Grenache, Viognier, and Marsanne are the best known.

Of the white varieties, the following are grown:

Northern and Southern Rhône

- Marsanne—produces wine with power and longevity in the north, but rather burly wines in the south, where it is used for blending.
- Roussanne—the most important grape for good white Hermitage, very aromatic with scents of honey, flowers, coffee, and nuts. It has a low yield and high susceptibility to diseases. As a result, it is slowly disappearing in the north.

Northern Rhône

- Viognier—produces exotic wine with scents of tropical fruits, honeysuckle, peaches, and apricots.

Southern Rhône

- Bourboulenc—provides body and a scent of roses.
- Clairette Blanche—provides fruitiness, deep colour, and high alcohol.
- Grenache Blanc—provides high yields that are low in acidity and high in sugar.
- Picardan—used for blends.
- Picpoul—produces a rather thin wine used for blends and Vermouth.

Of the red varieties, the following are grown:

Northern and Southern Rhône

- Syrah—the most famous red grape of the Rhône, providing backbone, deep colour, and longevity. Its complex, peppery nose, with aromas of hickory wood, dried fruits, and black currants, is at its best in the Côte Rôtie, Hermitage, and Cornas.

Southern Rhône

- Cinsault—with a generous yield, produces fruity wines low in tannin.
- Counoise—an underestimated grape, grown in small quantities. It has great finesse and provides deep, rich, fruity flavours with scents of red berries and smoked meat.
- Grenache—loves heat and dominates the south. Prolific and rich in sugar, it produces a fruity wine that varies in quality depending on the yield.
- Mourvèdre—a late-ripening grape, high in colour, with good acidity. It is used mainly to add complexity to the bouquet, with scents of leather, truffle, mushrooms, and humus.
- Muscardin—provides great structure, perfume, and alcohol to the wine.
- Terret Noir—grown in very small quantities, it gives high acidity, which cuts the richness of other grapes.
- Vaccarèse—used to add complexity to the nose with scents of tobacco, tar, licorice, and bell peppers.

The average annual production of the Côtes du Rhône region is 3 000 000 hL of red, rosé, and white wines. The greatest characteristics of the Rhône wines are their strength, deep colour, distinctive bouquet, and ability to age. The best, such as Côte Rôtie and Hermitage, require cellar ageing to show their full potential. Others, such as Tavel and Vacqueyras, disclose their qualities at an earlier stage.

The Rhône valley region has many appellations, including the following:

- Côtes du Rhône—a general and broad appellation for wines made anywhere inside the north and south delimitation of the Rhône AOC.
- Côtes du Rhône-Villages—comprised of sixteen villages. These wines include the naturally sweet wines of Rasteau made from Grenache and Beaumes-de-Venise. (The village of Beaumes-de-Venise, south of Gigondas, is well known for producing one of the greatest sweet Muscat wines.)

Northern Rhône—West Bank

- Côte Rôtie—above the town of Ampuis the steep, terraced vineyards are planted with 80% Syrah and 20% Viognier, and are divided into Côte Blonde with its sandy topsoil, and Côte Brune with its iron-bearing clay topsoil. The wines are rich, fragrant, and full bodied.
- Condrieu—planted with Viognier, this AOC produces peach-blossom-and-honey-scented wines that should be consumed young.
- Château Grillet—the smallest AOC of France, it is entirely owned by the Neyret-Gachet family. Made of Viognier, the wine is cask aged for eighteen months in a huge cellar, and requires five years to mature.
- St-Joseph—this AOC yields red wine made from Syrah grapes, plus up to 10% Marsanne and Roussanne. The wine, known as *vin de Mauves*, was once very famous. A small amount of white wines from Marsanne and Roussanne are also produced.
- Cornas—this deep red, tannic wine made from Syrah needs at least three years to mature, but ages well. The low yield of 35 hL/ha produces a strong, dark wine with a licorice, roasted chestnut, and truffle nose.
- St-Péray—this area of limestone soil makes a small amount of white wines from Roussanne and Marsanne and a larger amount of sparkling wines using the méthode Champenoise.

The name of both Hermitage AOCs comes from a little chapel at the top of the hill, and the hermit who once lived there.

Northern Rhône—East Bank and Outlying Areas

- Crozes-Hermitage—located around the town of Tain-l'Hermitage, the majority of its red wines are made from Syrah, and its whites from Roussanne and Marsanne. Quality varies depending on the location of the vineyards. The northern hills are more suited to grape growing than the alluvial land in the south.
- Hermitage—on a hill rising 162 m above the town of Tain l'Hermitage, these terraced vineyards are divided into small parcels whose names sometimes appear on the wine label. Most of the wines are red. The red made from Syrah (with 15% Roussanne and Marsanne) is dark, ages well, and has intense scents of black currant and pepper, becoming chocolatey with age. Styles depend on ageing and whether the wine has been oaked. The whites range from bland to herbal, and have scents of wet slate and minerals.
- Clairette de Die—on the Drôme River, this AOC produces a tiny amount of white table wine, plus sparkling wines made from Clairette Blanche and Muscat à Petits Grains, either by the méthode Champenoise or by using arrested fermentation.
- Châtillon-en-Diois—also on the Drôme, this AOC makes red and rosé wines from Gamay, Syrah, and Pinot Noir, and white wines from Chardonnay and Aligoté.

Southern Rhône—West Bank

- Lirac—this area produces mainly reds (from a minimum of 40% Grenache with additions of Carignan, Syrah, Mourvèdre, and Cinsault) and rosés, but also some whites (from Bourboulenc, Clairette Blanche, Picpoul, Calitor, and Macabeo).
- Tavel—this produces one of the best rosés of France. Made from Grenache, Cinsault, Syrah, Mourvèdre, Bourboulenc, Clairette Blanche, and Picpoul grown on pebbly clay soil, the wine turns to an onionskin colour with age.

Southern Rhône—East Bank and Outlying Areas

- Gigondas—the red and rosé wines are made from 80% (maximum) Grenache plus Syrah, Mourvèdre, and Cinsault. The clay soil produces powerful, rich, spicy wines that take four to five years to mature.

- Châteauneuf-du-Pape—this is the best-known name from the Rhône. The majority (95%) of its wines are red. The AOC permits thirteen grape varieties: Grenache, Syrah, Mourvèdre, Counoise, Muscardin, Vaccarèse, Cinsault, and Terret Noir for red, as well as Clairette Blanche, Picpoul, Bourboulenc, Roussanne, and Picardan for white. The soils vary noticeably between Château Beaucastel in the north, Domaine de Chante Perdrix in the south near the river, and Château La Gardine near Châteauneuf-du-Pape.

  As an example of the blend of grapes in Châteauneuf-du-Pape, Château Beaucastel uses only six grapes, including 30% Grenache, 30% Mourvèdre, and 20% Syrah. Château Rayas uses exclusively Grenache, and Château des Fines Roches use 60% Grenache, 15% Syrah, 10% Mourvèdre, 5% Cinsault, and some Clairette and Muscardin. These proportions change with the vintage.

In addition, the village of Vacqueyras, Côtes du Ventoux, and Côtes du Lubéron each have AOC status. Côtes du Vivarais wines have VDQS status.

PROVENCE

Located in the south of France on the Mediterranean coast, this region and former province is the birthplace of French vineyards. The region has 116 000 ha of vineyards and produces 4 000 000 hL of wine annually.

GEOGRAPHY

Spanning the eastern portion of the Mediterranean coast (from Nice in the east, to the mouth of the Rhône River near Arles, and extending inland), this region is natural for grape growing. The cool winds from the Mediterranean prevent the summer temperatures from soaring. The winters are warm and the hills in the north protect the vines from the mistral wind. Various rocks and gravel make up the soil in this hilly landscape. The area includes the city of Marseilles, as well as Cassis, Toulon, and Nice.

HISTORY

Around 650 B.C., Phocaeans settled on the coast and planted vines. The Greeks and the Romans continued the practice.

Until recently, the wines of the area have been considered mainly easy-drinking summer wines for tourists and café patrons. However, their quality and prestige has increased, thanks to the hard work and dedication of a few growers (such as the brothers Ott), and investments from outside.

GRAPE VARIETIES

Most of the grapes are the same as those cultivated in the Rhône, with the addition of Cabernet Sauvignon, Sauvignon, and Sémillon.

WINE PRODUCTION

Traditionally, this region produced mainly rosés that were cheap and which complemented most foods. Now the wines are generally much better and some are excellent. This improvement has resulted in new AOCs being awarded, such as Coteaux du Varois in 1993.

The AOCs for the region include the following:

- Côtes de Provence—extends along the rocky coast from St-Tropez to Toulon and inland to the hills. It produces 800 000 hL of rosé, red, and white wines. These wines range in quality from good to excellent. The main grapes are (for reds) Grenache, Cinsault, Carignan, and Mourvèdre, with some Cabernet Sauvignon and Syrah; and (for whites) Clairette, Ugni Blanc, Rolle, and Sémillon.
- Bellet—a small vineyard north of Nice (where most of this expensive AOC wine is consumed). It is made in red, white, and rosé. The red grapes are Braquet (Brachetto), Folle Noire (Fuella Nera), Cinsault, and Grenache. The white grapes are Rolle, Chardonnay, Roussanne, Ugni Blanc, Clairette, and Bourboulenc.
- Bandol—a coastal town surrounded by vineyards on limestone soil. Most of its wines are red, and half of all red grapes planted are Mourvèdre. The remaining reds planted are Grenache, Cinsault, Calitor, Carignan, Syrah, and Tibouren. Bandol's rosés are best. The few white wines are made from Bourboulenc, Ugni Blanc, Clairette, and Sauvignon grapes.

- Cassis—located on the coast between Bandol and Marseilles. The best wines are white, made from a blend of Marsanne, Ugni Blanc, Clairette, Doucillon (Grenache Blanc), and Sauvignon. Reds and rosés are made from Grenache, Cinsault, Mourvèdre, and Carignan.
- Palette—a tiny appellation east of Aix-en-Provence consisting of two properties: Château Simone and Domaine de la Cremade. The red wines (made from old vines growing on limestone) are quite tannic and need ageing. They are made from mainly Mourvèdre, Grenache, and Cinsault grapes. Rosé and white wines are made from Clairette, Grenache Blanc, Sémillon, Muscat, and Ugni Blanc.
- Coteaux d'Aix-en-Provence and Coteaux d'Aix-en-Provence les Beaux-de-Provence—promoted to AOC in 1985. The latter is a parcel of the former. This large area consists of chalky soils and produces red, rosé, and white wines made from the classic southern grapes (those used also in the Rhône), with Cabernet Sauvignon gaining in importance.
- Coteaux du Varois VDQS—takes its name from the département of Var. It is a large area around the town of Brignol, making mainly red and rosé wines. The red grapes are Cinsault, Grenache, Mourvèdre, Carignan, Alicante, and Aramon, with an increasing amount of Syrah and Cabernet Sauvignon. The white grapes are Grenache Blanc, Ugni Blanc, Clairette, and Malvasia.
- Coteaux de Pierrevert—a small VDQS area that extends along the Durance River and includes the town of Manosque. It produces red, white, and rosé wines.

CORSICA

In the past, Corsica (the "island of beauty") sent its wine to the mainland for blending. Most of these wines were strong, full of colour, and cheap. The few good ones were drunk by the locals. Located southeast of Provence in the Mediterranean Sea, the island is actually closer to Italy than to mainland France.

In 1768 Corsica was acquired by France from Genoa, in time for Napoleon Bonaparte to be born a French citizen in 1769.

GEOGRAPHY

The island is 180 km long, 80 km wide, and mountainous. Its vineyards cover 9100 ha. The climate is generally Mediterranean with hot to scorching (in the south) summers, and mild winters; the rain falls in the spring and autumn. The soil is mainly granite on the west side of the island, schist in the north and the Cap Corse, chalk and clay in Patrimonio, and marly sand on the east coast. Most of the vineyards are found clinging to sheer slopes, or in numerous small coastal valleys.

GRAPE VARIETIES

The primary grape varieties used for red and rosé wines are Nielluccio (probably related to Sangiovese), Sciarcarello, Grenache, Cinsault, Carignan, Syrah, Aramon, and some Merlot and Cabernet Franc. For the whites, Vermentino (Malvasia), Russula Bianca (Ugni Blanc), and Muscat are used, along with some Biancone, Sauvignon, and Chardonnay.

WINE PRODUCTION

The style of Corsica's wines is somewhat similar to that of the Rhône and Provence.

The town of Sartène, in the south, produces the most wines; the best wines come from Patrimonio AOC, on the northern Cap Corse. There are a few other interesting wines from Tallano and Domaine Peraldi near Ajaccio. Of all Corsica's wines, the reds are best—with a heady perfume, deep in colour, and high in alcohol.

Corsica's AOCs include the following:

- Vin de Corse—the basic general appellation for 40 000 hL of red, rosé, and white wines. The most interesting are Vin de Corse Calvi, Vin de Corse Coteaux du Cap Corse, Vin de Corse Figari, Vin de Corse Sartène, and Vin de Corse Porto-Vecchio.
- Ajaccio—produces some nice red wines with a minimum of 40% Sciaccarello, complemented with Carignan, Grenache, and Cinsault.
- Patrimonio—produces a small quantity of wines, most of which are red. The reds are made from a minimum of 90% Nielluccio grapes. This variety is cultivated on limestone soils.

LANGUEDOC-ROUSSILLON

Actually two connected regions, Languedoc and Roussillon combined form the largest and third-oldest wine region of France. Known also as the Midi, its 380 000 ha extend from the Spanish frontier to the Rhône River near Arles, and inland from the Mediterranean coast. It represents about 38% of the total area of land under vine in France.

GEOGRAPHY

Here, the Mediterranean climate is constant; the main preoccupation is water. The majority of the production is table wine, with the focus on vin de pays and on vin doux naturel (VDN), that is, naturally sweet wine.

GRAPE VARIETIES

A multitude of grapes are used, but Carignan, Cinsault, and Grenache dominate.

WINE PRODUCTION

Languedoc-Roussillon's AOCs include the following, roughly from west to east:

- Côtes du Roussillon and Côtes du Roussillon-Villages—wines made in the eastern Pyrenees near the Spanish border. The deep red wines and fruity rosé wines are made mainly from Carignan, Grenache, and Syrah grapes. The greenish-white wines are made from Macabeo and Tourbat grapes. Côtes du Roussillon-Villages (from twenty-eight villages) must have a minimum of 12% alcohol and a maximum yield of 45 hL/ha.
- Collioure—red wine made from Grenache. It is the oldest wine of the seaport of Banyuls.
- Corbières—has a history of winemaking over 2000 years old. The 13 000 ha of vineyards are divided into four zones and eleven terroirs, producing red, rosé, and white wines. The four zones are: Corbières Littorales (with chalk hills), Corbières Centrales (with sandy soil), Hautes Corbières (with steep chalk hills and schist soils), and Corbière d'Alaric (on the chalky soil to the north bordering the AOC Minervois).
- Fitou—two enclaves located south of the AOC Corbières. The first enclave surrounds the town of Fitou near the sea and the second lies further inland, south of the town of Durban-Corbières. The fleshy red wine is made mostly from Carignan and Grenache, and comes from nine communes. It is aged for a minimum of nine months before being bottled.
- Minervois—located inland, east of Carcassonne. Its 18 000 ha are divided into five zones of chalky vineyards that produce mainly lithe red wines from Mourvèdre, Grenache Noir, Carignan, and Syrah. The small amount of rosé and white wines are made from Macabeo, Bourboulenc, Marsanne, and Roussanne.
- Blanquette de Limoux—the oldest sparkling wine of France, made from Mauzac, Chardonnay, and Chenin Blanc grapes.

Minervois, originally a stronghold of the Tenth Roman Legion, was named for the goddess Minerva.

BLANQUETTE DE LIMOUX

To make this sparkling wine, the grapes are partly pressed to produce 100 L of juice for every 150 kg of grapes. The wine undergoes a second fermentation in the bottle after the addition of the liqueur de tirage.

Liqueur de tirage is a combination of wine, yeasts, and cane sugar.

- Coteaux du Languedoc—a large area north of Montpellier. It includes the AOC St-Chinian and Faugères. Twelve vineyard names may be added to the AOC.
- St-Chinian and Faugères—promoted to AOC in 1982. The wines are red and rosé.
- Costières de Nîmes—covers 2500 ha of stony soil, from the ancient delta of the Rhône River. These are typical southern wines made from Cinsault, Carignan, and Grenache, with a token amount of Mourvèdre, Syrah, Ugni Blanc, and Clairette Blanche.
- Clairette de Bellegarde—an appellation for the white wine produced in the southeast of the Costières de Nîmes.

In addition, Languedoc-Roussillon is well-known for its vins doux naturels (VDN). These wines are made from Grenache, Macabeo, Malvasia, and Muscat grapes. The VDN of Languedoc-Roussillon include the following:

- Banyuls and Maury—red sweet wines, made from Grenache, that age very well.
- Muscat de Rivesaltes, Muscat de Frontignan, Muscat de Mireval, Muscat de Lunel, and Muscat de St-Jean-de-Minervois—delicious, sweet, fruity Muscat wines best drunk when young.

THE SOUTHWEST

The region is a collection of scattered vineyards surrounded by Bordeaux, Languedoc-Roussillon, and the Pyrenean Mountains on the Spanish border.

WINE PRODUCTION

This area is influenced by an Atlantic climate but is protected from the cold, wet, westerly winds of winter by the Landes pine forest. The 30 000 ha of vineyards have a diversity of soil, climate, and grapes that makes it hard to generalize, but they can be divided into two basic groups:

- southern Pyrenean vineyards south of the town of Mont-de-Marsan and the River Midouze, encompassing the regions of the Basque country, Béarn, and Gascony.
- northern vineyards along rivers flowing toward the Gironde and the Bordeaux region from Albi, Rodez, and Périgueux.

Southern Pyrenean Vineyards

- Irouléguy—in the heart of the Basque country and at the extreme southwest of France, produces mainly red and rosé wines from Tannat, Fer, Cabernet Sauvignon, and Cabernet Franc grapes. A very small quantity of white wines is made from Gros-Manseng, Petit-Manseng, Courbu, Lauzat, Baroque, Sauvignon, and Sémillon grapes. The mountainous vineyards are located on terraced slopes at elevations between 200 m–400 m.
- Béarn—produces opulent red and excellent rosé wines made from Tannat (60%), Cabernet Franc, and Cabernet Sauvignon, plus Courbu, Fer, Manseng Noir, and Pinenc grapes. Also, a very small quantity of white wines is produced from Raffiat de Moncade, Gros and Petit Manseng, Courbu, Baroque, Lauzat, Sémillon, and Sauvignon grapes.
- Jurançon—at an altitude of 300 m, located south of the town of Pau, between the Gave d'Oloron and the Gave de Pau. The varied soils are composed of clay, chalk, sand, and rocks. The AOC is divided among twenty-five villages. The fame of Jurançon comes from the production of a very sweet wine made from Petit Manseng, Gros Manseng, and some Courbu, all of which are left on the trellised vines until the grapes shrivel and the sugars concentrate. The wine is rich with a good balance of acidity and has a nose of dry apricot, cinnamon, and cloves. A dry, more ordinary, white wine is also made.
- Madiran and Pacherenc du Vic-Bilh—both wines come from the same 1300 ha of vineyards northeast of the town of Pau. Madiran is red, made from various proportions of Tannat, Cabernet Sauvignon, Cabernet Franc, and Fer. The more Tannat in the blend, the darker and more tannic the wine. The white Pacherenc du Vic-Bilh is dry or sweet, made in small quantity from Ruffiat, Gros and Petit Manseng, Courbu, Sauvignon, and Sémillon.
- Tursan and Côtes de St-Mont are two connected VDQS areas of Madiran.

Northern Vineyards

- Côtes du Frontonnais—located just north of Toulouse, between the Tarn and Garonne rivers. The soil is composed of red pebbles (rich in iron), and decomposed clay from the moraine of an ancient glacier. The wines are red or rosé made from Négrette (70%) plus some Cabernet Sauvignon, Cabernet Franc, Fer, Malbec, Syrah, Cinsault, and Gamay.

- Gaillac—located on the Tarn River, including the towns Gaillac and Albi. It produces red, rosé, white, and sparkling wines. The red and rosé wines are mainly made from Duras, Fer (or Braucol, as it is known locally), Syrah, Gamay, and Négrette grapes, plus Merlot, Cabernet Franc, Cabernet Sauvignon, Portugais Bleu, Jurançon Noir, and Mauzac Blanc. The dry or sweet white wines, which used to predominate, are made from Mauzac Blanc, Len de l'El, Ondenc, Sauvignon, Sémillon, and Muscadelle grapes.

  The sparkling varieties are made in the méthode gaillacoise. Gaillac Perle is a fizzy (i.e., with very light carbonation) wine that undergoes malolactic fermentation in the bottle.

The méthode gaillacoise involves a second fermentation in the bottle without the addition of yeast or sugar. The first fermentation is stopped by refrigeration, then the wine is bottled.

- Vins de Lavilledieu and Côtes du Brulhois—two VDQS areas south of Buzet along the Garonne, downstream of the town of Moissac.
- Buzet—on the south bank of the Garonne between Buzet and Nerac. Between 1973, when the area was awarded the AOC, and 1988, it was called Côtes de Buzet. Most of the wines are red, with some rosés and whites made in the tradition of Bordeaux.
- Cahors—centred on the town of Cahors and the Lot River, the vineyards grow on soil composed of clay, chalk, sand, alluvial deposits, gravels, and broken rocks. The wine is red, produced from 70% Malbec (known also as Auxerrois), either Tannat or Merlot, and some Jurançon Noir. The wine demands ageing and can become a complex wine of quality.

  Cahors has a long history of wine. The Popes at Avignon, the Tsars of Russia, and the Russian Orthodox Church favoured the "black wine of Cahors." Later, the area was ravaged first by phylloxera and, in 1956, by a killer frost. Now the vineyards have recovered from a mere 100 ha in 1958 to 4300 ha.
- Vins d'Estaing and Vins d'Entraygues et du Fel—two VDQS areas located on the Lot River at the foothills of the Massif Central, northeast of the Marcillac.

  Delimited in 1993, Côtes de Millau is an additional small VDQS.
- Vins de Marcillac—received the AOC in 1990. The 145 ha of picturesque vineyards on red sandstone soil reach elevations of 600 m. The red and rosé wines are made from 80% Fer Servadou (known locally as Mansois) and some Cabernet Sauvignon, Cabernet Franc, Merlot, and Jurançon Noir.

- Côtes du Marmandais—near the Bordeaux region and Entre-deux-Mers area, on the Garonne River. The wines here are made from Bordeaux grapes, plus some Syrah, Fer, and Arbouriou. Most of the wines are red.
- Côtes de Duras—on the fringe of the Bordeaux region between Entre-deux-Mers and Monbazillac. Using Bordeaux grapes, half of the production of this area is semisweet. This AOC also includes dry, white wines made from Sauvignon, Sémillon, Muscadelle, Chenin Blanc, Ugni Blanc, Mauzac, and Ondenc.
- Bergerac—immediately east of the Bordeaux region, extending along the Dordogne River from St-Emilion to the town of Bergerac, and north to the Isle River. The summers are warmer than in the Bordeaux region and the soil is comparable with that in St-Emilion. The best wines are red, made from Cabernet Sauvignon, Cabernet Franc, and Merlot, to which a little Malbec and Fer are sometimes added. Some rosé and white wines are also made from classic Bordeaux grapes, with the usual addition of Chenin Blanc and Ondenc to the white wines.
- Monbazillac—southeast of Bergerac, famous for its sweet, white wine made in the Sauternes style. The 2500 ha are mainly planted with Sémillon, which can develop a good noble rot in this area. Usually Sémillon accounts for 70%, with Sauvignon grapes added to correct the acidity. Some Muscadelle is needed when noble rot is minimal.
- Pécharmant—also in the Bergerac area. The vineyards grow on soil composed of limestone and clay covered by layers of gravel. This terrain is particularly suitable to the production of very good red wines from Bordeaux grapes. These offer more structure than those of the neighbouring areas, and embody the pride of this Dordogne River region.
- Rosette, Saussignac, and Montravel—located east and south of Bergerac. Essentially, they produce sweet white wines from the classic Sémillon, Sauvignon, and Muscadelle grapes, plus Ugni Blanc, Chenin Blanc, and Ondenc.

Bordeaux

The wines of Bordeaux have always commanded respect and are now the international benchmark for Cabernet Sauvignon, Merlot, and Sémillon.

With 100 000 ha of vineyards (95% AOC), some 13 000 producers, an annual production of 660 000 000 bottles of wine, and more than a quarter of all France's appellations, the Bordeaux region is a huge force in the French wine industry. However, with so many players and so much wine produced, the wines necessarily vary, and do so in a way that Bordeaux's classification cannot always reflect.

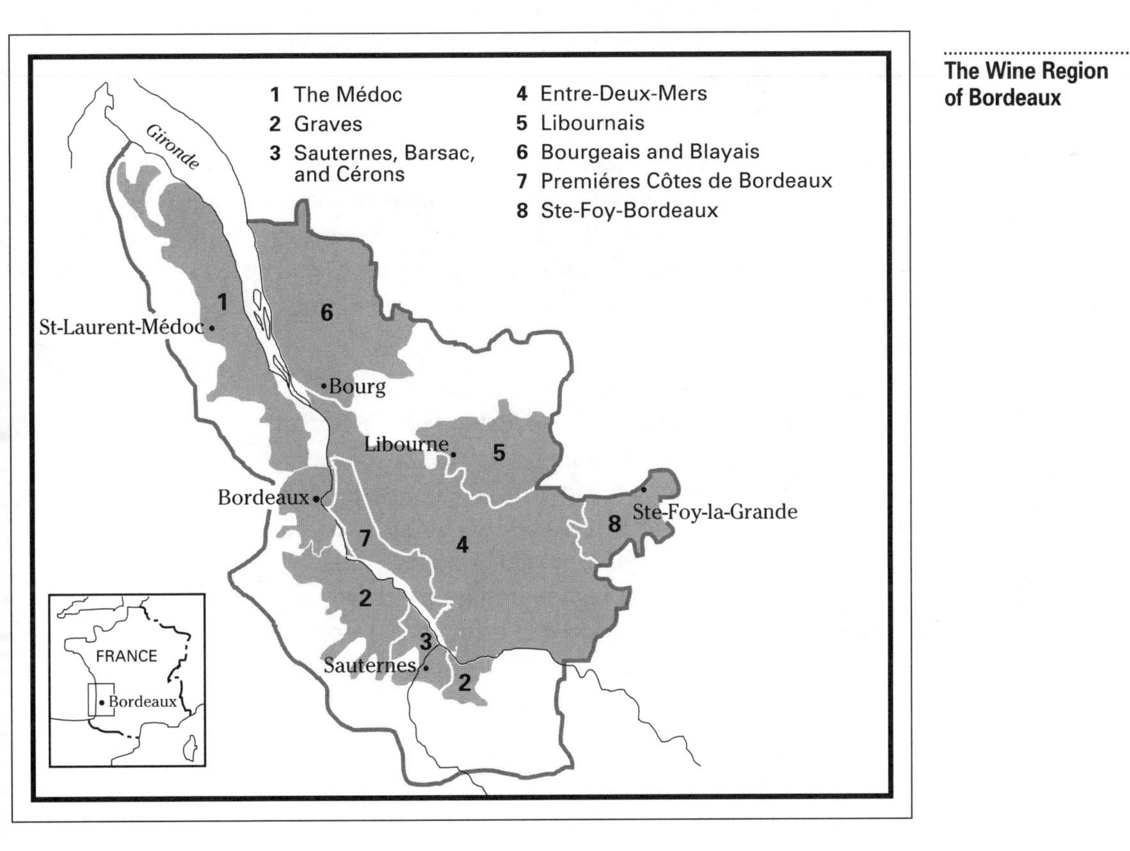

The Wine Region of Bordeaux

GEOGRAPHY

Bordeaux is one of the largest wine regions in the world. It is found inland from the Atlantic Ocean, hugging the Gironde estuary and portions of the rivers that flow into it—the Garonne and Dordogne. The coastal Atlantic climate is wet and mild. Geologically, this wine region is extremely unusual. The landscape was carved by a huge glacier running from the Pyrenees to the Atlantic, leaving a significant quantity of boulders and gravel over various soils. Each subregion or district has its own climate and particular soil.

HISTORY

Although the earliest efforts at viticulture in the region cannot be confirmed, it seems that the Romans arrived to find plots of land planted with vines by the Bituriges, the local tribe. Just as they did elsewhere, the Romans planted their own vines and helped to improve winemaking. In A.D. 379, the Latin poet Ausonius wrote of the grapes he grew in Bordeaux. (His property is now called Château Ausone, in St-Emilion.) The province was named Aquitana, which later became Aquitaine; the town of Bordeaux was known as Burdigala.

Long after the fall of the Roman Empire in A.D. 476, the harbour of Burdigala was still busy with shipments of wine. In 1152, Henry Plantagenet of England married Eleanor of Aquitaine and acquired Eleanor's territories—most of southwestern France. The region's wines benefited from both popularity in England and tax exemptions.

In 1453, the region reverted to French rule. Bordeaux remained important in the wine industry worldwide, and became increasingly so in France. The wine merchants of Bordeaux were among the first to sell château-bottled wine, to create glass bottle factories, to use the practice of topping up ullage, and to buy wine "in future," that is, to buy on speculation and before production. However, the great power of Bordeaux's wine merchants decreased following the French Revolution (1789–1802), when the First Republic abolished many privileges.

Today the Bordeaux region is very significant in international wine trade. Every two years, the city of Bordeaux holds VinExpo, the largest wine fair in the world.

Bordeaux-England Connection

The connection between Bordeaux and England can be traced up to the present time:

- The "vin clairet" left Bordeaux every spring and autumn on ships that were assessed in terms of tonneaux, that is, the number of the traditional Bordeaux wooden barrels they could hold in their hulls. Today, clairet has become claret (the English name for red Bordeaux), and tonnage (units of 100 cubic feet or 252 imperial gallons) is a standard measure of the volume of a merchant ship.
- The English connection is still evident in such names as Château Talbot (for John Talbot, fourth Earl of Shrewsbury, marshal of the English army), Château Haut Brion (for O'Brian), and Château d'Olivier (from Oliver, a friend of Edward the Black Prince).

GRAPE VARIETIES

The majority of grape varieties grown in the Bordeaux region are shown below. (Also given is the local name for the grape.)

White Grapes

- Sauvignon (Blanc Doux)—used in the region's dry white wines and blended in sweet wines to balance with acidity. The bunch is tight and the berries small and thick skinned. It is susceptible to powdery mildew and fungal disease.
- Sémillon (Crucillant)—widely planted in Bordeaux and very vigorous in gravelly soil. This thin-skinned grape is low in acidity, but can reach a high sugar content. Susceptible to noble rot, it is the main contributor to the rich, sweet wines of Sauternes, Barsac, and Sainte-Croix-du-Mont. When used for dry wine, it tends to produce rather soft wines.
- Muscadelle (Guépu)—susceptible to moulds and parasites but producing very sweet grapes (up to 400 g/L) with a Muscat aroma and a very strong nose. The berries are large with a very thin skin that tends to burst near maturity, thus attracting bees and wasps.

Chenin Blanc (Pineau de la Loire), Colombard (Pied Tendre), Folle Blanche (Enrageat), Mauzac Blanc (Blanc-Lafitte), Merlot Blanc, Ondenc (Blanquette), and Ugni Blanc (St-Emilion or Muscadet) are also used.

Red Grapes

- Cabernet Sauvignon (Petit Cabernet)—very important in Graves and the Médoc. The vine has average yields but loves sandy and clay soils. The small berries have a thick skin resistant to mould, but are late in ripening. The grape is susceptible to fungal disease, powdery mildew, black spot, and dead arm.
- Cabernet Franc (Bouchet, Gros Cabernet)—dominant in St-Emilion and second in Graves and the Médoc. This early-ripening grape is paler, less tannic, and less scented than Cabernet Sauvignon. The wine matures early, with a fruity bouquet reminiscent of green olives. It ages well, but is susceptible to dead arm and powdery mildew.
- Merlot (Sémillon Rouge)—dominant in the wines of Libourne, early in ripening, rich in colour and sugar, low in acid and tannin. It produces supple wines that mature early and have a deep colour and red-berry nose. It is susceptible to grey rot, coulure, and mildew.

- Malbec (Cot Rouge)—used in small quantities for blending. This prolific, early-ripening grape is deep in colour, high in acidity, and low in tannin. It will produce wines for early drinking and it is very susceptible to coulure.
- Petit Verdot (Carmelin)—ripens late in the season and has difficulty reaching maturity in Bordeaux. When not fully ripe, the grapes make a wine that is very tannic and high in acidity. Fully ripe Petit Verdot grapes create a wine with deep colour, high sugar and tannin levels, the ability to age well, and a complex bouquet. It is a difficult grape to grow and has a very low yield. It is susceptible to powdery mildew, various infections, and degeneration, but resistant to mould.

Carmenère (Grande Vidure), Bequignol (Fer), and Bouchalès (Grapput or Prolongeau) are also used.

WINE PRODUCTION

Most of the region's wines are red, and most of these are blends. Each blend depends on the varieties used, their availability and quality, the terroir, and the winemaker's particular style.

Many of Bordeaux's wines are classified. In the Médoc, the idea of a growth (cru) is fundamentally linked to ownership of vineyards and estates. While there has been a list that ranked Bordeaux estates based on the demand for their wines and their reputations since 1700, the growth classification (cru classé) system really began in 1885. For the Exposition Universelle in Paris that year, the Bordeaux wine brokers established five classes of Médoc and Sauternes wines. Their classifications were based on the average selling prices of the various estates' wines over the previous 100 years. The quality of the wines and the maintenance of the vineyards were additional considerations. The central belief was that the best wines fetched the highest prices. This resulted in the classification of sixty châteaux in the Médoc and one in Graves: four First Growths (Premier Crus), sixteen Second (Deuxièmes), thirteen Third (Troisièmes), ten Fourth (Quatrièmes) and eighteen Fifth (Cinquièmes). This was revised in 1973 with the promotion of Château Mouton Rothschild from Second Growth to First. Twenty-seven Sauternes were also classified.

Not everybody is satisfied with the classification system, which does not necessarily reflect the quality of a wine. Since 1885, many estates have changed hands, and newly acquired vineyards have become part of classified growths.

In addition to the crus classés, Bordeaux wines may be classified or labelled as follows:

- Cru Bourgeois—in the fifteenth century the wealthy bourgeois of Bordeaux were allowed to purchase land from the landlords. They acquired some of the best land and planted vineyards. The wine produced was labelled *Cru Bourgeois*. Classifications were drawn up in 1932 and revised in 1966 and 1978: Cru Bourgeois, Cru Grand Bourgeois, and Cru Bourgeois Exceptionnel. New EU regulations recognize only the term *Cru Bourgeois*. Though overshadowed by the crus classés, they are good wines of excellent value.
- Bordeaux Supérieur—wine made anywhere in Bordeaux with a higher minimum natural alcohol than for Cru Bourgeois, and from a maximum yield of 40 hL/ha instead of 50 hL/ha.
- Château—in Bordeaux the term is synonymous with "growth." By definition, it is a vineyard or group of vineyards (not necessarily of one holding) consisting of an estate with a wine shed or building.
- Mise en Bouteille au Château—indicates that the wine was bottled by the proprietor on the property where it was made.

The Médoc

The Médoc was initially all the land north of Bordeaux and west of the Gironde, as far as the Atlantic coast. The area was covered with forest, marsh, lakes, and grassland, but no vineyards, until the eighteenth century. In that century the name Médoc came to mean only the land north of the Jalle de Blanquefort.

The Médoc covers an area of more than 10 000 ha on the south bank of the Gironde estuary. It has a warm and humid climate, with no temperature extremes. The vineyards are planted on ridges made of gravel from the Pyrenees and pebbles from the Gironde, mixed with sand and layers of clay. This provides good drainage during the wet season and stabilizes the night temperature by releasing the heat stored during the day.

The grapes are red with three predominant varieties: Cabernet Sauvignon, Cabernet Franc, and Merlot, plus some Petit-Verdot and Malbec.

Crus classés represent about 25% of the total wine production in the Médoc.

The Médoc has eight AOCs—Médoc, Haut-Médoc, and six commune appellations. They are all given below (in order from north to south).

- Médoc—extends from St-Vivien in the north to St-Seurin de Cadourne in the south.
- Haut-Médoc—extends from St-Seurin de Cadourne in the north to the Jalle de Blanquefort near the city of Bordeaux in the south. Some fine Châteaux with this appellation are Châteaux Cantemerle, Châteaux La Lagune, and Châteaux La Tour-Carnet.
- St-Estèphe—located along the Gironde on a soil heavy with clay. The wines are traditionally full bodied, very aromatic, with good acidity and firm tannin. They are slower to mature but age very well.
- Pauillac—separated from St-Estèphe by the Jalle du Breuil and from St-Julien by the Juillac brook. The soil varies. For example, Château Latour has greater thickness of clay and Mouton has more sand and pebbles. The wines are full bodied with good tannin, rich flavour, and a typical bouquet of black currants and cedar. They need time to develop, but age very well. Pauillac has four Premier Crus.
- St-Julien—located in the centre of the Haut-Médoc. This AOC has vineyards on the gravel hills of St-Julien and Beychevelle that slope to the Gironde. The wines are extremely rich in flavour and have a delicate nose, medium tannin, and great finesse. They mature faster than those of St-Estèphe and Pauillac.
- Listrac—inland, 34 km northwest of the city of Bordeaux and 5 km from the Gironde. At 43 m, it is the highest point of the Haut-Médoc. The wines are fruity, robust, and somewhat leathery.
- Moulis—between Listrac and Margaux, this is a small appellation located on a gravelly plateau. The wines are full bodied with good colour and bouquet. They age well.
- Margaux—has soil containing more limestone than is found in the nearby city of Bordeaux. Consequently, the wines are softer and have great finesse and a delicate nose. Of the six commune appellations, Margaux wines are the fastest to mature.

Graves

This area is the continuation of the Médoc south of the city of Bordeaux, on the south bank of the Garonne River. Graves covers a stretch of vineyards 60 km long and 15 km–20 km wide. It is the oldest wine district of Bordeaux: for example, Château Pape-Clément was created in 1300 by Bertrand de Got, who became Pope Clément V in 1305. Graves was producing wines when the Médoc was mostly covered with timber and swamp.

In Châteaux Haut-Brion, the gravel layer can be 15 m thick.

The soil is composed of a thick layer of gravel mixed with sand and patches of clay. The climate is slightly warmer than in the Médoc during the summer, and the vineyards near the Garonne are more subject to mould during the misty mornings of the late season. The grape varieties are the same as those for the Médoc, with the addition of Sémillon, Sauvignon, and Muscadelle.

Graves produces a great amount of white wines, although the reds are the best, in the Médoc style with a light touch overall. The reds in the south have a greater percentage of Merlot, making them ready to drink at an earlier stage. The white wines can be dry or sweet. The dry white wines from Graves are the only ones from Bordeaux that can be said to age with grace.

The northern part of Graves was awarded a new appellation in 1986, under the name of Pessac-Léognan. This area of about 900 ha includes fifty-five estates producing wines from a lower yield per hectare than in Graves. When mature, the wines exhibit roundness, with a nose of red berries, tobacco leaves, and a mineral smell. In 1959, Graves was classified in a way similar to the Médoc. The result was one Premier Grand Cru, twelve Crus Classé red, and ten Crus Classé white. All of these are in Pessac-Léognan.

Sauternes, Barsac, and Cérons

Sauternes, Barsac, and Cérons are AOCs that form three separate enclaves in the southern part of Graves. As well, Sauternes is the luscious sweet white wine made from the five communes of Sauternes, Barsac, Bommes, Preignac, and Fargues, located on the south side of the Garonne River.

The Ciron River, a small tributary of the Garonne, separates the AOC Sauternes from the AOC Barsac. Wines from Barsac may use the Sauternes appellation or the Barsac.

Wines in this area are made from Sémillon, with the addition of some Sauvignon and Muscadelle grapes.

Sauternes
In the autumn, when the cold water of the Ciron meets the warm water of the Garonne, a morning fog shrouds the vineyards. These are ideal conditions for noble rot, which spreads quickly among the vines and which easily penetrates the thin skin of the Sémillon grape. As the mist disappears late in the morning, noble rot lives on the water of the grape, and combined with the sun, shrivels them. The botrytized berries are hand picked over a period of months in successive passes by patient harvesters.

The production of Sauternes, Barsac, and Cérons wines depends on the area's proximity to a large body of water, for the development of noble rot.

Sauternes AOC can produce some of the greatest sweet wines in the world. At their best they are very rich with a great balance of acidity and an amber colour. Aromas vary from dry apricot and honey, to incense. These wines age very well. According to the 1855 classification of Sauternes, there is one First Great Growth (Premier Cru Supèrieur), eleven First Growths (Premier Crus) including two Barsac, and fourteen Second Growths (Deuxièmes Crus), including eight Barsac.

Cérons produces sweet white wines labelled *Cérons* and dry white wines labelled *Graves*.

Entre-Deux-Mers

Deriving its name from salty tidal waters, Entre-Deux-Mers ("between two seas") forms a triangle of 3000 ha between the Garonne River to the south and the Dordogne River to the north. The land is flat, with vineyards located here and there among crop fields and wooded areas. The soil is high in limestone.

The wines of this region are mainly white and dry, and are made from Sauvignon, Muscadelle, and Sémillon (70%), plus Merlot Blanc (20%), and Colombard, Mauzac, and Ugni Blanc (10% maximum).

The AOCs for the area include the following:

- Entre-Deux-Mers—the large AOC between the rivers Garonne and Dordogne.
- Bordeaux or Bordeaux Supérieur—for red wines.
- Premieres Côtes de Bordeaux—the slopes down to the Garonne, facing Graves, produce the best red wines of the area. The dry white wines are sold under the Entre-Deux-Mers label and the sweet white wines as Premières Côtes.

- Cadillac, Loupiac, and Ste-Croix-du-Mont—areas within Premières Côtes de Bordeaux that face the Sauternes vineyards. They produce sweet white wines in the Sauternes style, but much thinner.
- Côtes de Bordeaux St-Macaire—terminates the stretch along the Garonne; produces white wines.
- Graves de Vayres—lies on the south side of the Dordogne; produces red and white wines. Ste-Foy-de-Bordeaux, to the far east of the area, produces red and white wines.

Libournais

This district, located on the north side of the Dordogne, takes its name from the main town of Libourne. The climate is similar to that of the Médoc, but drier and warmer in summer, and colder in winter. Spring and winter rain and frost can be of concern on the plateau. The soil is mostly clay-sandstone with patches of sand and gravel, and outcrops of limestone.

In contrast to its counterparts Graves and the Médoc (where Cabernet Sauvignon dominates the blend), Merlot and Cabernet Franc dominate the blend here. The Libournais wines are sturdy reds with less tannin, a higher alcohol content, deep colour, and a truffle aroma.

The AOCs for the area include the following:

- St-Emilion—produces only red wines. St-Emilion is the area's basic appellation. Its wines were delimited in 1936, first classified in 1955, and have been reclassified every ten years since. As of 1985, there were two Premier Grands Crus Classés A, nine Premier Grands Crus Classés B, sixty-three Grands Crus Classés, and sixty-four Grands Crus.

  Generally, the wines of St-Emilion have a deep colour, a dense fruitiness, and a penetrating bouquet. They range from those of the *côte* (the steep south-facing hill of the town St-Emilion, with limestone or clay and sandstone soils), to those of the *plateau* (north of St-Emilion, with alluvial sand on a clay and sandstone base). Château Cheval Blanc and Château Figeac are on the gravel and sand mixture brought by the Isle River.

  There are also appellations for satellite villages separated from St-Emilion by the Barbánne River. They are St-Georges St-Emilion, Montagne St-Emilion, Parsac St-Emilion, Lussac St-Emilion, and Puisseguin St-Emilion. The soil contains more clay than limestone and produces wine similar to that of St-Emilion. However, it has less depth and is sold at a lower price.

- Côtes de Castillon and Côtes de Francs—located east between St-Emilion and its satellites. The vineyards of the Côtes de Castillon (named after the town of Castillon-la-Bataille) mainly produce red wines; the small amount of white wine is labelled *AOC Bordeaux*. Côtes de Francs produces mainly red wines.
- Pomerol—produces some of the greatest red wines of Bordeaux. Most of the vineyards are small. The plateau of Pomerol (tucked between the town of Libourne, the St-Emilion area, and the Barbánne River) consists of sandy soils combined with gravel. Of the grapes used, Merlot dominates, followed by Cabernet Franc, Cabernet Sauvignon, Malbec, and Carmenère. The deep red wines have a firm structure and an intense fruity aroma.
- Lalande-de-Pomerol—across the Barbánne River from Pomerol and roughly the same size. It includes the village of Néac. Here, also, Merlot prevails and yields wines of intense colour and a plummy taste, although they are leaner than those of Pomerol and less costly.
- Fronsac—west of Libourne, situated between the Dordogne and Isle rivers. This area has a glorious past dating back to Charlemagne, and is regaining some recognition. The wines are now made from a high percentage of Merlot combined with the other traditional Bordeaux grapes. They are softer than in the past, when they had the reputation of being long-lasting but chewy.
- The tiny appellation of Canon-Fronsac (around the villages of Fronsac and St-Michel-de-Fronsac) produces wines that are firmer and of better quality than those of Fronsac.

Bourgeais and Blayais

On the northern side of the Gironde facing the Médoc, lie the quaint wooded hills of Bourgeais and Blayais. This large area has been exporting wine since medieval times from its harbours of Bourg and Blaye. When the Médoc was still covered with bog and forest, English ships were filling their holds with the *vin clairet* from this district, conveniently located on the estuary. Today the area is filling the need for less expensive Bordeaux wines, made to be drunk young.

The following AOCs are included in the Bourgeais and Blayais area:

- Côtes de Bourg—a small area, at the mouth of the Dordogne River. The (mostly) red wines have more character than those of Blaye.

- Blaye—produces a large amount of red wine. The modest quantity of dry white wine is entitled to the appellation Côtes de Blaye. Some better quality red and white wines are labelled Premiéres Côtes de Blaye.

LOIRE VALLEY

The longest river in France, the Loire runs 1010 km to the Atlantic from its source in the mountains of the Massif Central, not far from the Rhône. Because of the diversity of soils and climates, the vineyards produce a variety of wines, mostly white.

GEOGRAPHY

The main vineyards of the Loire valley extend from the Atlantic Ocean to the region of Pouilly-sur-Loire near Burgundy, and as far as Roanne in the Massif Central. From the Atlantic to the river's source, the main towns and cities of the wine region include Nantes, Angers, Saumur, Chinon, Tours, Blois, Orléans, and Gien. The climate is relatively mild to cool, with more precipitation near the Atlantic Ocean. Overall, the climate favours the cultivation of white grapes. Because the climate and soils (and thus the grape varieties and wines) vary so much, they will be discussed for each area within the region.

The region can be divided into: Pays Nantais; Anjou and Saumur; Touraine; Upper Loire.

HISTORY

The origin of this region's viticulture is hazy. Certainly the Romans planted vines before the fall of the Roman Empire in A.D. 476, and the wines of Anjou in particular were popular with the Loire natives in England (e.g., Henry of Plantagenet, king of England from 1154 to 1189). But significant expansion only occurred in the fourteenth, fifteenth, and sixteenth centuries, during the time of the Renaissance.

After the French royalty and nobility began building summer residences in this mild, bucolic country, they demanded its fresh, fruity wines.The Dutch also were serious customers, until the Seven Years War (1756–1763).

There followed a period of deterioration due to decreased demand, poor transportation, and (in the late nineteenth century) various pests. Today the wine business is on the increase, thanks to aggressive marketing and modern methods of vinification.

WINE PRODUCTION

The Pays Nantais

This area closest to the Atlantic is named for the city of Nantes at its centre. The two main grapes are Folle Blanche (Gros Plant) and Muscadet, whose real name is Melon de Bourgogne. The latter was brought to the area in 1709 after a destructive frost annihilated the vineyards.

The best known wine of the area is Muscadet. All Muscadet wines should be consumed young. The appellations include:

- Muscadet—covers 11 000 ha of vineyards west and south of Nantes. In turn, it is divided into three appellations: Muscadet, Muscadet Coteaux de la Loire, and Muscadet de Sèvre-et-Maine. The soil consists of a mixture of hard, primary rocks such as sandstone, granite, and mica schists. Muscadet wine is produced at a rate of 120 000 hL per year.
- Muscadet de Sèvre et Maine—east of Nantes, named after the two tributaries of the Loire. It produces 530 000 hL per year.
- Muscadet des Coteaux de la Loire—in the northeast portion of the district between Nantes and Ancenis. It has an annual production of 25 000 hL per year.
- Muscadet Primeur—regulated not to be sold before the third Thursday of November after the harvest. Like Beaujolais Nouveau, this wine is best when very fresh and young.

MUSCADET SUR LIE

Many Muscadet wines are labelled *sur lie*. Generally, this means that the wine has spent a winter in its cask on its lees and has been bottled directly without racking. This permits the wine to retain its freshness and fruitiness, and gives it a yeasty nose and light carbonation. To ensure that this slight carbonation is genuine, look for the words *Mise en bouteille à la propriété*, *...au château*, or *...au domaine*. These are the best representatives of the Muscadets.

- Gros Plant du Pays Nantais VDQS—mainly located between the Muscadet area and the ocean. Five other tiny areas that are part of the appellation skirt the other appellations around Nantes. It yields crisp white wines made from Folle Blanche and is sometimes bottled sur lie.
- Coteaux d'Ancenis VDQS—around the town of Ancenis. These wines are mainly red, with some rosé made from Gamay. The area produces a few Cabernets, as well as wines from Pinot Gris and Chenin Blanc.
- Fiefs Vendéens VDQS—five areas south of Nantes. The wines are fruity, light, and crisp. The reds are made from Pinot Noir and Gamay, and the whites from Chenin Blanc. They should be consumed young.

Anjou-Saumur

Anjou and Saumur are two neighbouring areas totalling 14 500 ha, upstream from Nantes. Angers and Saumur are the main towns. AOC Anjou is best known for its rosé made with Grolleau, Cabernet Franc, and Cabernet Sauvignon. Only the red wines of Saumur are authorized as AOC Saumur; whites from Saumur are AOC Anjou.

The wines of the two areas include the following:

Rosés

- Rosé d'Anjou—a semisweet wine produced anywhere within Anjou, from Grolleau grapes and some Cabernets.
- Cabernet d'Anjou—a semisweet-to-sweet rosé made anywhere within the region, from Cabernet Franc and Cabernet Sauvignon.
- Cabernet de Saumur—a dry rosé made from Cabernets in the area of Saumur.
- Rosé de la Loire—a dry rosé made from Cabernets (30% minimum), plus Gamay, Grolleau, and Cot (Malbec). It is also produced in Touraine.

Reds

- Anjou red—made from Cabernet Franc, Cabernet Sauvignon, the local Pineau d'Aunis, and Cot.
- Saumur and Saumur-Champigny—produced around the town of Saumur on the south banks of the Loire from Cabernet Franc and Cabernet Sauvignon. They are considered the best red wines from Anjou-Saumur.
- Anjou-Gamay—fruity red wines made from Gamay grapes and often sold as *Primeur.*
- Anjou-Villages—red wines made from Cabernet Franc and Cabernet Sauvignon from forty-six selected villages. Wines must be aged for one year before being released.

Whites

- Anjou—made from Chenin Blanc to which 20% Sauvignon and Chardonnay may be added.
- Quarts-de-Chaume, Bonnezeaux, and Coteaux du Layon—rich, sweet wines made from Chenin Blanc, often affected by noble rot.
- Savennières, Coteaux de la Loire, and Coteaux de l'Aubance—dry white wines made from Chenin Blanc.

Sparkling

- Saumur—rosé and white sparkling wines made from a second fermentation (in the bottle) of wines stored in caves cut in the chalk. Another sparkling wine is Crémant de la Loire, made similarly, but with less carbonation.

Touraine

This central district of the Loire valley covers 10 000 ha of vineyards scattered on both sides of the Loire. It centres on the town of Tours and includes the communities of Vouvray, Mountlouis, and Amboise along the Loire, and Mesland north of the Loire. Two tributaries of the Loire, the Cher and the Indre, flow through the area. The soil consists of chalk, silica, and clay combined with silica.

The grapes used for whites are Sauvignon and Chenin Blanc (also called Pineau de la Loire); the reds are Cabernet Franc, Gamay, Cot, Grolleau, and Pineau d'Aunis.

The area's appellations include the following:

- Touraine—produces still and sparkling white wines.
- Touraine-Villages—wines from the areas of Amboise, Azay-le-Rideau, and Mesland.
- Chinon, Bourgueil, and St-Nicolas-de-Bourgueil—the most typical red wines from the Loire Valley, made mainly from Cabernet Franc. They develop a particular nose of raspberry, violets, and dill pickle. The wines should be consumed cool and young, but can age moderately.
- Vouvray and Montlouis—dry to sweet, still and sparkling wines from Chenin Blanc. Vouvray (on the north bank of the Loire) is very famous for its wines, as well as its spectacular cellars and galleries cut in the chalky cliffs.
- Cheverny—a former VDQS recently promoted to AOC status. It makes interesting crisp white wines from Sauvignon, Chenin Blanc, and Chardonnay. Some red and rosé wines are produced from Gamay, Cabernet Franc, Pinot Noir, and Cot (Malbec).
- Jasnières—white wines from Chenin Blanc.

- Coteaux du Loir—produces red, dry rosé, and white wines.
- Valençay, Vins du Thouarsais, Coteaux du Vendômois, Orléanais, and Haut-Poitou—all have VDQS status.

Upper Loire

From the city of Orléans, the Loire River turns south. Further upstream are the famous districts of Sancerre and Pouilly-sur-Loire. In the area's chalky soil mixed with broken clay and volcanic rocks, the Sauvignon develops its characteristic qualities: herbaceous aromas of gooseberry as well as wet flint. The Sauvignon wines from this area may be imitated, but never equalled.

The area's appellations include the following:

- Sancerre—on the west side of the Loire. It has 1600 ha of mostly Sauvignon grapes, with a little Pinot Noir for red and rosé wines. The appellation includes fourteen towns and villages, one of which is Sancerre. The town of Sancerre is located on a volcanic core and the area is split by a seismic fault, with various soil compositions on either side. The red wines made from Pinot Noir have a tendency to be rather acidic, but some have great balance and character.
- Pouilly-Fumé—made from Sauvignon in the area surrounding Pouilly-sur-Loire, on the east side of the river, facing Sancerre. The best vineyards are on chalky soil mixed with broken flint.
- Pouilly-sur-Loire—a wine from the same area as Pouilly-Fumé, made from the Chasselas grape.
- Menetou-Salon—west of Sancerre toward Bourges, the area makes very good Sauvignons of a lighter style and at an affordable price. Some red and rosé wines are also produced here.
- Further west of Menetou-Salon, Reuilly (on the Cher River) and Quincy produce Sauvignon of a lighter style than Menetou-Salon.
- Coteaux du Giennois (centred on the town of Gien) and Châteaumeillant (which is southwest of Nevers), both have VDQS status.
- Further in the Massif Central, Côtes du Forez and Côtes Roannaises (both on the Loire), and Côtes d'Auvergne and St-Pourcin-sur-Sioule (both on Allier, a Loire tributary) also have VDQS status.

Chapter 11

Champagne

Champagne is both the name of a French province and the name of a sparkling wine made there, according to a distinct method developed more than three centuries ago. Before that time, the wine of the Champagne region was red or white, but always a still wine. The region centres on Reims and covers an area of about 34 000 ha, of which 28 000 ha produce an average 2 000 000 hL of wine annually. Champagne production is the most important production of sparkling wine in France, and the most imitated in the world.

assemblage (ass-som-BLAZH)—harmonious blend of wines forming a cuvée
cuvée (kew-VAY)—the first 2050 L of juice extracted during the first pressing of 4 t of grapes; the blend of young wines destined to become Champagne
disgorgement—removing the sediment brought to the bottle neck by remuage
dosage (doe-ZAHZH)—a mixture of cane sugar and wine added after disgorgement
liqueur de tirage (lee-CUR de tee-RAHZH)—combination of wine, yeasts, and cane sugar used in the making of sparkling wines
pupitres (pew-PEET-ruh)—perforated wood panels or racks in which the bottles are placed neck down for the remuage
rebêche (ruh-BESH)—last pressing
remuage (ruh-mew-AHZH)—process of moving the sediment toward the bottle's neck so that the sediment can be removed
taille (tie)—"cut," that is, the later juice that is pressed from the cake of pomace (marc) after it is loosened with wooden spades

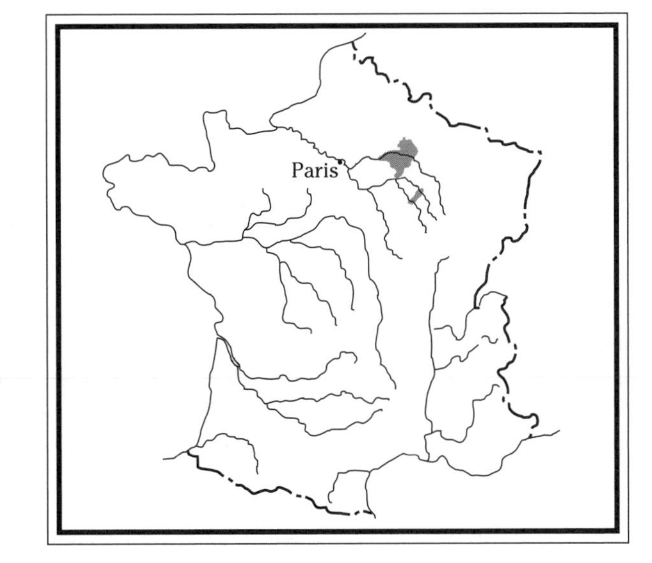

The Wine Region of Champagne

GEOGRAPHY

Champagne, the most northerly wine region of France, is located roughly 145 km to the northeast of Paris. The main towns are Reims and Epernay, and the main rivers are the Marne and the Vesle.

The climate is barely adequate for viticulture. Spring frosts are common, summer nights are cool, and winter comes early. But the summers are (fortunately) long, and provide an average of 1870 hours of sunshine. This long growing season enables the grapes to accumulate a significant concentration of minerals from the soil, and to develop an appropriate sugar-acid balance for the production of sparkling wines. The best vineyards are located on slopes between 80 m–200 m. Below 100 m, the heavier cold air stagnates during the winter and increases the danger of frost. (Because of this, the Marne valley is planted mainly with the more resistant Pinot Meunier.)

The entire region rests on the chalky bed of an ancient sea. The sedimentary soil is composed of two types of chalk: micaster (unique to the region of Champagne and made of fossilized sea urchins) and belemnites (made from mollusc shells). Both provide good drainage but retain moisture. On the tops of the hills are layers of sandstone and clay mixed with lignite (brown coal). This topsoil is a scant metre deep. The chalky subsoil, which is more than 200 m deep, provides ideal conditions of moisture and steady temperature for storing and maturing the 450 000 000 bottles of Champagne in the 200 km of galleries and cellars beneath the towns of Reims and Epernay.

HISTORY

The oldest relics of vines found in the region of Champagne are fossilized vine leaves discovered near the town of Sézanne. These 60-million-year-old fossils have been identified as Vitis sezanennsis, and would have flourished in a subtropical climate. Historical records of the area date back to 52 B.C., when the Romans conquered the region. As they did elsewhere, the Romans improved upon the region's existing vineyards and developed new ones. They built roads and developed the town of Durocotorum (now Reims). For this project they needed stones, which they took from quarries dug out of the limestone subsoil. These quarries later proved to be critical in the making of sparkling wine.

The sixteenth to eighteenth centuries are particularly significant in the region's wine history. Under Henri IV of France (1553–1610), the practice of making pale wine from red grapes began. Dom Pérignon (1639–1715) became a monk, took charge of the cellars of the Benedictine Abbey of Hautvillers, and laid down the principles of making the sparkling wine now known as Champagne.

DOM PÉRIGNON

Dom Pérignon did not invent or create Champagne; rather he developed some practices, perfected others, and systematized more. He enlarged the vineyards of the abbey, improved pruning methods, identified terroirs, perfected the assemblage (selection) of grapes from particular vineyards, oversaw the excavation of large cellars cut into the chalky soil, and supervised winemaking. During his time in charge of the abbey's cellars, Pérignon was confronted by an annoying problem. Every spring, wine stored in the cellars became fizzy and burst the barrels. Since no one truly understood the process of fermentation, it was a phenomenon that Pérignon was unable to cure. His innovation (and major contribution to Champagne) was to perfect and improve this natural tendency of the wine to sparkle. And he imported stronger bottles from England and corks from Spain, to withstand the pressure of the sparkling wine.

In the late eighteenth century, techniques of assembling the red and white grapes began to improve, as did the blending of wines, disgorgement, and the use of pupitres for remuage. In 1836, a French pharmacist named André François calculated the exact amount of sugar (dosage) to be added per bottle to create a controlled carbonation. And in 1884, local producers organized their first professional association.

By 1890, refrigeration baths were developed to help the disgorgement. The total export of Champagne had risen to 21 000 000 bottles.

Unfortunately, at this time phylloxera appeared in the vineyards. This was followed by an unpopular delimitation of the vineyards in 1908, and violent demonstrations against it in 1910 and 1911. By 1919, however, the region's wine industry seemed to be getting back on track: classification of the vineyards began. The best communes were ranked according to the quality of their productions. The commune awarded the highest evaluation (100%) received 100% of the market price, a 70% evaluation received a price of 70% of the market price, and so on. This ranking system continues, but, since 1985, ranges from 100% to 80%.

LAW AND REGULATIONS

The Comité Interprofessionnel du Vin de Champagne (CIVC) is an organization created by the French government for the control and regulation of this unique wine. It is responsible for legislation regarding the growing of grapes and winemaking techniques, the ranking of the communes, the pricing of wine, and the promotion of Champagne.

To be called a Champagne:

- the wine must be made from Chardonnay, Pinot Noir, and/or Pinot Meunier grapes grown in the départements of Aisne, Aube, and Marne, plus a few communes in Seine-et-Marne and Haute-Marne.
- the wine must be made using the méthode Champenoise and must not exceed 13% alcohol.
- the wine must be made in the AOC Champagne region.
- the wine must be aged for a minimum of one year in the bottle before distribution.
- the word *Champagne* must be printed on the cork and on the label.
- the grape yield must not exceed 13 t/ha and the pressing not exceed 102 L from 160 kg of grapes.

The méthode Champenoise is described later in this chapter.

In addition, to be designated vintage Champagne, a wine must:

- be aged for a minimum of three years in the bottle before distribution.
- show the vintage year on the label.
- have a minimum of 11% alcohol.
- not exceed 80% of the harvest.

Champagne is the only AOC wine that does not need to state its AOC on the label.

Champagne labels indicate the origin of a wine with these abbreviations:

BOB—buyer's own brand

CM (coopérative-manipulant)—wine made by a co-operative

MA (marque auxilliaire)—wine made as a second label or a Champagne according to the specifications of the buyer and labelled under the name of that buyer (e.g., a restaurant or a supermarket)

NM (négociant-manipulant)—wine labelled with the name of the merchant who made the wine but bought the grapes from another grower

RD (récemment degorgé)—used for Champagne kept a long period on its lees and corked only recently

RM (récoltant-manipulant)—wine made by a single producer from grapes grown by the producer

SR (société des recoltants)—wine made by a group of growers

The following are specific AOCs within Champagne:

- AOC Coteaux Champenois—still, red, rosé, or white wines made from the same grapes, from the same area, and with the same restriction for yield and pressing as for AOC Champagne wines.
- AOC Rosé de Riceys—still rosés made from Pinot Noir harvested from specific vineyards in the commune of Riceys. The minimum level of alcohol is 10%. Restrictions for the yield and pressing are the same as for Champagne.

In addition, Champagne rosé is a VQPRD. It is made by blending white wines with 10%–15% red wine prior to the second fermentation. It is the only VQPRD rosé for which this mixing of red and white is authorized.

GRAPE VARIETIES

Three grape varieties are used to create Champagne: Pinot Noir (37% of the vineyards), Pinot Meunier (36% of the vineyards), and Chardonnay (27% of the vineyards). Several other grapes (L'Arbanne, Petit Meslier, Pinot Gris, and Pinot Blanc) were used, but have slowly disappeared from Champagne production.

- Pinot Noir favours warm (not hot) weather, and its early budding and flowering make it susceptible to spring frosts. Pinot Noir gives backbone and structure to Champagne.
- Pinot Meunier is more resistant to frost, flowers late, and has the ability to send out second shoots if damaged by frost. It yields abundantly and prefers rich, moist soils that are low in chalk. It produces thin wines that are short in acidity and do not age well, but are very fruity in their youth. Pinot Meunier contributes freshness and aroma at an early stage in the blend. It is not permitted in blends destined to be Premiers Crus, Grands Crus, and Grande Marque.
- Chardonnay (often wrongly called Pinot Chardonnay), is the only white grape used in the production of Champagne. Flowering quite late, it is sensitive to coulure and botrytis but produces well in soil made of clay or chalk. Chardonnay gives Champagne its finesse and elegance.

WINE PRODUCTION

The following are the various types of Champagne:

- Grande Marque—a prestige Champagne from a particular firm (e.g., the wine *La Grande Dame* from Veuve Clicquot-Ponsardin).
- Grand Cru—Champagne made from grapes coming from vineyards rated 100%.
- Premier Cru—Champagne made from grapes coming from villages rated 90%–99%.
- Blanc de Noir—white wine made from red grapes.
- Blanc de Blanc—white wine made from white grapes.
- Crémant—Champagne with only 3 atmospheres of pressure, instead of the usual 5–6 atmospheres for regular Champagne.

One atmosphere is equivalent to a pressure of about 101 kPa.

Méthode Champenoise

The wine Champagne undergoes the following process.

Harvesting

Only well-formed and unblemished grapes are kept. (Many fine producers harvest the grapes by hand.) The chosen grapes are placed in baskets or perforated plastic bins, to allow any juice to drip out. The containers are carefully and gently transported to the press houses, usually located within 15 km of the vineyards.

Pressing

The press is loaded with exactly 4 t of one grape variety (Chardonnay, Pinot Noir, or Pinot Meunier). A few wineries use horizontal presses, but the traditional cocquard is a vertical press made of oak. It is only 80 cm high, and can be square or circular, with a surface of 9 m^2. A pressure of 20 t/m^2 to 40 t/m^2 is applied.

Since 1992, regulations require that the pressing procedure be as follows. The cuvée is pressed in three stages: the first pressing yields 1025 L, the second 615 L, and the third 410 L, for a total of 2050 L. The taille consists of the fourth and fifth pressings, which together yield 500 L, for a grand total of 2550 L grape juice.

A sixth pressing of up to 200 L can be made, but this juice can be used only for distillation, not for the vinification of Champagne. This last pressing is called the rebêche.

All the pressings are done very quickly to prevent colour bleeding from the skins. This is particularly important since the (mainly) red grapes are used to produce white juice.

First Fermentation

The juices are identified according to the grape, origin, and level of pressing, and transported separately to the winery. There they rest before being decanted. After transfer into stainless steel vats (with capacities of 60 hL to 1000 hL), the juice undergoes its first fermentation.

A few wineries continue to ferment some wine in wood for added character.

During the fermentation, the temperature of the must cannot exceed 20°C. Most wines are allowed to undergo malolactic fermentation. Other wines, intended for long cellaring, are not. These last mentioned wines are more difficult to make: if the malolactic fermentation happens spontaneously during the second fermentation, the wine will cloud and take on an undesirable flavour.

Elaboration of the Cuvée

This is the most critical phase in the making of Champagne. A group of experts tastes up to seventy young wines from the three grapes, and from various vineyards. A small proportion of wines from previous years can also be included.

The aim is to create a blend (cuvée) that fits with the style of the firm. To do so, the tasters must anticipate what flavour the wine will have after the second fermentation and cellaring. It is an exceedingly complex task, and mistakes cannot be corrected later!

Second Fermentation

After the cuvée is created, it is stored in large tanks until the end of the winter. In the spring, the liqueur de tirage is added, and the wine is then bottled. The liqueur de tirage is a combination of older wine, yeasts, and cane sugar in the proportion of 500 g sugar to 1 L wine. The yeasts will ferment in the bottle to form carbon dioxide and raise the alcohol by 1.3%.

The bottles are then transported to cellars (cut into the moist chalk of the region) for the second fermentation. Million of bottles are laid on their sides. Thin strips of wood between the layers cushion the weight and prevent the rows from collapsing if any bottle bursts. The temperature must be a constant 9°C–10°C. This long, slow fermentation produces Champagne's fine bubbles.

Remuage

When the second fermentation is complete, the dead yeasts that have settled on the sides of the bottles must be removed from the wine, so that it will not be cloudy when served. Remuage is the traditionally slow process of removing the sediment. The bottles are placed neck first in perforated pupitres (wooden racks about 1.5 m high and 75 cm wide, holding 60 bottles per side) in a near horizontal position. Every other day, cellar staff will riddle each bottle (i.e., give it a shake), and change its slant. Eventually, the bottles are in a near-vertical position, neck down. The process brings the sediment in the bottle down to the neck. Once this is done, the bottles are stored sur pointe (upside down).

A cheaper alternative to riddling by hand is the gyropalette—a computerized machine that jiggles and rotates bottles in a metal cage.

Disgorgement

Disgorgement removes the sediments brought to the bottle's neck by remuage. Traditionally, a worker holds the bottle in a near-horizontal position, removes the stopper so that the pressure in the bottle expels the sediment, and tilts the bottle more upright so as not to lose too much wine. The bottle is then temporarily stopped up, to limit the loss of carbon dioxide, and placed on a rotating plate, ready to receive the final dosage.

Non-vintage Champagnes are usually disgorged mechanically. The bottles are held upside down by a machine, and the necks immersed in a glycol bath refrigerated to –18°C. This freezes the wine and sediment in the neck to the consistency of sherbet. A machine removes the stopper and the sediment is expelled.

Dosage

Prior to corking, most Champagne receives the addition of a dosage, or liqueur d'expédition. This is a mixture of wine and cane sugar in the proportion of 500 g–750 g sugar diluted in 1 L wine. (Producers decide upon a dosage based on the preferences of the intended market.) The dosage is indicated on the label.

The following terms on a label indicate the level of dosage. Dosage is expressed as percentage of the volume of wine in the bottle.

brut—up to 1%

extra sec (extra-dry)—from 1% to 3%

sec (dry)—from 3% to 5%

demi-sec (semisweet)—from 5% to 8%

doux (sweet)—from 8% to 15%

Some Champagne is sold without dosage and may be called *Brut Zero, Brut de Brut, Ultra Brut, Brut Sauvage, Brut Total, Brut Nature*, depending on the winery.

Finally, the wine is corked and fitted with a wire cage over the cork. If the wine is to be released right away, it will be labelled and dressed with foil at the neck. These final stages can be done mechanically, or by hand.

Champagne does not need to be aged, since it is at its best and ready to drink as soon as it is released for sale. Premium and vintage Champagnes, however, will keep for ten years in a good cellar. Some consumers appreciate older vintages.

WINE REGIONS

Montagne de Reims

This area extends south of Reims from the villages of Gueux to Ambonnay near the Marne River. Its vineyards on the north slopes of the Montagne de Reims are mainly planted with Pinot Noir. (The mountain is actually only 287 m high.) The area around the village of Bouzy makes the best still red wine of the Champagne region.

Vallée de la Marne

This is a succession of vineyards trailing for 70 km along the Marne Valley, from the village of Charly in the west and past Mareuil-sur-Ay in the east. The soil contains some chalk but more clay. The vineyards are planted with Pinot Meunier and Pinot Noir.

Côte des Blancs

The area takes its name from the vineyards mainly planted with the white (blanc) Chardonnay grape. The Côte des Blancs starts south of Epernay and extends to Bergères-les-Vertus. This region is warmer than the Montagne de Reims and the Vallée de la Marne, and the soil is chalkier.

Côte de Sézanne

This is the newest district of Champagne (1960s). The vineyards from the village of Allemant to Villenauxe-la-Grande are planted with Chardonnay, on south-facing slopes.

Bar-sur-Aubois and Bar-Séquanais

The Aube area is the most southerly part of Champagne and includes Bar-sur-Aubois and Bar-Séquanais. The vineyards are mainly planted with Pinot Noir, on a soil made of Kimmeridge marl and clay—similar to that of Chablis, only 60 km away. The hotter summer here can better ripen the Pinot Noir, which has a tendency to yield wines with a stronger character. In theory, no wine should dominate a cuvée and the wine from Aube has the habit of leaving its signature fragrance in the blend.

This area produces Rosé de Riceys, one of the rarest rosés of France. The wines are made, not by mixing red and white wine, but by bleeding the colour from the skin of the Pinot Noir. The AOC was granted in 1971.

There are three types of Rosé de Riceys:

- those made in tanks without wood contact—a very fruity wine that should be consumed within two years.
- those aged in wood for six months—a wine with more body that should be drunk within two years.
- those aged in wood for at least two years and in the bottle for five years, to develop its full character.

ITALY

Ancient Italy was called Oenotria—the land of wine. Today it produces a stunning 59 300 000 hL—in other words, one-fifth of the world's wine production—in a land half the size of France. Thus Italy is the largest wine producer in the world, and second only to Spain in terms of the land dedicated to vineyards. The average annual consumption is about 63 L per capita.

GEOGRAPHY

Typically described as having the shape of a boot, mainland Italy extends 1200 km south from the Alps into the Mediterranean, with the Apennines forming a spine along the peninsula. Calabria is the toe of the boot, with Sicily at its tip in the warm Mediterranean waters. Italy is bordered on the north by France, Switzerland, Austria, and Slovenia. The Adriatic, Ionian, Tyrrhenian, and Ligurian seas surround the remainder.

In general, the climate is Mediterranean, although in northern Italy it might be better described as continental. It does change from north to south, but altitude tends to play a greater role than does latitude. Nearly 80% of the land is hilly or mountainous, thus providing ideal exposures and slopes for climbing vineyards. There is a wide a range of soils.

La Dolce Vita

Italy brings to mind images of the arts, fashion, sun, la dolce vita ("the sweet life"), and of course, wine. As with most countries, stereotypes abound (e.g., cuisine is always made with olive oil, spaghetti with tomato sauce and Parmesan cheese is the essential meal, Chianti comes only in wicker flasks) but the reality is far more varied and complex. If one might generalize, however, wine is woven into the fabric of Italy as possibly nowhere else. Stop in at a local vineyard—they are everywhere throughout Italy—and hear how the grower speaks of the wine. Taste it, freshly drawn from the barrel and served in a stemless goblet. See how gently your host holds the precious liquid and talks of its robe, its body, its nose. You may well think this wine were a beloved child.

Wine Regions of Italy

abboccato (ah-bo-KAH-toh)—medium dry
amabile (a-MA-bee-lay)—semisweet
bianco (BYUNK-oh)—white
cantina (cun-TEEN-a)—cellar or winery
cantina sociale (so-CHYAH-lay)—co-operative winery
classico (CLAH-see-ko)—geographical term applied to quality wine. It shows that the wine comes from the central, traditional, and better part of a production zone.
colli (COAL-lee)—hill, slope
dolce (DOLE-chay)—sweet
fattoria (fah-toh-REE-ya)—farm or estate consisting of several poderi
frizzante (free-TSUN-tay)—semi-sparkling, rather than spumante
imbottigliato all'origine (eem-boat-teel-YAH-toe allo-REE-jee-nay)—estate bottled
imbottigliato nella zona di produzione (nell-ah ZONE-ah dee pro-doot-SYOH-nay)—bottled in the area of production
liquoroso (lee-kwo-ROE-soh)—refers to wine which has a high alcohol content, but is not necessarily fortified
metodo classico (may-TOH-doh CLAH-see-ko)—term on sparkling wine labels to indicate that it has been made using the méthode Champenoise. Current EU rules forbid member countries from labelling their sparkling wines with the term *méthode Champenoise.*
novello (no-VEL-loh)—the first wine of the harvest. It is the equivalent of the French *vin nouveau* or *primeur.*
passito (pah-SEE-toh)—produced from semi-dried grapes. Most passito is sweet.
podere (poh-DAY-ray)—plot, small farm, or estate
poggio (PUDGE-yoh)—hill
produttore (pro-doo-TOR-ee)—producer
recioto (ray-CHYO-toh)—special strong wine made from semi-dried grapes and containing some residual sugar
recioto amarone (ah-ma-ROE-nay)—recioto wine that is dry and higher in alcohol than recioto
riserva (ree-ZAIR-vah)—quality wine aged longer in cask
riserva speciale (spate-SYAH-lay)—aged longer than riserva
rosato (roh-ZAH-toh)—rosé
rosso (ROE-so)—red
secco (SECK-ko)—dry, when applied to still wines; semisweet, when applied to sparkling wines
spumante (spoo-MUN-tay)—sparkling
stravecchio (strah-VECK-yo)—very old
superiore (soo-pair-YORE-ay)—quality wine with a higher alcohol content and (sometimes) longer ageing
tenuta ((ten-OO-tah)—one of the holdings of a farm or estate
vecchio (VECK-yo)—old
vendemmia (ven-DEM-yah)—vintage

Novello cannot be sold before November 6 and must be bottled before December 31. Residual sugar is limited to 10 g/L, minimum alcohol is 11%, and at least 30% of the wine must be made by carbonic maceration.

HISTORY

Long before the Phoenicians reached Apulia on the southeastern coast around 1000 B.C., the tribes of the peninsula were cultivating wild strains of Vitis vinifera and making wine. Because the Phoenicians were traders not conquerors, they recorded the winemaking they saw, traded with the local inhabitants, and established commercial ports in Sicily and on mainland Italy.

Wine was also a part of daily life in the north as far back as 750 B.C. In what today is Tuscany, Etruscans made wine and trained vines onto pergolas or trees, in a method known as *etrusco*, or *alberate*. Eventually, they spread their knowledge north into the Po Valley and south as far as the Bay of Naples.

By far the greatest influence in Italy's early history of winemaking came from the Greeks. They established trading colonies in southern Italy (Naples in 750 B.C.) and Sicily (in 735 B.C.). They also introduced new varieties from their homeland, including Aglianico, Muscat, and Malvasia. As well, they introduced a second method (known as the *greco* method) of pruning and training vines—in the short, freestanding, bush or goblet shape.

In time, the Romans developed new viticultural and winemaking practices. They developed cane pruning (training vines horizontally along reed canes), divided vineyards into categories of growths according to quality, and established winemaking techniques that would not be improved until the seventeenth century.

During the Middle Ages (A.D. 476 to the fifteenth century), winemaking was the domain of the Church and well-documented within it, but monastic vineyards were maintained to produce wine for the Mass, not the masses. However, the Renaissance (fourteenth to sixteenth centuries) expanded the vineyards, albeit slowly.

Wine traders in Genoa and Venice shipped wine all over the Mediterranean and as far as England. By the nineteenth century, early attempts to improve the quality of the wines by strictly selecting sites and vines brought new vigour to the industry. Trade increased and Italian winemakers learned techniques used abroad. In Piedmont, they started to use French barriques (in place of large chestnut vats) for Barolo. The town of Asti started to make sparkling wines. In Sicily, an Englishman named John Woodhouse developed a fortified wine from grapes grown around Marsala.

The nineteenth century also brought powdery mildew and phylloxera. Hundreds of indigenous vines disappeared. Replanting took time and resulted in familiar vines being grafted onto American rootstocks, plus the introduction of other varieties such as Cabernet Sauvignon, Merlot, and Chardonnay.

In the twentieth century, the industry has thrived. Soon after World War II (1939–1945), Italy became the world's largest producer of wine. In 1963, Italy established a system modelled on the French appelation of origin. Italy's best wines began to receive recognition in the international market. Today, new generations of winemakers travel regularly to research the market, inspect the latest equipment, and identify emerging trends. Marketing is aggressive and wine exports reached 12 500 000 hL by 1993.

The challenge for the industry now is to make sense of Italy's confusion of grapes and wines.

LAWS AND REGULATIONS

Historically, Italy's wine industry has seemed somewhat chaotic, with a great number of producers making wine from countless grape varieties. Thus, Denominazione di Origine Controllata (DOC) laws enacted in 1963 were widely welcomed. The regulations were modelled on the French AOC system, but with added restrictions (e.g., prohibiting chaptalization, requiring the declassification of an entire crop if a vineyard exceeds its maximum yield, reducing the amount of sulphur used, and prohibiting the production of more than one wine in any DOC zone). In 1992, the laws were revised to offer a simpler, more precise, and yet broader framework than those of 1963 laws. The 1992 laws, known as the Goria laws, adhere to the principles of the EU framework.

DOC History

The DOC laws of 1963 honoured the traditions of winemaking. However, their strength became their weakness: the regulations defended tradition rather than quality. This became apparent when some new wines (that had to be listed only as *vino da tavola*) achieved higher quality and price than the DOC wines of the same area. The 1992 laws (named for the then prime minister Goria) are seen as significant improvement. They permit the use of vineyards, estate, or commune names instead of the name of the DOC zone.

Within the larger EU requirements of a two-tier system (table wine and quality wine), there are four basic categories in the new classification:

Vino da Tavola

Any Italian wine of any colour. The label cannot state the grape variety, vintage, or locality.

Vino da Tavola con Indicazione Geografica Tipica

Equivalent to the French vin de pays and German landwein. It must come from a specific area not classified as a DOC. The label can state the grape variety.

Denominazione di Origine Controllata (DOC)

Wine from specific vineyards and made from authorized grapes in a delimited geographical zone. The DOC may apply to an entire zone, to a commune, or to a specific vineyard within the zone. The wine must be made according to well-defined rules as to yield, grapes, method, and length of ageing. It must also pass a chemical analysis, and attain quality standards for colour, nose, and taste as appraised by a select panel of tasters.

Denominazione di Origine Controllata e Garantita (DOCG)

Wine that meets the requirements of DOC as well as stricter controls and tasting requirements. Any wine judged not to achieve excellence is declassified.

All wines that are exported have the seal of the Instituto Nazionale Enologico (INE) on the top of the bottle.

GRAPE VARIETIES

Italy has innumerable varieties of native grapes. Many are clones or subvarieties that, over the centuries, have mutated and adapted to a particular area. Over 1000 types have been counted. Grape growers typically practise monoculture, some with modern varieties, some reviving traditional and oft-forgotten ancient vines.

Monoculture is the cultivation of a single product.

About 400 varieties are authorized for cultivation in DOC zones. The best and most important include:

- Aglianico—imported by the Greeks to southern Italy around the eighth century B.C. It is one of the best red varieties of the south and does particularly well in Basilicata (in the Aglianico del Vulture wine) and in Campania (in the Taurasi wine).
- Barbera—a native of Piedmont. This red variety is the second most-planted variety, after Sangiovese.
- Malvasia—an ancient grape brought from the Greek city of Monemvasia and famous in Ancient Roman times as Apianae. It has a lighter colour in the northern part of Italy and somewhat darker skin in the south, much like Muscat, a grape to which it is related. Wine made from Malvasia can be either light or dark, sweet or dry.
- Montepulciano—used extensively for red wine in central and southern Italy. The name should not be confused with Vino Nobile di Montepulciano, a wine produced from Prugnolo Gentile, a clone of Sangiovese.
- Nebbiolo—one the best red grapes of Italy, it is used for Barolo, Barbaresco, Gattinara, and others. It is also known as Spanna, Chiavennasca, Picutener, Picotendro, and Pugnet.
- Sangiovese—the most widely planted red variety of Italy. Its origin is thought to be native Etruscan. It has a multitude of clones, the two main ones being Sangiovese Grosso (also called Brunello di Montalcino, Prugnolo di Montepulciano, or Prugnolo Gentile) with medium-sized berries, and Sangiovese Piccolo (also called Sangioveto in the Chianti region), which is the small-berried variety.
- Trebbiano—known as Saint Emilion in Cognac and Ugni Blanc elsewhere in France. It is a prolific white grape and the most planted of all varieties in Italy. The grape was planted by Etruscans and has been popular in Italy ever since.

WINE REGIONS

Italy has 20 wine regions, 94 provinces, approximately 250 regulated wine zones, 8090 communes, and more than 2 000 000 producers. Among them they create a huge variety of wines.

VALLE D'AOSTA

This mountainous region on the northwest border has the smallest wine production in Italy, the highest vineyards, and the highest per capita consumption. It is the source of a very few rare, subtle wines made from unique local varieties. The region is bilingual; the second language is French.

Valle d'Aosta (DOC 1986) has seven subzones including:

- Blanc de Morgex et La Salle—a peculiar white wine made from ungrafted Blanc de Valdigne grown as high as 1400 m. Phylloxera has never reached this altitude.
- Donnaz (DOC 1971)—a red wine made from Nebbiolo.
- Enfer d'Arvier (DOC 1972)—a red wine made from Petit Rouge with a yearly production of only 6000 bottles.

The other zones are Arnad-Monjovet, Chambave, Nus, and Torette, all enjoyed locally and by visiting tourists. Very little is exported.

PIEDMONT

Located at the foot of the Alps and bordering France, Piemonte (Piedmont) has more DOC wines than any other Italian region, with thirty-eight geographical DOC areas and forty-two distinct types of wine. This is the land where the Nebbiolo grape reigns supreme, clinging to sun-kissed hillsides and drinking from the morning dew. Other important varieties grown include Barbera, Moscato (Muscat), and Cortese. Nearly half of all the Piedmontese wines are made from Barbera. Piedmont is also the birthplace of Vermouth.

Spanna is a popular Piedmont vino da tavola, produced mainly in the area of Novara from Nebbiolo grapes. In good vintages (and with sufficient barrel ageing) this wine offers exceptional value.

VERMOUTH

In 1786, Antonio Benedetto Carpano of Turin decided to flavour the local white wine with aromatic herbs and spices, reviving a drink from Ancient Roman times. The beverage was first named Punt e Mes ("point and a half"). It eventually became known as Vermouth, from the German *wermut* (wormwood), a plant used in concocting the drink.

Piedmont includes the following DOCs:

- Barolo (DOC 1966, DOCG 1980)—a robust and generous red wine made from Nebbiolo's subvarieties Michet, Lampia, and Rosé. The wine is named after the small village of Barolo located a few kilometres southwest of Alba. Vineyards are situated on the upper portions of steep south-facing hills surrounding the area. It is one of the greatest wines of Italy and was the first to receive the DOCG status. Deep in colour, with assertive, complex aromas of black pepper, tar, truffles, leather, and tobacco, this wine needs long ageing to develop its full personality.

  Barolo can be subdivided into two distinct districts: the valley of Serralunga around the villages of Serralunga d'Alba, Castiglione Falletto, and Monteforte d'Alba (more robust and long-lived wines with dense flavours and a lengthy finish), and the valley of Barolo surrounding the villages of Barolo and La Morra (softer and less powerful wines which can acquire remarkable fragrance, subtlety, and finesse).

TWO STYLES OF BAROLO

There are two schools of thought in the making of Barolo. Each school produces a distinct style—both good, and certainly different.

The old school practises long fermentation of up to two months on the skins followed by extended ageing (up to eight years) in large chestnut or oak vats called *botti*. These wines are massive, with tremendous amounts of extract and tannins. They are made to last for decades.

The new school believes that the wine best develops its qualities in the bottle rather than in wood, where excess oakiness tends to mask the wine's true character. This style of wine is fermented on the skins for ten days only and is aged in small Bordeaux casks for the two-year minimum required by law. This Barolo is more elegant and ready to be drunk early on, although it too can age admirably to develop harmonious subtleties.

- Barbaresco (DOC 1966, DOCG 1980)—made from the same grapes as Barolo. Named for the village of Barbaresco north of Alba, the area encompasses the townships of Neive, Barbaresco, Treiso, and Alba. The vineyards cover the rolling Langhe Hills (200 m–350 m) to get maximum sun exposure. The climate is drier and warmer than in Barolo, and the grapes generally ripen earlier.

Long regarded as the baby brother of Barolo, Barbaresco is slightly less firm and reaches maturity at an earlier age. Its main characteristics are softer, thinner body, higher acidity, and intense aroma of violets. The wine is made according to the same regulations as Barolo, with two exceptions: it can be released after two years, with only one year in wood, and the minimum alcohol level is 12.5% rather than Barolo's 13%.

- Gattinara (DOC 1967, DOCG 1990)—made from Spanna (Nebbiolo) with a possible addition of up to 10% Bonarda di Gattinara. On the south and east side of the village of Gattinara, the meagre soil (laced with broken porphyry and quartz) produces a unique terroir in which Nebbiolo yields an aromatic wine, high in acidity and with firm tannins. In general, the wine needs a decade to soften up but can develop a wonderful bouquet, charm, and elegance. In great years, Gattinara can equal Barolo's depth and sophistication.
- Ghemme (DOC 1969)—located on the south-facing slopes of Ghemme, directly across the River Sesia from Gattinara. It is comparable to Gattinara, but lighter. It is made from Spanna (60% to 85%), Vespolina (10% to 30%), and the rare Bonarda Novarese (15%).
- Nebbiolo d'Alba (DOC 1970)—a red wine from vineyards on both sides of the Tanaro River, between Barolo and Barbaresco and to the east. The wine comes in a variety of styles: dry, sweet, and sparkling. The dry version makes a pleasant wine ready to drink after two or three years of ageing.
- Barbera d'Alba (DOC 1970)—the Barbera grape is believed to be a native variety of Piedmont and widely planted there. Around Alba, it produces a deeper red wine than in any other part of Piedmont, perhaps because regulations allow up to 15% Nebbiolo to be added. Barbera d'Alba wine matures well and can develop an intense yet delicate aroma. This wine must be aged for at least one year in wood.
- Barbera d'Asti (DOC 1970)—produced from the same grape variety as Barbera d'Alba, but from the larger area surrounding the town of Asti. The moist and heavier clay soil here produces a full-bodied wine with a zesty aroma and supple tannins. Modern temperature control techniques and early malolactic fermentation result in a softer and fruitier wine than in the past. As with Barbera d'Alba, this wine needs a minimum one year in wood.

- Barbera dei Colli Tortonesi (DOC 1973), Barbera del Monferrato (DOC 1970)—produced from Barbera. The former is 80%–100% Barbera with possible additions of Freisa, Bonarda Piemontese, and Dolcetto. The latter may contain up to 25% Freisa, Grignolino, and Dolcetto. Both wines are light in body and should be consumed young.
- Gavi or Cortese di Gavi (DOC 1974)—the most popular white wine of Piedmont, from Cortese grapes grown on the eastern part of the Monferrato Hills in a balmy climate. The wine has a clean, crisp taste and a candy nose. Gavi is also made in a spumante version.
- Moscato d'Asti and Asti Spumante (DOC 1967, DOCG 1994)—both made from Moscato Bianco (sometimes called Moscato Canelli), the most important white grape of Piedmont. This intensely aromatic grape has flourished in the region since the seventeenth century. Asti Spumante is at its best fresh and young. It suffers too easily from warm storage and should therefore be kept cool at all times.

The local spumantes and frizzantes have established themselves as among the world's best-known Muscat wines.

ASTI

Asti producers take great precautions to preserve the refined aroma and delicate sweetness of the Moscato (Muscat) grape. The grapes are harvested early to preserve acidity, gently crushed to prevent any bitterness, and thoroughly centrifuged and filtered. The must is stored in stainless steel tanks at near-freezing temperatures and fermented in small batches according to market needs.

Unlike other sparkling wines, Asti is made in one go, without a second fermentation. The must is inoculated with yeast, sealed in a stainless steel tank, and fermented to the desired level of alcohol and sweetness. Natural carbon dioxide is retained. The fermentation is arrested by refrigerating; the wine is then filtered, bottled, and immediately shipped. The alcohol level is 7%–9%, with natural sweetness between 3%–5%. Some Asti is fermented totally dry (brut) and has a higher alcohol level. Moscato d'Asti is a frizzante with only 5% alcohol and residual sugar of 5%–7%.

Other DOC wines of Piedmont include Carema, Erbaluce di Caluso, Lessona, Bramaterra, Boca, Sizzano, Fara, Grignolino del Monferrato Casalese, Rubino di Cantavenna, Gabiano, Colli Tortonesi, Dolcetto di Ovada, Cortese dell'Alto Monferrato, Brachetto d'Acqui, Dolcetto d'Acqui, Freisa d'Asti, Freisa di Chieri, Malvasia di Casorzo d'Asti,

Grignolino d'Asti, Malvasia di Castelnuovo Don Bosco, Ruchè di Castagnole Monferrato, Dolcetto d'Asti, Roero, Roero Arneis, Dolcetto d'Alba, Dolcetto di Diano d'Alba, Dolcetto di Dogliani, Dolcetto delle Langhe Monregalesi, and Loazzolo.

LIGURIA

This narrow strip of land is the Italian Riviera and extends from the French border to near La Spezia at the border with Tuscany. All along the coast of the Gulf of Genoa, terraced vineyards tumble into the Mediterranean Sea like a giant staircase. Over 100 vine varieties are cultivated here.

The region is divided into two main territories—the western riviera (Riviera di Ponente) and the eastern (Riviera di Levante).

Riviera di Ponente

Riviera di Ponente extends from the French border to Arenzano west of Genoa. It includes the following DOCs:

- Dolceaqua (DOC 1972)—a pleasant, fruity, red wine from the Rossese grape.
- Riviera Ligure di Ponente (DOC 1988)—a larger area where Pigato, Vermentino, and Ormeasco grapes are among those cultivated. Ormeasco (called Dolcetto in Piedmont) yields a dark-coloured wine with a pronounced red-berry aroma. Ormeasco is also turned into a dry rosé called Sciacchetrà, a local abbreviation for *schiacciare e trarre* (crushing and bleeding the juice from the skins).

Riviera di Levante

Riviera di Levante is tucked into the area southeast of La Spezia. This tiny area has two DOC zones:

- Cinqueterre (DOC 1973)—literally, five lands or five villages. The wine is a dry white, made from a combination of 60% Bosco and up to 40% Albarola or Vermentino.
- Colli di Luni (DOC 1989)—a red wine made from up to 70% Sangiovese. At least 15% Canaiolo, Pollera Nera, and Ciliegiolo Nero are added, and other red varieties can make up to 25%. Cabernet can make up no more than 10%. A white is also produced using 35% Vermentino, 25%–40% Trebbiano Toscano, and 30% other whites. Wine labelled Vermentino is made from 90% Vermentino.

Lombardy

At the foot of the Alps and south of Switzerland, Lombardy (Lombardia) makes a wide variety of wines—red, white, and rosé, and both still and sparkling. With nearly nine million inhabitants, it is the most populated province. The land is rich and very fertile. Even the difficult Nebbiolo has adapted well to the rugged mountainous region. The following are three key DOCs:

- Oltrepò Pavese (DOC 1970)—a large area east of Voghera. It is the most productive in all Lombardy with an output of 1 300 000 hL, of which 260 000 hL are classified as DOC. The red wines are made from Barbera, Bonarda, Croatina, Malvasia, Pinot Nero (Pinot Noir), Ughetta, and others. Some of the best-known wines are Buttafuoco, Barbacarlo, and Sangue di Giuda. The whites are produced mainly from Chardonnay, Riesling Italico, Renano, Müller-Thurgau, Pinot Grigio (Pinot Gris), Moscato (Muscat), Cortese, and Malvasia. The area also makes a good sparkling wine.
- Valtellina (DOC 1968)—a long, narrow band north of the Adda River from Tirano to Ardenno. The vineyards are terraced and extend up to elevations of 800 m on south-facing hills. A simple, red Valtellina is made from 70% Nebbiolo (locally called Chiavennasca) and 30% Pinot Nero (Pinot Noir), Merlot, Pignola Valtelinese, Rossola, and Brugnola. Valtellina Superiore contains 95% Nebbiolo from around the towns of Inferno, Grumello, Sassella, and Valgella. It must have a minimum 12% alcohol level and age for two years in wood—not the one year required for regular Valtellina. Sfursat (also known as Sfurzato and Sfurzat) is produced from 70% Nebbiolo left to dry in the sun to a raisinlike state. The result is a powerful red wine of 14.5% alcohol and a Port-type character.
- Franciacorta (DOC 1967)—located between the town of Brescia and Lake Iseo. Vineyards are planted on ancient glacial moraine and benefit from the lake, which moderates the climate. A multitude of grapes produce red, white, and rosé wines. It is the sparkling wine, however, that gets the most attention. This is a metodo classico wine blended from Chardonnay, Pinot Bianco (Pinot Blanc), Pinot Grigio (Pinot Gris), and Pinot Nero (Pinot Noir). A rosato spumante is also made from these grapes. Red wine with a bright, ruby colour and a pleasant, assertive bouquet is made from Cabernet Franc,

Barbera, Nebbiolo, and Merlot. A pale, greenish-white wine made from Chardonnay and Pinot Bianco has a refreshing, crisp acidity. The Lake Garda area is also well known for Chiaretto, a rosé.

Other Lombardy DOCs include Valcalepio, Cellatica, Botticino, Capriano del Colle, Riviera del Garda Bresciano, Lugana, Tocai di San Martino della Battaglia, Colli Morenici Montovani del Garda, Lambrusco Mantovano, and San Colombano al Lambro.

TRENTINO-ALTO ADIGE

This is the most northern region of Italy and, as one would expect, produces Italy's best white wines. Fully 55% of the region's production is DOC wine, the highest percentage for any Italian region. The vineyards are located in Alpine valleys where the cool, sunny climate is perfect for aromatic, fresh, crisp wines. Reds tend to be light in colour yet full of fruitiness.

Trentino and Alto Adige are two small provinces joined together for administrative purposes.

Trentino

Trentino is south of Alto Adige. Vineyards edge the banks of the Adige and Isarco rivers. Because the climate is warmer than in Alto Adige, the land is more productive. A new consortium in the area is producing an increasing amount of metodo classico sparkling wines called Spumante Trento Classico.

The DOCs include the following:

- Trentino (DOC 1971)—the largest appellation, covering red, white, rosé, sparkling wines. A delicious Vino Santo is made from Nosiola. Most wines are varietals from a wide range of grapes including Cabernet, Chardonnay, Lagrein, Lambrusco, Muscat, Müller-Thurgau, Pinot, Riesling, and Traminer Aromatico (Gewürztraminer).
- Sorni (DOC 1979) and Casteller (DOC 1974)—primarily Schiava-based red wines.
- Teroldego Rotaliano (DOC 1971)—a pretty violet-red wine with a charming raspberry and almond aroma. It is best when aged about six years.
- Valdadige/Etschtaler (DOC 1975)—covers thirty-eight communes and describes all wines grown in the Adige valley as far as Veneto.

Alto Adige

Alto Adige was called South Tyrol until it was ceded from Austria after World War I (1914–1918). The region is autonomous and fiercely bilingual. Places and labels often have German names, and Austria's influence is reflected in the style of the wines. Many wines are labelled after the variety, and Pinot Grigio (Pinot Gris, Ruländer) excels here as does Traminer Aromatico (Gewürztraminer). The only arable land is along the banks of the Adige and Isarco rivers. Vines are grown on south-facing slopes between 250 m–1000 m.

The DOCs include:

- Santa Maddalena/St Magdalener (DOC 1971)—one of the best reds of the region. It is made from Schiava grapes and has an intense ruby colour, with a pleasing bouquet and bitter almond aftertaste.
- Lago di Caldaro/Kalterersee (DOC 1970)—a subtle and harmonious wine with a bouquet typical of the Schiava grape. Its beautiful garnet colour when young turns to brick red after three years.

Other DOCs of Alto Adige are Colli di Bolzano/Bozner Leiten, Terlano/Terlaner, Valle Isarco/Eisacktaler, Casteller, Alto Adige/Südtiroler, and Meranese di Collina/Meraner Hügel.

FRIULI-VENEZIA GIULIA

Friuli-Venezia Giulia is located in the northeast corner of Italy along the Austrian and (former) Yugoslav borders. Austrian, Slavic, and Hungarian influences are evident in the wines and cuisine. Friuli is a corruption of the Roman *Forum Julii.*

To the north, mountains cover 43% of the land. Most vineyards are planted in the plains that roll toward the Adriatic. The growing season is mild and humid with plenty of sunshine. As in Alto Adige, this is an ideal environment for lively, fruity, white wines. The reds are also fruity and aromatic because they rarely undergo malolactic fermentation. Most wines of the region are labelled with the name of the grape variety.

The DOCs include:

- Collio (DOC 1968)—along the border, from Gorizia to the little village of Mernico. This DOC produces Tocai Friulano and some of the best Pinot Grigio (Pinot Gris).

The area also grows good Gewürztraminer, Cabernet Franc, and Chardonnay, although Chardonnay is not yet recognized by the DOC. Collio Bianco is made from Ribolla Gialla (45%–55%), Malvasia Istriana (20%–30%), and Tocai (15%–25%). This wine is best consumed young to appreciate its full, fresh, subtle bouquet.

- Colli Orientali del Friuli (DOC 1970)—the northern extension of Collio DOC. The appellation grows seventeen varieties. Rosato is a rosé made from 90% Merlot. Ramandolo is a dessert wine made from 90% Verduzzo. This is also the home of the famous dessert wine, Picolit. Regrettably, this grape variety suffers badly from cryptogamic (i.e., carried by spores) vine diseases, e.g., downy mildew and coulure, and is slowly disappearing.
- Grave del Friuli (DOC 1970)—in the Adriatic basin from Veneto to Isonzo. Merlot blankets nearly half of the area. Cabernet is heavily planted as well. The best wines, however, are the whites from Chardonnay, Pinot Grigio (Pinot Gris), Sauvignon Blanc, and Tocai Friulano.
- Carso (DOC 1986)—the source of the region's most outlandish red wines. Carso and Terrano del Carso are tart, purple wines with a grassy aroma. They need years before they can be drunk. Malvasia del Carso is a delicious white with hints of honey and almonds.

Other Friuli-Venezia Giulia DOCs are Isonzo, Aquileia, and Latisana.

VENETO

Veneto produces 8 000 000 hL of wine, so ranks third after Apulia and Sicily in production, but leads in DOC output with 1 700 000 hL. The region is split between the mountains in the north from Belluno to Lago di Garda (Lake Garda) and the fertile plains of the south.

The DOC authorizes 83 varieties grown on more than 160 000 individual sites, yet most wine is made from Bardolino, Valpolicella, and Soave—what might be called the Veronese triumvirate. Modernization and the implementation of more efficient practices have brought about the consistent high quality of these admittedly mass-market wines.

The DOCs of this region include the following:

- Valpolicella/Recioto della Valpolicella/Amarone (DOC 1968)—northeast of Verona, with vineyards decorating every slope. The particularly mild climate influenced by Lake Garda and the Lessini Mountains creates ideal sites in the hills. Red wine is made from 40%–70% Corvina Veronese, which gives aroma and richness, 20%–40% Rondinella for colour and tannin, 5%–25% Molinara for roundness, and up to 15% Negrara Trentina, Rossignola, Barbera, and/or Sangiovese. Because of the options that makers of Valpolicella have, these wines can range greatly from one producer to another. There is a light, fruity Valpolicella with a bright ruby colour, ready for immediate enjoyment; Valpolicella Classico with a rich bouquet of banana and ripe cherry; Valpolicella della Valpantena, lighter bodied with more subtle fruit; and Recioto della Valpolicella, which is a sweet, weighty, red wine made from passito grapes. Great Valpolicella are some of the best Italian red wines.

VALPOLICELLA WINES

To make Recioto della Valpolicella, the grapes are dried (to concentrate sugars, aromatic substances, and tannin) on screen shelves in dry, well-ventilated lofts. The grapes lose at least 30% of their original weight. They are allowed to be affected by noble rot (Corvina grapes are particularly susceptible). The drying process takes three months. This is followed in January by pressing. Then the cold winter weather results in a cold maceration followed by slow alcoholic fermentation. The wine emerges rich and sweet after forty to fifty days. Recioto della Valpolicella can be still or spumante. The label indicates the alcohol level with two numbers, e.g., 14% + 2%, meaning that 14% is the actual alcohol level and 2% additional would have resulted had the residual sugar been fermented completely. So, the higher the second number, the higher the sweetness.

Amarone, until recently labelled *Recioto della Valpolicella Amarone*, is made by totally fermenting the semi-dried grapes, and ageing the wine in barrels. The alcohol can reach a lofty 16% or 17%. The wine is bottled four years after the vintage. Amarone is an opulent wine with a deep garnet hue; aromas of bitter almond, chocolate, and spices; penetrating flavours of fruit jam, dried dates, and cherry preserve; and a long, satisfying, bitter aftertaste.

Valpolicella Ripasso is made by placing the new fermented wine into vats containing the lees of the previous year's recioto. The result is a richer wine, with a bitter almond nose, and ageing potential.

- Bardolino (DOC 1968)—named for the village of Bardolino, with vineyards between Valpolicella and the shore of Lake Garda. The wine is a blend of 35%–65% Corvina Veronese, 10%–40% Rondinella, 10%–20% Molinara, up to 10% Negrara, and up to 15% Rossignola, Barbera, Sangiovese, and Garganega. Pale ruby, fruity, and light bodied, it is always at its best very young. Chiaretto is the rosé and can be still or spumante.
- Soave and Soave Classico (DOC 1968)—between Vicenza and Verona (Classico is the northeast portion). Production is huge considering the acreage. In some vineyards, vines are planted 13 000/ha. Garganega, Trebbiano di Soave, and up to 30% Trebbiano Toscano are combined to give a simple, easy drinking, rather ordinary wine. Recioto di Soave is a sweet version made from semi-dried grapes in still, spumante, and liquoroso styles.
- Prosecco di Conegliano-Valdobbiadene (DOC 1971)—produces sparkling wine. Prosecco grapes are used, often with the addition of Verdiso, Pinot Bianco (Pinot Blanc), Pinot Grigio (Pinot Gris), and up to 15% Chardonnay. Still wines are produced in dry, amabile, dolce, or frizzante versions.
- Piave (DOC 1971)—produces good Cabernet, Merlot, and Pinot wines for early drinking.

Other Veneto DOCs are Bianco di Custoza, Lessini Durello, Breganze, Gambellara, Colli Berici, Colli Euganei, Montello e Colli Asolani, and Lison-Pramaggiore.

Emilia-Romagna

Emilia-Romagna is south of Veneto and Lombardy, north of Tuscany, stretching from the Adriatic to Liguria. The land is flat and fertile. Winters are cold, damp, and foggy but summers can be scorching.

The region combines two distinct areas and their traditions. Emilia extends west of the capital Bologna to Piacenza; it takes its name from the Via Emilia, built by the Romans in A.D. 187. Romagna extends east from Bologna to the Adriatic Sea and was created as a Roman dominion. Emilians are fond of the native Lambrusco grape and consume vast amounts of its wine in various forms—red, white, or rosé, dry or sweet, still or fizzy. Their fresh, zesty wines wash down rich Bolognese cuisine. In contrast, the Romagnans prefer Sangiovese, Albana, and Trebbiano wines.

The DOCs include the following:

- Lambrusco di Sorbara (DOC 1970), Lambrusco Salamino di Santa Croce (DOC 1970), Lambrusco Grasparossa di Castelvetro (DOC 1970), and Lambrusco Reggiano (DOC 1971)—wines made from subvarieties of Lambrusco around the town of Modena. Most of the wine is lightly frizzante, sweet, and made to be drunk young.
- Montuni del Reno (DOC 1988)—a white, spritzy wine usually made from the Montuni grape.
- Albana di Romagna (DOC 1967, DOCG 1987)—from the sweet and prolific Albana grape grown in the southeast triangle of the Romagna region, between the Apennine Hills and the Adriatic. The area produces a dry, white wine, pale in colour, with 13% alcohol, plus an unctuous amabile with a golden hue. The rich Scaccomatto is a botrytized wine, fermented on its lees. The area also produces spumante and frizzante wines.

Other DOCs in Emilia-Romagna are Colli Piacentini, Colli di Parma, Bianco di Scandiano, Colli Bolognesi-Monte San Pietro, Sangiovese di Romagna, Trebbiano di Romagna, Cagnina di Romagna, Pagadebit di Romagna, and Bosco Eliceo.

TUSCANY

Toscana (Tuscany), whose name reveals its Etruscan origins, is the second largest producer of DOC wines in Italy. Its coastline borders the Mediterranean Sea, and the region includes the island of Elba. Most of the land is hilly or mountainous, except for the flat valley of the Arno River. Chianti is the best-known wine of the region (and of Italy).

The DOCs include:

- Chianti (DOC 1967, DOCG 1984)—produced in a huge, fragmented area that takes in 5 provinces containing 103 communes. The area extends from Pistoia, northeast of Florence, to the Mount Cetona pass at Montepulciano in the south, from Arezzo to a few kilometres east of Livorno. The Chianti DOC overlaps many other DOCs and has seven DOC subzones: Chianti Classico, Colli Aretini, Colli Fiorentini, Colli Senesi, Colline Pisani, Montalbano, and Rufina.

  Traditionally, Chianti was made from 75%–90% Sangiovese, 5%–10% Canaiolo Nero, and 5%–10% of the two white grapes Trebbiano Toscano and Malvasia del Chianti. New regulations allow 10% Cabernet and Merlot.

Chianti

Chianti has changed quite frequently since it was first drunk by the Etruscans. In the fourteenth century, Tuscan natives Petrarch, Galileo, Boccaccio, Leonardo da Vinci, Dante, Michelangelo, and Machiavelli would have tasted quite a different wine. And this wine changed dramatically again during the nineteenth century when Baron Ricasoli of Castello Brolio redirected the production to more tart wines made for earlier drinking. More recent makers of Chianti have swung the pendulum back. For example, Piero Antinori created Tignanello (20% Cabernet) and Solaia (80% Cabernet), and the Frescobaldis established their Pomino and Capitolare. Today, products like Sassicaia and Vigorello are dubbed Super-Tuscans.

The wine must be aged for six months before release, and for three years if it is to be labelled Riserva. Wines made using the system of Governo all'uso del Chianti must state so on the label and be sold within a year of the harvest. By reducing the minimum of white grapes to 5%, modern Chianti Classico is fruitier, with deeper colour and better longevity than in the past. Rufina requires longer ageing to bring out its complexity.

- Brunello di Montalcino (DOC 1966, DOCG 1980)—both appellation and grape are named for the hilltop town Montalcino (50 km south of Siena) and the dark-brown (brunello) colour of this Sangiovese Grosso clone. The high, dry slopes around the town offer ideal exposures. Brunello must age three and a half years in oak or chestnut casks, plus six months in the bottle, before its release. A Riserva must age five years. The wine has a deep colour and unyielding tannin when young. With ten to fifteen years ageing, a wonderful complexity emerges: the bouquet hints at tar, leather, spices, and black truffle. A Riserva from a good vintage may need at least twenty to twenty-five years to show its full potential.

CAPITOLARE

Capitolare wines, previously known as Predicato, are classified as vino da tavola even though they fetch prices much higher than the local DOCs. There are four classifications:

Muschio—Chardonnay blends

Selvante—Sauvignon Blanc-based wines

Cardisco—red blends made mainly with Sangiovese

Biturica—red blends made mainly with Cabernet Sauvignon

- Rosso di Montalcino (DOC 1984)—can be made from the young vines of the Brunello di Montalcino vineyards, or by using the harvested fruit of lesser years. It is a lighter, fruitier, and more accessible version of Brunello, although it can age almost as well.
- Vino Nobile di Montepulciano (DOC 1966, DOCG 1981)—named for the commune of Montepulciano, not far from Montalcino. The appellation is divided into two territories: Argiano, Caggiole, Canneto, and Casalte east of Montepulciano; and Valiano east of the Val di Chiana. Like those of Montalcino, the vineyards are located on steep slopes (250 m–600 m). The wine is made from 60%–80% Prugnolo Gentile (a clone of Sangiovese), 10%–20% Canaiolo Nero, Mammolo, Grechetto, or Trebbiano, and up to 20% Malvasia. Aged two years in wood, the wine becomes dark and hard. In good years, and with some maturing, Vino Nobile di Montepulciano can be a robust wine full of aromas, sophisticated, with a bitter finish. When made from young vines (or from lighter vintages), it is usually labelled Rosso di Montepulciano.
- Carmignano (DOC 1983)—This tiny district west of Florence produces wines with the same basic grape content as Chianti, but with the addition of 10% Cabernet Sauvignon or Cabernet Franc. The result is a wine with deeper colour, and a more pronounced nose of black cherry and nuts. Red Carmignano is a DOCG.
- Vernaccia di San Gimignano (DOC 1966)—this tiny district surrounds the town of San Gimignano, between Florence and Siena. The white wine is produced from Vernaccia. Praised since the Renaissance, the wine has an interesting flavour combination of fruit, crispness, and honey. A spumante is produced with the addition of up to 15% Chardonnay, Pinot Bianco (Pinot Blanc), and Pinot Nero (Pinot Noir).

Other Tuscan DOCs include Bianco dell'Empolese, Bianco della Valdinievole, Galestro, Pomino, Val d'Arbia, Moscatello di Montalcino, Bianco Vergine Valdichiana, Morellino di Scansano, Bianco di Pitigliano, Parrina, Bianco Pisano di San Torpè, Bolgheri, Candia dei Colli Apuani, Colli dell'Etruria Centrale, Colli Lucchesi, Elba, Montecarlo, Montescudaio, and Val di Cornia.

MARCHES

The Marches (Marche) extends 175 km along the Adriatic to the mountains inland. It ranges from the beaches and sand dunes to the mountains (2500 m) at the western border with Umbria. The climate is generally stable, but summers can be hot and arid. Of the region's vineyards, 12% are dedicated to growing Verdicchio grapes. The varieties Sangiovese, Montepulciano, and Trebbiano make up most of the planted areas, although Merlot is gaining ground.

Verdicchio, typically bottled in amphora-shaped bottles, is the region's best-known wine.

The DOCs include the following:

- Verdicchio dei Castelli di Jesi (DOC 1968)—near the town of Jesi. The best classico vineyards are on either side of the Esino River between 200 m–500 m in the hills west of Jesi. Some wines are entirely Verdicchio, although up to 15% Malvasia and Trebbiano is permitted. Verdicchio sur lie is becoming increasingly popular. The area produces some sparkling wine, in bulk, or using the méthode Champenoise.
- Verdicchio di Matelica (DOC 1967)—a second, smaller zone hidden high up in an Apennine valley between Camerino and Fabriano. The local wine has more depth and weight than Verdicchio dei Castelli di Jesi.
- Rosso Piceno (DOC 1968)—the highest-volume red of the region. It is a blend of 60% Sangiovese and 40% Montepulciano, best drunk between age three and seven years.
- Rosso Conero (DOC 1967)—similar in style and longevity to Rosso Piceno, but sturdier and more tannic. It is made from a higher proportion (up to 85%) of Montepulciano.

Sur lie means that the wine has spent a winter in its cask on its lees, and has been bottled directly, that is, without racking.

Sangiovese dei Colli Pesaresi, Bianchello del Metauro, Lacrima di Morro d'Alba, Vernaccia di Serrapetrona, Falerio dei Colli Ascolani, and Bianco dei Colli Maceratesi are other Marches DOCs.

UMBRIA

One of the few landlocked regions of Italy, Umbria is a hilly place where Etruscans and their wine culture thrived long before the existence of Rome. Orvieto, once the most famous dry white wine of Italy, was the favourite of native painters Pinturicchio and Raphael. The abboccato version has regained some of its popularity as a dessert wine in recent years. Umbria's eminent wine is Torgiano Rosso Riserva, which was granted DOCG status. Merlot and Barbera are the most-planted red varieties, followed by Cabernet Sauvignon and Pinot Nero (Pinot Noir).

Umbria's DOCs include:

- Orvieto (DOC 1971)—around the town of Orvieto. The vineyards of Orvieto Classico surround the town, expand north and south to the regular DOC zone, and spill into Latium. The white wine is made from a blend of Procanico (the local name for Trebbiano Toscano) 40%–65%, Verdello 15%–25%, Grechetto 20%–30%, and Malvasia Toscana up to 20%. Traditionally, the glory of this wine was its sweet taste and golden colour; however, the exploitation of the vineyards and industrialized production after the DOC classification yielded a clean but rather dull wine. Today Orvieto seems to have regained a bit of its fame. A few growers manage to produce amabile with good colour and hints of noble rot.
- Torgiano (DOC 1969) and Torgiano Riserva (DOCG)—named for the town at the fork of the Tiber and Chiascio rivers. Torgiano Rosso is a distinguished wine made from Sangiovese (50%–70%), Canaiolo (15%–30%), Trebbiano (up to 10%), and Cilegiolo and Montepulciano (up to 10%). The DOCG Rubesco Riserva is from more select grapes from the same vineyards. What is not used in Rubesco Riserva goes into regular DOC Torgiano. A white Torgiano made from Trebbiano and Grechetto is pleasant, with balanced fruit and acid.

Umbria's other DOCs are Colli Altotiberini, Colli Amerini, Colli del Trasimeno, Colli Martani, Colli Perugini, and Montefalco.

Abruzzi

Located on the Adriatic south of Marches, the rugged hills of the region (also known as Abruzzo) produce mainly wines for everyday drinking. Two-thirds of the production is made by co-operatives. Only two DOCs are found on the coastal hills—the Montepulciano d'Abruzzo and the Trebbiano d'Abruzzo.

Molise

The region was a southern extension of Abruzzi until 1963. Almost all (90%) the vineyards are in parcels of less than half a hectare; only two properties in the entire region have more than 20 ha. Sangiovese, Barbera, Bombino Grosso, and Aglianico are the most important red varieties. Trebbiano leads the whites. Most of the wines are consumed locally. Within the region, Biferno (1983) and Pentro di Isernia (1984) have DOC status.

Latium

This is the region known also as Lazio, and is the site of Rome, city of seven hills. Much of the production is white wine based on Trebbiano and Malvasia grapes. With sixteen DOCs, Latium is the most productive region in central Italy. More than two hundred grape varieties are still used here but only a handful attract attention, e.g., Trebbiano, Malvasia, Moscato, and Cesanese. Cabernet and Merlot are slowly gaining ground. Most of the white wines are fresh and easy drinking when consumed young. The DOCs include the following:

- Aleatico di Gradoli (DOC 1972)—a wine similar to Port from the volcanic slopes on the northwestern side of Lake Bolsena at the northern tip of the region. Aleatico yields a very purple wine with a high sugar concentration.
- Est! Est!! Est!!! di Montefiascone (DOC 1966)—a rather plain, white wine despite the legends. The vineyards surround all of Lake Bolsena, overlapping the Aleatico zone. Est! Est!! Est!!! is made from Procanico (Trebbiano Toscano), Malvasia Bianca Toscana, and Rossetto (Trebbiano Giallo).

- Frascati (DOC 1966)—one of the most popular drinks in Rome's terraces and cafés. The light, white wine is made from Malvasia Bianca di Candia and Trebbiano Toscano (at least 70%), plus Malvasia del Lazio and Greco (up to 30%).

The other DOCs in Latium include Aprilia, Bianco Capena, Cerveteri, Cesanese del Piglio, Cesanese di Affile, Cesanese di Olevano Romano, Colli Albani, Colli Lanuvili, Cori, Marino, Montecompatri-Colonna, Velletri, and Zagarolo.

Campania

This sun-soaked region lies south of Latium, extending east from the Tyrrhenian Sea, and includes Naples. Campania's volcanic and alluvial soil is rich and fertile. The famous wine Falernium hailed from the volcanic slopes of Mount Vesuvius in Roman times. Some grape varieties, such as Coda di Volpe ("fox tail"), Falanghina, Aglianico, and Greco di Tufo are very old, of Greek origin. Asprinio is a legacy of Etruscan times. Fiano was cultivated by the Romans for its sugar-rich berries that attract bees.

The DOCs include:

- Fiano di Avellino (DOC 1978)—one of the most-honoured white wines of Italy. It combines Fiano (85%) with Greco, Coda di Volpe, and Trebbiano Toscano (up to 15%). Vineyards surround the tiny town of Avellino east of Naples. The terroir and the Fiano grape result in an almond-scented wine with cinnamon hints and poached pears on the palate.
- Greco di Tufo (DOC 1970)—a smaller area surrounding the village of Tufo and adjacent to the Fiano zone. It is regarded as equivalent in quality to Fiano di Avellino. The blend is 85% Greco di Tufo and up to 15% Coda di Volpe.
- Taurasi (DOC 1970)—around the village of Taurasi, to the east of Greco di Tufo and Fiano di Avellino. The wine is dubbed "the Barolo of the South" for its structure and ability to age, but has not yet achieved the distinction of DOCG. It is made from Aglianico (70% minimum) with additions of Sangiovese, Piedirosso, or Barbera.

Aglianico del Taburno, Capri, Castel San Lorenzo, Cilento, Falerno del Massico, Ischia, Solopaca, and Vesuvio are other DOCs of the area.

Apulia

On the southeast coast of Italy, Apulia is the heel of the boot that separates the Adriatic and Ionian seas. Puglia, as it is also known, is a dry and sunny land with more wines and more grapes than any other region of Italy. Production is around 12 000 000 hL, but the region's twenty-four DOCs make up only 1.6% of this.

The DOCs include:

- Aleatico di Puglia (DOC 1973)—a regional DOC covering the entire area. The sweet, red wine is either dolce or liquoroso. It is mainly made from Aleatico, with additions of Negroamaro, Malvasia Nera, and Primitivo (up to 15%).
- Castel del Monte (DOC 1971)—probably the best red wine of Apulia. The large production zone is located east of Bari around Castel del Monte. The grapes (e.g., Uva di Troia, Sangiovese, Montepulciano, Aglianico, Pinot Nero) grow on meagre, arid soils. A riserva can age well and will develop some intriguing and pleasant qualities.
- Moscato di Trani (DOC 1975)—a rich, sweet, aromatic, white wine made from Moscato Reale. It is hard to find.
- Primitivo di Manduria (DOC 1975)—a purple wine made from the Primitivo grape (alleged parent of the Californian Zinfandel). The growing zone is around Manduria on the east side of the Gulf of Taranto. Annual production is a minuscule 12 000 bottles.

The other Apulia DOCs are Alezio, Brindisi, Cacc'e Mmitte di Lucera, Copertino, Gioia del Colle, Gravina, Leverano, Lizzano, Locorontodo, Martina Franca, Matino, Nardò, Orta Nova, Ostuni, Rosso Barletta, Rosso di Cerignola, Salice Salentino, San Severo, Squinzano, and Rosso Canosa.

Basilicata

Basilicata, also called Lucania, is the instep of the Italian boot between Apulia, Campania, and Calabria. It is an impoverished land with only one classified wine, the great DOC (1971) Aglianico del Vulture. The vineyards are located on the slopes of the extinct volcano, Mont Vulture, at an altitude of 600 m. The deeply coloured, hearty, red wine can improve with age to become complex, smooth, and fleshy, with a charming aftertaste. The region also produces some sweet, red, sparkling wine and amabiles.

CALABRIA

This mountainous region forms the toe of the Italian boot. The land is poor with hard winters, scorching summers, and an insufficient water supply. It takes a hardy Calabrian farmer to manage this rugged land.

Eight districts have achieved DOC classification. The following are among them:

- Cirò (DOC 1969)—red, white, or rosé. Red wine is made using Gaglioppo. To produce rosé, Gaglioppo is blended with Trebbiano Toscano and Greco Bianco. In white wines, Greco Bianco dominates, with as much as 10% Trebbiano Toscano added.
- Greco di Bianco (DOC 1980)—sweet, white wine produced in the hills west of the coastal town of Bianco. The wine has a rich copper tone and a herbaceous nose.

The other DOCs are Donnici, Lamezia, Melissa, Pollino, Sant'Anna Isola Capo Rizzuto, and Savuto.

SICILY

Sicily is the largest island in the Mediterranean Sea. The wine region extends to include several nearby islands. Viticulture in this region goes back to the fifth century B.C., when Greek and other Mediterranean civilizations regularly visited. Now Sicily is the second-largest producer of wine in Italy, yet only 2.5% of its annual 11 100 000 hL have DOC status. Four-fifths of the production comes from co-operatives, and producers tied to the old traditions are reluctant to accept DOC regulations.

The island is mountainous, with only 15% of the land lying below elevations of 100 m, and the soil is volcanic. The vineyards are slowly being modernized with the installation of irrigation and the trellising of vines to yield lighter, fruitier wines.

Most of the wines are white, which is somewhat surprising for a southern region. Bianco d'Alcamo leads the way with 2.5 million bottles annually. Corvo-Duca di Salaparuta and the Tasca d'Almerita family's Regaleali are vini da tavola with consistent high quality. The letter Q is printed on the labels of Sicily's best wines.

Sicily's DOCs include the following:

- Malvasia delle Lipari (DOC 1974)—from the Aeolian islands, including Lipari to the northeast of Sicily. The wine lost ground in the last decade but has regained some of its lustre. Passito, dolce, and liquoroso wines are made using Malvasia delle Lipari, together with a small portion of Corinto Nero.
- Moscato di Pantelleria or Passito di Pantelleria (DOC 1971)—from a small parched island halfway between Sicily and the coast of Tunisia. Shrivelled Zibibbo grapes are turned into an intensely sweet wine with a pronounced Muscat flavour and a deep golden colour.
- Marsala (DOC 1969)—one of the most underestimated fortified wines. The zone of production covers the western tip of Sicily from Marsala to beyond Trapani. The grapes for white Marsala (Oro and Ambra) are Grillo, Catarrato, Ansonica (Inzolia), and Damaschino. For red Marsala (Rubino), the same grapes are used, plus Perricone, Nero d'Avola (Calabrese), and Nerello Mascalese.

Other DOCs include Bianco d'Alcamo, Cerasuolo di Vittoria, Etna, Faro, Moscato di Noto, and Moscato di Siracusa.

MARSALA

Marsala was created by John Woodhouse of England, in 1773. He adapted the traditional Spanish method of making Sherry to the local grapes. His wine became popular in Victorian England. Over time, Marsala developed a reputation as a cheap wine, useful mainly in the kitchen for making Veal Scaloppine and Zabaglione. A revival started after the 1971 DOC classification, and subsequent DOC reform in 1986. The result has been a significant increase in quality and production (220 000 hL), placing Marsala in the ranks of the ten top DOCs.

Marsala is classified as follows:

- Fine—basic Marsala, which can be red or white but must have 17% alcohol and be aged for at least one year. The wine, which may be called Italy Particular (IP), is secco, semi-secco, or dolce.
- Superiore—aged in wood for at least two years and with a minimum of 18% alcohol. Riserva must be aged in wood for at least four years. Superiore has a variety of names, such as Superiore Old Marsala (SOM), London Particular (LP), and Garibaldi Dolce (GD).
- Vergine or Soleras—dry wine, aged in wood for five years, with a minimum 18% alcohol, and no concentrated must. Stravecchio or Riserva must have a minimum of ten years in wood.

Sardinia

The second largest island of Italy, Sardegna (Sardinia) is a colourful island west of the mainland. Its history of successive invasions by Carthaginians, Romans, Vandals, Byzantines, Genoese, and Spaniards has left it rich in obscure old species of vines with names like Giro, Cannonau, Torbato, Monica, and Nuragus. The Spanish stayed the longest, and left the biggest imprint with the Sherry-type Vernaccia di Oristano and Nasco di Cagliari.

Drought is a constant problem and many vineyards are irrigated. Most wine is produced in the southern Campidano region and Oristano in the west. The wine is mostly white and produced by co-operatives.

There are eighteen DOCs, of which the following are particularly noteworthy:

- Vernaccia di Oristano (DOC 1971)—Sardinia's most distinctive wine. Vernaccia grapes are pruned low to the sandy, hot soil where they accumulate a tremendous amount of sugar. The wine is aged in small casks kept in small brick shelters. The barrels are only partly filled so a film develops on the surface (as in fino Sherry). The wine reaches a natural alcoholic strength of 15% and has the characteristic Sherry nose.
- Cannonau di Sardegna (DOC 1972)—a powerful red wine produced mainly from Cannonau grapes, plus Boval Grande, Boval Sardo, Pascale di Cagliari, and Monica (up to 10%), and Vernaccia di San Gimignano (up to 5%). The DOC applies to the entire island. The wine can be red, rosé, superiore, or liquoroso. Cannonau is best in the form of Anghelu Ruju ("red angel"), a Port-type wine made from grapes dried on cane mats.

Other DOCs are Arborea, Carignano del Sulcis, Giro di Cagliari, Malvasia di Cagliari, Monica di Cagliari, Moscato di Cagliari, Nugarus di Cagliari, Malvasia di Bosa, Mandrolisai, Moscato di Sorso-Sennori, Vermentino di Sardegna, Monica di Sardegna, Moscato di Sardegna, Campidano di Terralba, Monica di Cagliari, Nasco di Cagliari, and Vermentino di Gallura.

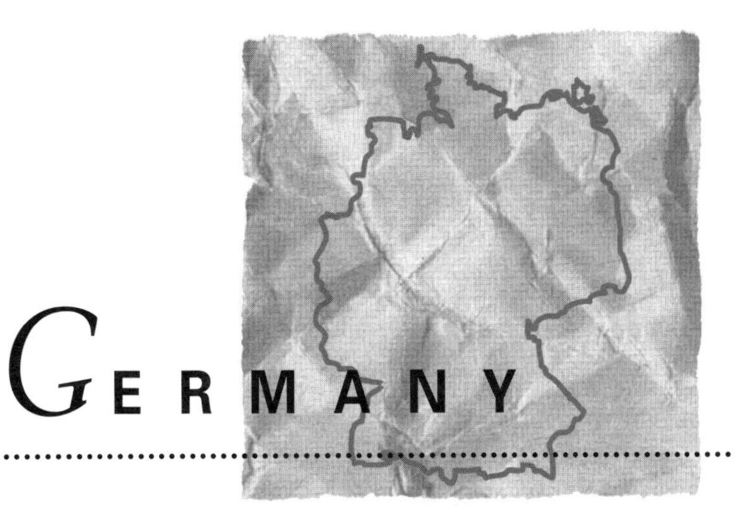

GERMANY

Germany has an annual production around 10 200 000 hL and a consumption rate of 23 L per capita. The country has the well-deserved reputation for a variety of distinctions: the quality of its sweet wines, the northern extremity of its wine regions and necessary dependence on white grapes, its sparkling wine consumption and production, and its system of appellations and classifications. Fruitiness, a judicious balance of sweetness to complement the crisp acidity, and subtle aftertastes are the key factors in the great German wines. And its naturally sweet Rieslings have a reputation so high that they have become the standard of excellence worldwide.

abfüllung (UP-fill-oong)—bottling
anbaugebiet (UN-bow-ga-beet)—wine region; the broadest permitted designation for quality wine, e.g., Rheinhessen
Amtliche Prüfnummer (UMT-li*CH*-uh PREWF-noo-mer)—the official quality control test number given to quality wines
bereich (buh-RYE*CH*)—district or subregion (within an anbaugebiet) including many wine-growing villages and taking the name of the most prominent. The plural is bereiche.
bocksbeutel (BUX-boy-tul)—flagon-shaped bottle from Franken
edelfäule (AY-dul-foy-luh)—noble rot
einzellage (INE-tsel-lah-guh)—individual vineyard site
eiswein (ICE-vine)—icewine
erzeugerabfüllung (air-TSOY-ger-up-fill-ung)—estate bottled
gemeinde (guh-MINE-duh)—community or parish
grosslage (GROSS-lah-guh)—general or collective vineyard site
kellerei (kell-er-EYE)—wine cellar
perlwein (PAIRL-vine)—slightly sparkling table wine made from red or white grapes with natural or injected carbonation
rotling (ROTE-ling)—rosé wine made by blending red and white grapes, or their pulp
schaumwein (SHOWM-vine)—sparkling wine
schillerwein (SHILL-er-vine)—rotling quality wine. The name derives from the German schillern, "to change colour."
sekt (zeckt)—quality sparkling wine
spritzig (SHPRIT-si*CH*)—slightly sparkling
süssreserve (ZEWS-ray-zair-vah)—a small quantity of grape juice held back before fermentation, to be added later as a sweetener
trocken (TRUCK-in)—dry
weinberg (VINE-bairg)—vineyard
weingut (VINE-goot)—wine estate
weinkellerei (VINE-kell-er-eye)—commercial wine cellar
weissherbst (VICE-hairbst)—rosé quality wine made from a single red grape variety
winzergenossenschaft (VINTS-er-guh-nuss-en-shufft)—growers' co-operative

GEOGRAPHY

German vineyards are located between 47°N–52°N. There are thirteen regions in all, eleven in the southwest corner of the country (mostly in the valleys of the Mosel and Rhine rivers, and their tributaries). Two other regions are found in the former East Germany—one where the Unstrut and Saale rivers meet near Liepzig, and another on the Elbe River near Dresden, close to the Czech border.

The Mosel is known as Moselle in France; the Rhine is known as the Rhein in Germany, a fact evident in many wine names.

With the exception of Baden in the south, all of Germany's wine-growing regions are in what the European Union classifies as Zone A (a climatically defined zone that is more northerly than most of the EU's wine regions). The German climate is continental, with cold winters and hot summers. The southeast regions benefit from a somewhat kinder climate: earlier springs and longer, milder autumns. Germany's wine regions are primarily suitable for white grape varieties.

WINE REGIONS OF GERMANY

Ahr
Mittelrhein
Mosel-Saar-Ruwer
Rheingau
Nahe
Rheinhessen
Pfalz
Hessische Bergstrasse
Franken
Württemberg
Baden
Saale-Unstrut
Sachsen

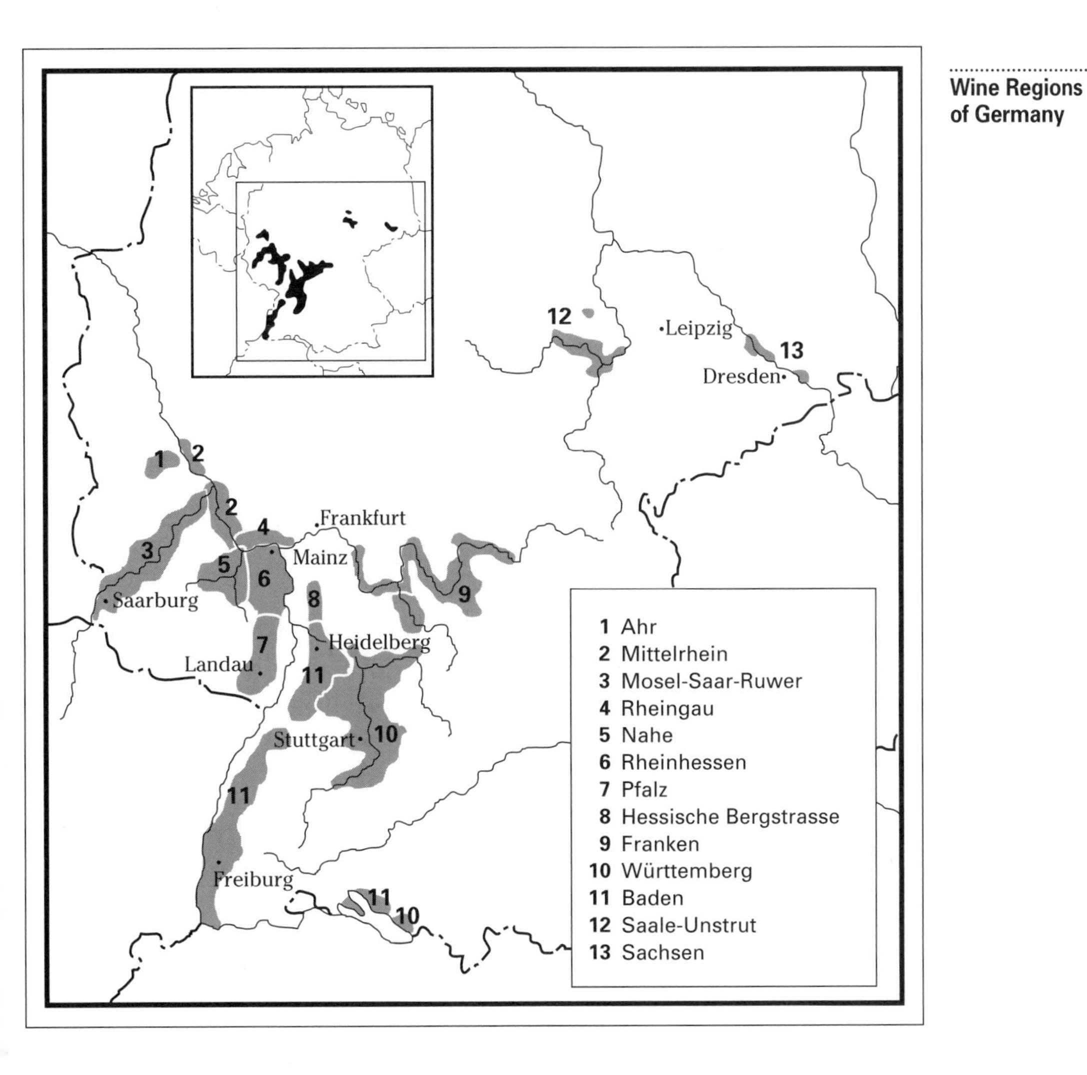

Wine Regions of Germany

HISTORY

Much of Germany's wine-growing area was part of the Roman Empire after Julius Caesar crossed the Rhine in 55 B.C. Vineyards and the wine trade flourished until the collapse of the Roman Empire in A.D. 476.

Viticulture was stalled again until vineyards were replanted under Charlemagne, king of the Franks from 768 to 814. One early spring morning, from his castle in Ingelheim (on the Rhine, near Mainz), he observed snow melting on the opposite (south-facing) bank of the Rhine. Thus, he ordered the planting of new vineyards, on the place known today as Johannisberg. This was repeated on other south-facing slopes—their names (e.g., Kaiserstuhl, Kaiserpfalz) are proof of their origin. Charlemagne also encouraged viticulture by fostering the proliferation of monasteries and abbeys. Their well-tended vineyards provided wine for the Mass (as well as for the elation of abbots). Over time, Charlemagne enacted laws to regulate winemaking (e.g., grapes could not be trodden by foot but only in a wine press), hygiene (animal skins could not be used to store wine), and selling (winemakers could announce sales of wine by hanging green branches above their doors).

Eventually, the Church controlled many of the best vineyards in the deep valleys along the Mosel, the Rhine, and their tributaries. In 1135, the archbishop of Mayence (Mainz) granted land to Cistercian monks, who then founded Kloster Eberbach. During the twelfth and thirteenth centuries, this became the largest wine-growing enterprise in the world, with 200 other establishments between Worms and Köln (Cologne). By the sixteenth century, the archbishop of Mainz controlled almost all the Rheingau. Meanwhile, Benedictine monks controlled Johannisberg, laid out new vineyards around Trier, and expanded those of Berkastel, Kues, and Trittenheim.

By the late nineteenth century, the golden wines of the Rhine were prized in England, particularly by Queen Victoria (1837–1901). She was especially fond of the sweet wine of Hochheim. Since that time, Rhine wines have been known in England as Hock wines.

LAWS AND REGULATIONS

The German system of proven quality and origin was introduced in 1971. Under the European Union's system, which recognizes two categories of wine (table and quality), Germany has the following categories and subcategories.

Table Wine

- Tafelwein—table wine.
- Landwein—country wine.

Quality Wine

- Qualitätswein bestimmter Anbaugebiete (QbA)—quality wine from a specific region.
- Qualitätswein mit Prädikat (QmP)—quality wine with distinction.

Tafelwein

Tafelwein is the lowest-ranked wine. It can be a blend of wines from anywhere within the EU, provided it is bottled in Germany. Deutscher tafelwein (German table wine) must be made from grapes grown in Germany. Within this category, some wines of very good quality are produced, often from declassified wines, or wines that were not submitted for official classification.

Landwein

Landwein is wine from a specific place (a similar category to France's vin de pays). For a wine to qualify as a Deutscher landwein, it must come entirely from one of the Landwein regions, and not include any concentrated grape must.

QbA

Qualitätswein bestimmter Anbaugebiete (quality wine from a specific region) is the lower rank of the quality wines. For a wine to qualify as a QbA wine, it must come entirely from one of the thirteen delimited wine regions, and be submitted for an official taste test.

QbA wines can be enriched or chaptalized to obtain a higher alcohol level. If enriched with a süssreserve, the süssreserve must be produced in the same region as the wine, be of no less than QbA ripeness, and may not exceed 25% of the final wine. Neither acid nor water may be added, but partial de-acidification is permitted. QbA wines must state the name of the region on the label. Showing the name of a bereich, grosslage, or einzellage is authorized.

QmP

Qualitätswein mit Prädikat (quality wine with distinction) is the highest level of wine. Like QbA, QmP wines must each come from one of the thirteen delimited regions and undergo a taste test. The name of the region and the bereich must be clearly stated on all QmP wine labels. Names of the community, vineyard, or site of origin are also allowed.

The big difference between the QbA wines and the QmP is that QmP wines are made from riper grapes and may not be chaptalized.

Within the QmP category there are six prädikats (levels of distinction) to indicate the degree of ripeness and sugar content of the grapes at harvest. Each QmP wine undergoes quality testing, which determines its prädikat. From lowest to highest, the levels are as follows:

- Kabinett—(originating from the French word cabinet) indicates that the wine is good enough to set aside in the wine cabinet for a later day. Light and delicate, Kabinett wines can be dry or semi-dry. At minimum, the must weights are between 70°–81° Oeschle.

DEGREE OESCHLE

This system was invented by Christian Ferdinand Oeschle in the mid-nineteenth century to measure the sugar in must. A hydrometer is used to measure the density of the must, by comparing the weight of 1 L must to the weight of 1 L water. A must that is 1095 g on the hydrometer is equivalent to 95° Oeschle. By using a chart, a winemaker can deduce that the potential alcohol in the future wine with 95° Oeschle will be 13% if all the sugars are fermented.

- Spätlese—"late harvest." The grapes are left on the vine to ripen until a few weeks after the normal harvest. Spätlese wines can be sweeter than Kabinett but are increasingly made in a trocken (dry) style. The legal minimum must weights are between 76°–91° Oeschle.
- Auslese—"selected harvest," or "selected late picked." The wine is produced from carefully selected, fully ripe, whole bunches of grapes that have been left to ripen after the normal harvest date, and which can be affected by noble rot. Auslese wines are generally sweet and rich but can also be made in a dry style. The legal minimum must weights are between 83°–101° Oeschle.
- Beerenauslese (BA)—"late-harvested individual berries." Beerenauslese wines are made from individually selected overripe grapes, often affected by noble rot and thus very sweet. The legal minimum must weights run from 110°–128° Oeschle. If the wines were to be fermented to complete dryness, they would reach at least 18% alcohol by volume. They are traditionally very sweet, luscious wines, but (increasingly) semi-dry and dry wines are produced.

- Eiswein—made from grapes that have been naturally frozen on the vine and pressed at temperatures of no more than –8°C. The grapes must be of Beerenauslese Oeschle levels. Eiswein is extremely concentrated and delicious, providing the balance of acidity is right. Riesling grapes are the most suitable because of their high natural acidity. From other grapes, the wines are often fat and syrupy without a satisfying finish.
- Trockenbeerenauslese (TBA)—"late-harvested individual berries that have shrivelled to near dryness." TBA wines are made from grapes that have been dried and shrivelled by noble rot. They are at the zenith of German wines. Luscious, concentrated, and rich, they are very expensive and quite scarce, produced only in exceptional years. The legal minimum weight of the must is 150° Oeschle.

Other Categories

In addition to the above categories and subcategories, German wines may be described as follows:

- Hock is any table wine made in the Rhine region from Riesling or Silvaner grapes.
- Moseltaler is a white QbA wine produced from Riesling, Müller-Thurgau, Elbling, and Kerner, or any blend thereof, within the Mosel-Saar-Ruwer district. No grape variety may be indicated on the label. Residual sugar is limited to 15 g/L–30 g/L and acidity must be at least 7 g/L.
- Schaumwein (sparkling wine) is regulated by the same labelling laws as still wine. These wines are produced by secondary fermentation in bulk and have a minimum alcohol content of 9.5%. Schaumwein can not bear a vintage nor the name of the grape variety on the label. Perlwein is the same as Schaumwein but slightly less sparkling (with values between 1 and 2.5 atmospheres).

One atmosphere is equivalent to a pressure of about 101 kPa.

For sparkling wine (schaumwein), the quality version (sekt), and the less fizzy version (perlwein), the following can indicate residual sugar:

extra brut or extra herb	up to 6 g/L
brut or herb	less than 15 g/L
extra dry or extra trocken	12 g/L–20 g/L
dry or trocken	17 g/L–35 g/L
halbtrocken	3g/L–50 g/L
mild	more than 50 g/L

- Sekt wines have higher production and cellaring standards than do schaumweins. Many sekt wines are based on Italian, French, or other non-German wines because German base wines are often too costly. The higher quality sekts are mainly made from Riesling and Weissburgunder (Pinot Blanc).

  Three production methods are used: méthode Champenoise, transfer method, and charmat method. Descriptions of bottle fermentation are permitted on the label.

- Liebfraumilch is blended wine produced using white grapes—primarily Riesling, Silvaner, Müller-Thurgau, and Kerner—which reach a minimum of 60° Oeschle and which originate in the Rheingau, Pfalz, Rheinhessen, or Nahe districts. The wine should be of QbA quality, and sweet, with a minimum of 18 g/L residual sugar. It must pass analytical tests and taste tests to receive an AP number (see next page).

LIEBFRAUMILCH

Historically, Liebfraumilch was the tasty product of grapes grown in the vineyards surrounding the Gothic monastery called the *Liebfrauenkirche* (literally, the Church of the Mother of Love, i.e., the Blessed Virgin Mary) in Worms. The history of the wine's name is a muddle of many scrumptious stories. Here is our version:

> *The name originated in the vineyards around the Liebfrauenkirche. The wine, of course, was made by the monks (Minch in the Old German language), and so was called Liebfrauenminch. It was so good and sweet that the name was changed to Liebfrauenmilch, or milk from the breast of the Mother of Love.*

As the demand became greater than the output of the small vineyard, grapes from adjacent vineyards and from additional plantings were used in the production. Now nearly half of the German wines exported in bottles are labelled Liebfraumilch, and at least half of them come from Rheinhessen. Wine from the famous original vineyard is called Liebfrauenstift Kirchenstück.

Labelling Requirements

German wine labels are required to present the following information:

- category of quality—e.g., Tafelwein, Tafelwein landwein, Qualitätswein (QbA or QmP).
- type of wine—e.g., red, white, rosé, rotling.
- origin—e.g., German, specified region for QbA, specified region and bereich for QmP.
- liquid content.
- bottler's (or shipper's) name and residence.
- alcohol content.
- AP number.

All information must be easy to read and not misleading in format or content. Lettering and print size are regulated.

Labels may give information regarding vintage, sweetness, seals and awards, grape varieties, and village or site, but this is optional.

Quality Control and Awards

Amtliche Prüfnummer (AP number) is the official quality control test number given after chemical analysis and a taste test. All quality wines must bear this number on the label. The test must be made following bottling and prior to the wines being offered for sale.

AP NUMBER

Consider, for example, the following AP number:

A.P. Nr. 4 371 123 8 96

4—number of the examination board

371—number of the community where the estate is situated

123—number of the bottler

8—bottler's current number

96—year of examination (not the year of the vintage)

Various seals and awards may also appear on German wine bottles. Distinctions are awarded to wines that surpass the minimum requirements for AP numbers, and so denote higher quality levels. The Deutsches Weinsiegel ("German wine seal") is awarded by the Deutsche Landwirschaft Gesellschaft (DLG), the German Agricultural Society. There are three different seals:

- the red seal, for wine with some sweetness.
- the lime green seal, for medium dry win.
- the yellow seal, for dry wine.

The yellow seal may also show precise information about the calories and the amount of unfermented sugars—of particular interest to diabetics.

Wines may receive other awards from regional and national wine competitions. The regional wine award labels are allocated by the local Chamber of Agriculture and come in gold, silver, and bronze. The DLG national wine award labels are called Grosser Preis ("Great Prize") and come in gold, silver, and bronze.

GRAPE VARIETIES

German authorities permit over 200 grape varieties in the production of quality wines. Almost 90% of German vineyards are planted with white varieties. The greatest wines are produced from Riesling grapes.

Kerner is a Trollinger/Riesling cross developed in 1969 in the Württemberg region, and named after Julius Kerner, a nineteenth century writer of drinking songs.

DOMINANT WHITE GRAPES OF GERMANY	
Müller-Thurgau	28.8%
Riesling	19.4%
Silvaner	9.8%
Kerner	6.3%
Scheurebe	4.5%
Bacchus	3.5%
Grauburgunder (Pinot Gris)	3.5%
Morio-Muscat	3%
Faber	2.3%
Elbling	
Gewürztraminer	
Gutedel (Chasselas)	
Huxelrebe	
Weissburgunder (Pinot Blanc)	

DOMINANT RED GRAPES OF GERMANY	
Spätburgunder (Pinot Noir)	4.6%
Portugieser	3.4%
Trollinger	2.5%
Dornfelder	
Lemberger	
Müllerrebe (Schwarzriesling or Pinot Meunier)	

Percents are approximate. Grape varieties below 1% are not given a percent.

WINE PRODUCTION

A leader in the production of white wines, particularly sweet wines, Germany is also well known for its production and consumption of sparkling wines: it has the highest per capita consumption of sparkling wines (5 L); annual production of sparkling wine is about 41 000 000 cases.

WINE REGIONS

AHR

The River Ahr flows into the Rhine River just south of Bonn. The steep valley is lined with vineyards on stony, slate soil.

Ahr is one of the most northerly of the Rhine grape-growing regions and, surprisingly, is famous more for its red wines than its whites. The red wines are much lighter than those grown in the warmer regions of Europe. Spätburgunder (Pinot Noir) is the red grape of choice here; more than half (246 ha) of the total 518 ha in this tiny region are planted with this variety. Other red varieties, such as Portugieser, Dornfelder, and Frühburgunder, produce mildly interesting wines with a somewhat lighter character. The best area for reds is around Heimersheim and Heppingen in the southern part of the district.

The dominant white variety is Riesling, which is usually selected for all the second-best sites—i.e., those not planted with reds. Though they are graceful and scented, the whites of Ahr are not of the same calibre as those produced in the Mosel and Rhine.

The region has one bereich—Walporzheim-Ahrtal.

MITTELRHEIN

The region begins about 5 km south of Bonn and runs south 110 km to Bingen, along one of the most well-known tourist areas of the Rhine. The steeply terraced vineyards, dotted with medieval ruins and castles, cascade in a green swell toward the banks of the Rhine River. Compared to around 300 hours on the highly mechanized plains, annual labour costs here run around 1200 hours per ha—vineyards with 60° pitch are dangerous and difficult to cultivate!

The higher slopes do have the benefit of a few critical additional hours of sunshine that result in some very nice wines, but many vineyards have been abandoned because of the hard labour and low profits.

Fully one-third of the district produces sparkling wine blends or sekts. Riesling accounts for 74% of all plantings.

There are three bereiche—Siebengebirge, Rheinburgengau, and Bacharach.

Mosel-Saar-Ruwer

Named for the three rivers its vineyards follow, this region has been well known for its excellent wines since early Roman times. Trier in the southwest has traditionally been an important wine trade centre. The 12 980 ha of vineyards extend from Koblenz on the Rhine, along the Mosel, and up Ruwer and Saar, all the way to the borders with France and Luxembourg. The best vineyards are located on steep, south-facing slopes in the river valleys.

The region is divided into the Lower Mosel, Middle Mosel, Upper Mosel, and the Saar-Ruwer; the best vineyards are in the Lower and Middle Mosel.

The predominant grape grown here is Riesling (50%), followed by Müller-Thurgau (25%). Other grapes such as Optima and Kerner are also used.

Slate subsoils produce racy wines with good acidity, a distinctive, green-apple character, with spicy, flinty, mineral overtones. The region also produces very good sparkling wines.

There are five bereiche—Zell-Mosel, Bernkastel, Obermosel, Saar-Ruwer, and Moseltor.

Rheingau

This is one of the most prestigious wine regions of Germany. From Lorchhausen to Russelsheim, the south-facing lower slopes on the Taunus Mountains are planted with 90% Riesling, and produce some of the most esteemed wines in the world. Schloss Vollrads, Schloss Rheinhartshausen, and Schloss Johannisberg are synonymous with the best.

The last of these is so famous that the true Riesling grape is called the Johannisberg Riesling (occasionally shortened to Jo'berg) in many parts of the world. The wine's most common characteristics are its wonderful balance, distinctive firm fruitiness, and seductive scent.

Vines are grown by around 1500 growers on 3119 ha. In some years as much as 10% of the crop is used for the production of sekt.

The only bereich is Johannisberg.

Nahe

The Nahe River flows northeast into the Rhine, joining it near Rüdesheim. It is in this area, southwest of Bingen, that the Nahe region lies. There are the 4635 ha of vineyards extending upstream as far as Martinstein and spreading along the valleys of the Guldenbach, Gräfenbach, Glan, and Alsenz rivers. The region has a fairly mild climate, relatively free from winter frosts.

The best wines are produced from Riesling grapes, which cover 25% of the planted area. Müller-Thurgau makes up another 25%, followed by Silvaner (12.1%), Morio-Muskat, and Weissburgunder (Pinot Blanc). Red varieties are currently on the increase, with Spätburgunder (Pinot Noir), Portugieser, and Dornfelder grapes making up the bulk of the newer red varieties.

Up to 22% of the region's QbA and QmP wines are produced in a dry style. Overall, the Nahe style is much heavier than that of the Rheingau and the Mosel due to the warmer climate and heavier soil. Nahe wines also have a clean, crisp acidity.

There is one bereich—Nahegau.

Rheinhessen

This is the largest wine-producing region of Germany, with 23 834 ha under vine. Located immediately south of the Rheingau, the region forms a triangle between the towns of Bingen, Mainz, and Worms, with the Rhine River as a natural border to the north and east.

The ever-popular Müller-Thurgau makes up 23% of all plantings, followed by Silvaner at around 14%, Kerner and Scheurebe at 8% each, Riesling and Bacchus at 7% each, Faber at 6%, and Portugieser at 5%. Kerner is currently on the increase, as it is one of the favourite varieties in the making of Liebfraumilch.

With the Taunus Mountains sheltering the region from the cold north winds, and the rich soil, the region's wines are assertive and robust. Almost two-thirds of the total production is sold in bulk or is bottled outside the region.

There are three bereiche—Bingen, Nierstein, and Wonnegau.

RHEINHESSEN SILVANER

In 1986, a new type of wine emerged from a sea of plonk, with the goal of reviving the sagging reputation of the wine business. Rheinhessen Silvaner is made from Silvaner (in France, Sylvaner) grapes. The wine is made from hand-picked grapes, cannot exceed the maximum yield of 55 hL/ha, is dry, and must pass two quality control tasting panels. Over fifty producers sell under a common label (black and orange with the letters "RS" on it). Thus far, it has been a huge success.

PFALZ

The Pfalz region runs along the west bank of the Rhine, from Worms in the north all the way to the French border in the south. The English name for the region is the Palatinate, and until recently it was called the Rheinpfalz in Germany. The 80 km by 10 km region encompasses 21 417 ha of vineyards. It is the second-largest region after Rheinhessen.

Pfalz can be considered a continuation of the Alsace region of France, as the soils are virtually identical. The climate is dry and mild with a balmy average of 15°C during the growing season.

The region produces a large quantity of decent, fairly ordinary wines along with a handful of better ones. Müller-Thurgau covers the most ground (22%), followed closely by Riesling (20%). The white Kerner, Silvaner, and Scheurebe and the red Portugieser, Dornfelder, and Spätburgunder (Pinot Noir) are gaining ground.

The relatively warm climate best suits the production of fruity, red wines. Many other grape types are planted, specifically for use in bulk wines, blends, and sekt.

There are two bereiche—Mittelhaardt-Deutsche Weinstrasse and Südliche Weinstrasse.

HESSISCHE BERGSTRASSE

Hessische Bergstrasse lies across the Rhine from Worms and is composed of two basic areas. The larger stretches north-south from Seeheim to Heppenheim to make the bereich of Starkenburg.

About 20 km northeast is the bereich of Umstadt. With only 380 ha, this wine region is one of the smallest. Its small, steep vineyards are cultivated by around 1000 growers who work part time and sell their produce to the two local co-operatives. The soil is composed of disintegrated granite combined with loess and sandstone. The wines are delicious Rieslings similar to those of the Rheingau. However, they are difficult to find since production is quite small and very much appreciated locally. Late Harvest Ruländer (Pinot Gris) wines from this virtually frost-free area are especially noteworthy for their great character.

There are two bereiche—Umstadt and Starkenburg.

FRANKEN

Located along the River Main and its tributaries, the region meanders through northern Bavaria, from about 80 km east of the Rheingau at Michelbach to Bamberg. The principal city of Würzburg is the centre of the local wine trade and the source of the renowned Würzburger Stein.

Franken's growing season is short with unpredictable weather in the spring and autumn. Its hot summers and bitter winters are characteristic of the continental climate. Although the soil varies from place to place, it consists largely of limestone with some silt and loess deposits.

Silvaner, Weissburgunder (Pinot Blanc), Bacchus, and Müller-Thurgau are the most-widely-planted varieties in the region's 5920 ha, followed by Riesling, Scheurebe, and Rieslaner (a local cross of Silvaner and Riesling).

Half of all the wines here are produced by co-operatives. Steinwein (from Würzburger Stein) became the traditional generic name for Franken wines and is usually sold in flattish flagons (similar to Portugal's Mateus) called *bocksbeutel*. As much as 70% of Franken wines are made either in the trocken or halbtrocken style, with residual sugar levels of less than 4 g/L as compared with 9 g/L throughout the rest of Germany. The wines are firmer (having an average of 11.5% alcohol), with characteristic earthiness and dried fruit flavours. The local Silvaner is unique and particularly good.

There are three bereiche—Mainviereck, Maindreieck, and Steigerwald.

Württemberg

This region follows the Neckar River (from Reutlingen in the south to Neckarzimmern in the northwest) and spills out as far as Schäffersheim in the northeast, along the rivers Jagst and Kocher to the Tauber at Bad Mergentheim. Thus the area connects the regions of Hessische Bergstrasse and Franken. The 10 720 ha of vineyards are divided among 15 500 proprietors, and co-operatives are responsible for 80% of all wine production.

Half the vineyards are planted with the red grapes Trollinger and Lemberger; another quarter are dedicated to Riesling. The other grapes planted are Schwarzriesling (Pinot Meunier), Kerner, Müller-Thurgau, and a new variety called Samtrot, which is a mutation of Schwarzriesling.

Most of Württemberg's wine is rather ordinary. The very small amount of higher quality wine is rarely found outside the area. Locals love their light, pale, red wines, soft in tannin and usually sweetened with süssreserve. A local speciality is Schillerwein, produced by pressing and fermenting red and white grapes together.

The four bereiche are—Kocher-Jagst-Tauber, Württembergisch Unterland, Remstal-Stuttgart, and Oberer Neckar.

Baden

This region is an assembly of districts stretching 400 km along the east side of the Rhine (opposite the French wine region of Alsace).

It includes vineyards on the north side of the Bodensee (Lake Constance) on the Swiss border. The Black Forest lies on the region's east; the major cities are Baden and Freiburg. The region of Baden covers a total area of 16 417 ha, with considerable climatic variation. It is the sunniest region in all of Germany and the only one lying in Zone B (as defined by the EU). Grapes are generally high in sugar and low in acid, thus produce wines that are fuller than most German wines. Chaptalization is permitted to a maximum of 2.5% alcohol.

Nearly 200 grape varieties are cultivated in this prolific region, but six dominate. One-third of the area is planted with Müller-Thurgau, a quarter with Spätburgunder (Pinot Noir), and significant amounts with Riesling, Gutedel (Chasselas), and Silvaner.

Most wines are soft and made in the trocken style.

This forward-looking region is growing a rare collection of Chardonnay clones, and is ageing red wines in French and Slovenian oak on an experimental basis.

There are eight bereiche—Tauberfranken, Ortenau, Badische Bergstrasse/Kraichgau, Breisgau, Kaiserstuhl, Tuniberg, Markgräflerland, and Bodensee.

Saale-Unstrut

The most northerly region of Germany lies where the Saale and Unstrut rivers meet, around the town of Freyburg. Its 390 ha of vineyards are located mainly on south-facing slopes along the Unstrut. The climate is typically continental with short, hot, sunny summers and bitterly cold winters, when temperatures can plummet to –34°C. Early frosts in autumn and late ones in spring regularly challenge grape growers. This is not a region for Late Harvest wines. The soil is composed primarily of limestone and sandstone.

The area is planted mainly with Müller-Thurgau (44%). Silvaner covers 11% of vineyards, followed by Riesling, which is saved for the best and steepest sites. Gewürztraminer and Weissburgunder (Pinot Blanc) are other white varieties planted. Red varieties include Portugieser, Bacchus, and Faber grapes.

There are two bereiche—Schloss Neuenburg, and Thüringen.

Sachsen

Also known as Saxony, this region is the most easterly of Germany's vineyards and lies along the Elbe River. The 310 ha of grape plantings stretch out around the town of Meissen and on the outskirts of Dresden. Winters here are so cold it is not uncommon to lose 90% of the crop in some years. Because warm weather comes only in June, only perfectly sheltered sites with southern exposures can expect to produce mature grapes. The yield is a low 10 hL/ha. The soil is mainly volcanic, composed of reddish sandstone and loess. The best sites are planted on granite that is overlaid with alluvial sandy deposits or limestone loam.

The cold-hardy Müller-Thurgau dominates the plantings (40%), followed by Weissburgunder (Pinot Blanc), Ruländer (Pinot Gris), and some Morio Muskat. The growing season is much too short for Riesling, so instead Goldriesling (a cross of Riesling and Muscat Précoce de Courteiller) is planted.

There are three bereiche—Meissen, Dresden, and Elstertal.

AUSTRIA

Like the waltz, classical music, and old-world charm, wine is part of Austria's culture, ever present in daily life. Wine is offered to guests at the slightest excuse—whether it be a national festivity, a business dinner, the greeting of a friend, or a family picnic. Austrians generally like their wine fresh, crisp, and dry. Wine production is 3 600 000 hL; wine consumption is 31 L per capita.

Wine Regions of Austria

1 Niederösterreich
2 Wien
3 Burgenland
4 Steiermark

Hollabrunn
Krems
Linz
Tulln
Baden
Neusiedl
Salsburg
Eisenstadt
Güssing
Graz
Deutschlandsberg

GEOGRAPHY

Austria is a landlocked country in central Europe. The wine regions lie in the eastern part of the country, in the warm and dry Pannonian basin. With roughly the same latitude as Burgundy, Austria's wine regions stretch out in temperate climate zones.

Altitudes and inclination of the slopes are key factors in determining the quality of the wines. Most vineyards are located at elevations between 200 m–400 m. The highest ones, in the Sausal area in Steiermark, reach 560 m.

The Danube River (in the northeastern wine regions) and Neusiedlersee (a shallow lake surrounded by sandy marshes) create distinct mesoclimates.

The yearly precipitation averages between 400 mm–800 mm, depending on the location. The summers are warm and sunny; the long autumn days are balmy.

Austria's soils vary from place to place and can be divided into two main types. Light, sandy soils containing loess warm up quickly and, with proper irrigation, yield delicate, fruity wines. Heavier clay soils retain more moisture and tend to provide larger yields of full-bodied wines.

HISTORY

Grape seeds found in Celtic tombs indicate that grapes were available in what is now Austria as far back as 700 B.C.

The Romans, under the leadership of Emperor Probus (A.D. 276–282), planted many vineyards in the northern province they called Noricum.

Later, grape growing came to a standstill with the invasions of the Franks, Slavs, and Avars in the fifth century. Under Charlemagne (742–814), viticulture surged ahead.

In 1526, the first Trockenbeerenauslese was recorded, made in the Donnerskirchen Mountains. But grape growing slumped again in the next century, during the Thirty Years War (1618–1648), and because heavy taxes and exorbitant wine prices caused an increase in the consumption of beer.

By 1784, taxes were again lowered, and the Holy Roman Emperor Joseph II (1741–1790) gave the wine industry new momentum by allowing winemakers to sell wine from their estates, a privilege that exists to this day.

In the nineteenth century, Austrian vineyards were ravaged by a series of calamities. Like the Four Horsemen of the Apocalypse, powdery mildew, downy mildew, phylloxera, and brutal freeze-ups devastated the vineyards. Several wine schools were established to combat the onslaught. Today, the most renowned is the Vocational School of Viniculture in Klosterneuburg.

The collapse of the Austro-Hungarian Empire (1918), as well as World Wars I and II, left the wine industry in disarray. Disintegration of the old structures made room for the introduction of new techniques and modern equipment. A highly publicized scandal over the addition of diethylene glycol to wines disgraced the industry in 1985 and crippled the export market. Today, under new rules of strict legislative control to guarantee the highest level of quality, Austria is regaining international acknowledgment.

LAWS AND REGULATIONS

Austria belongs to the European Union, and as such follows its wine policies. In general, Austria's wine laws resemble Germany's. They are based on origin, yield per hectare, quality categories, and quality control. There are five wine categories, largely determined by the sugar content of the must, which in Austria is measured using the Klosterneuburger Mostwaage (KMW) scale.

KLOSTERNEUBURGER MOSTWAAGE (KMW)

Klosterneuburger Mostwaage ("Klosterneuburg Must Weight") measures the amount of sugar in the must by the percentage of weight.

The scale was developed in 1869 by August Wilhelm Freiherr von Babo (1827–1894), the first director of the wine institute in Klosterneuburg. Roughly speaking, 1° KMW equals 5° Oechsle.

The five categories are as follows:

Table Wine

- Tafelwein—minimum of 10.6° KMW. No specification of origin, variety, or vintage date is permitted on its label.
- Landwein—minimum of 14° KMW. Also, the wine has to originate from a single wine region.

Quality Wine

- Qualitätswein—minimum of 15° KMW. It may be chaptalized with up to 4.25 kg sugar for each 100 L, to give a maximum 19° KMW for white wine and 20° KMW for red. Alcohol content must be a maximum of 9% for white and 8.5% for red.
- Kabinett—minimum 17° KMW, a maximum alcohol content of 12.7%, and no chaptalizing allowed. The wine must be bottled in Austria.
- Prädikatswein—bottled in Austria and with the most stringent regulations. There are seven types of prädikatswein. Neither chaptalization nor the addition of süssreserve are permitted in any subcategory. With the exception of Spätlese, these wines cannot be sold before May 1 following the harvest.

The seven subcategories of prädikatswein are as follows:

- Spätlese—minimum of 19° KMW.
- Auslese—minimum 21° KMW. All faulty or unripe grapes must be excluded.
- Beerenauslese (BA)—minimum of 25° KMW, produced from overripe or botrytized grapes.
- Strohwein—minimum 25° KMW, produced from raisined grapes. In Austria, this means that the overripe grapes are stored and air-dried on straw or reed mats for at least three months.
- Eiswein—minimum 25° KMW, produced from grapes that are harvested and pressed while still naturally frozen.
- Ausbruch—minimum 27° KMW, produced exclusively from overripe grapes that are naturally shrivelled and affected by noble rot.
- Trockenbeerenauslese (TBA)—minimum 30° KMW, produced from overripe grapes that are naturally shrivelled and affected by noble rot.

The maximum yield for tafelwein, qualitatswein, and prädikatswein is restricted to 9 t/ha and 6750 L/ha. Qualitatswein and prädikatswein must pass both a chemical analysis and a taste test. A prüfnummer (proof number) is indicated on a red and white stripe affixed to the neck of the bottle.

Other Types of Wine

- Bergwein—wine made from grapes grown on slopes with gradients greater than 26%.
- Gemischten satz—wine made from a blend of grapes all grown in the same vineyard.

SWEET OR DRY?

Many Austrian wines are fermented dry; the indication of, for example, Beerenauslese does not necessarily mean that the wine is sweet. The following terminology better indicates the level of sweetness of the wine:

extra trocken (extra dry)	up to 4 g/L residual sugar
trocken (dry)	up to 9 g/L residual sugar if the total acidity is less than the residual sugar by a maximum of 2 g/L. For example, a wine with 8 g/L residual sugar must have at least 6 g/L acidity in order to be called *trocken*.
halbtrocken (half dry)	up to 12 g/L residual sugar
lieblich (medium sweet)	up to 45 g/L residual sugar
süsse (sweet)	over 45 g/L residual sugar

GRAPE VARIETIES

Of the thirty or so grape varieties grown, twenty are white (weiss) and ten are red (roter). Not surprisingly, the soil and the climate are definitely kinder to the whites. The following lists the main grape varieties, their percentage of all plantings (unless negligible), and the types of wines they produce.

White Grapes

- Grüner Veltliner (37%)—fresh, with a peppery aroma.
- Welschriesling (9%)—wines develop a spicy bouquet, with a subtle green-apple aroma and racy acidity.
- Müller-Thurgau (7.8%)—delicate, similar to Muscat, scented, fruity, soft, and elegant.
- Weissburgunder (5%)—full bodied, with soft acidity and a nutty aroma; also called Pinot Blanc or Klevner.
- Riesling (2.6%)—delicate, with peachy overtones, lithe texture, and lively acidity; also known as Rhine Riesling and White Riesling.
- Neuburger (2.6%)—spicy, rich, with fruity acidity.
- Traminer (1%)—soft acidity, full body, and an aromatic, flowery bouquet reminiscent of dried rose petals; also called Gewürztraminer or Roter Traminer.
- Muscat Ottonel (1%)—distinct Muscat smell and flavour, soft, light, and charming.
- Bouvier (0.9%)—soft, with hints of sweetness, and a scent of Muscat.
- Chardonnay (0.6%)—green apple aroma, good acidity, and sound body; locally called Morillon or Feinburgunder.
- Sauvignon Blanc (0.3%)—grassy, with good acidity, and the aroma of elderberry.
- Muskateller(0.2%)—intense Muscat aroma, fruity acidity.
- Rotgipfler—fresh and spicy, with a distinctive tang.
- Zierfandler—spicy, with crisp, fruity acidity.

Red Grapes

- Blauer Zweigelt (8%)—full bodied, spicy, with rich tannins, and satisfying acidity; a cross of Saint Laurent and Blaufränkisch that is also called Rotburger.
- Blaufränkisch (5%)—fresh and racy, with good acidity.
- Blauer Portugieser (5%)—fruity, with soft acidity, and a mild flavour.
- Saint Laurent (0.9%)—flavourful, with low acidity, and good tannin.
- Blauer Burgunder (0.7%)—soft, rich flavour, a fruity bouquet, and medium tannins; also called Blauburgunder or Pinot Noir.
- Cabernet Sauvignon (0.6%)—black currant nose, firm tannins, and a grassy, herbaceous character when young, or unripe, or overwatered.
- Merlot—used mainly for blending.
- Pinot Noir—the real grape from Burgundy that is soft, low in tannins, light in colour, with a strawberry aroma.

WINE REGIONS

Austria has about 55 000 ha of vineyards and can be divided into four regions containing a total of sixteen districts. Roughly from north (the Danube plain) to south (Styria), the regions are as follows.

NIEDERÖSTERREICH

The Niederösterreich (or Lower Austria) region covers 32 000 ha of land, distributed among the following eight zones or districts.

- Weinviertel—the 18 004 ha contain mostly loess, loam, rock, and black soils. Dominant varieties include Grüner Veltliner, Welschriesling, Rhine Riesling, Weissburgunder, Feinburgunder, Blauer Portugieser, and Zweigelt. Particularly good eiswein originates here.
- Kamptal—the 4189 ha located here are found on primary volcanic rock, loess, and clay. The main grapes are Grüner Veltliner, Riesling, Feinburgunder, Zweigelt, Blauburgunder, Cabernet Sauvignon, and Merlot.
- Donauland—most of the 2814 ha are located on loess soils. Grüner Veltliner dominates, but Riesling and Weissburgunder are also planted. Frühroter Veltliner is a local speciality.
- Kremstal—this subregion is between Kamptal and Wachau. Its 2438 ha of vineyards are planted on primary volcanic rock and loess soils. Grüner Veltliner and Riesling dominate, followed by Feinburgunder. Some red wines are produced, as are good sparkling wines.
- Wachau—all 1448 ha are located mainly on primary volcanic rock soils. Grüner Veltliner and Riesling are the main grapes, followed by Neuburger, Feinburgunder, Müller-Thurgau, and Weissburgunder.

WACHAU'S WINES

In Wachau, the wine association has created three distinctive categories:

Steinfeder—light wine with a maximum alcohol content of 10.7%

Federspiel—wine with a minimum of 17°KMW and a maximum alcohol content of 11.9%

Smaragd—wine with a minimum 18°KMW

- Traisental—there are 696 ha of vineyards located over sand, gravel, and clay soils. The area is planted mainly with Grüner Veltliner, with some Welschriesling, Rhine Riesling, Feinburgunder, and red varieties.
- Carnuntum—the 995 ha of vines grow on gravel, loam, sand, and loess. Primary varieties are Grüner Veltliner, Welschriesling, Weissburgunder, Feinburgunder, Zweigelt, and Saint Laurent.
- Thermenregion—these 2814 ha benefit from generally stony and somewhat heavy loam soils. The main varieties are Neuburger and Weissburgunder. One-third of the area is planted with red Portugieser, Zweigelt, Blauburgunder, and Cabernet Sauvignon. The famous Gumpoldskirchener wine is made here from Spätrot (Zierfandler) and Rotgipfler.

Wien

The region named for Austria's capital, Wien (Vienna), has 731 ha planted with vines on slate, gravel, sand, and loess soils. The most important grape cultivated here is the Grüner Veltliner, followed by Neuburger, Traminer, Weissburgunder, Chardonnay, Riesling, Blauburgunder, Zweigelt, Portugieser, and Cabernet Sauvignon.

Most wines consumed locally in Wien as heurigen wine are also gemischten satz wines, that is, made from a blend of grapes all grown in the same vineyard.

HEURIGE

Heurige has two meanings—new wine from the most recent vintage, or the taverns where winemakers sell their own wine. Officially, a wine becomes "heurigen" on St Martin's Day (November 11), and may be sold as such until December 31 of the following year.

Burgenland

The Burgenland region totals approximately 19 000 ha, divided among the following four zones:

- Neusiedlersee—with 10 837 ha planted on loess, black soil, gravel, and sand, this area makes good sweet wines and eiswein.

The main varieties are Welschriesling and Weissburgunder. Local specialities include Bouvier, Traminer, and Muscat Ottonel. Red varieties (Zweigelt, Saint Laurent, Blaufränkisch, Cabernet Sauvignon, and Pinot Noir) contribute to 15% of all plantings.

- Neusiedlersee-Hügelland—there are 6264 ha planted over loess, sand, black soil, and loam. Welschriesling, Weissburgunder, Neuburger, Blaufränkisch, and Zweigelt dominate. Specialities include Sauvignon, Furmint, Chardonnay, and Cabernet Sauvignon, many of which benefit from noble rot.
- Mittelburgenland—all 2107 ha are on a flat loamy soil. As much as 95% is planted with red varietals. The main grape is Blaufränkisch, but some Zweigelt, Welschriesling, and Weissburgunder are also grown.
- Südburgenland—only 457 ha of vines are planted on thick, ferruginous, clay soil. Blaufränkisch, Zweigelt, and Welschriesling are the main varieties, with small amounts of Muscat Ottonel, Riesling, and Weissburgunder.

Steiermark

In the southeast of Austria, Steiermark has 3000 ha under vine, divided among three zones. One of the grapes, Blauer Wildbacher, can be planted in no other Austrian region.

- Süd-Oststeiermark—there are 1205 ha planted over volcanic basalt or sandy loam soils. The main variety is Welschriesling, followed by Weissburgunder and Traminer. Local specialties are Ruländer, Riesling, and Morillon (Chardonnay). Reds include Blauburgunder, Zweigelt, Saint Laurent, and Blauer Wildbacher.
- Weststeiermark—this small zone of only 480 ha is planted on gneiss and slate soils. The main variety is Blauer Wildbacher, but some Welschriesling, Weissburgunder, and Zweigelt are also planted. A good rosé under the name of Schilcher is made from Blauer Wildbacher.
- Südsteiermark—the most southerly region has 1902 ha on rich soil composed of slate, sand, marl, and limestone. The main grapes are Welschriesling, Sauvignon Blanc, and Morillon (Chardonnay).

HUNGARY

Hungary has a winemaking tradition that dates back to Roman times. It is the smallest of the eastern European wine-producing countries (and the only one that is landlocked), yet it produces a staggering variety of wines. Although its production ranks tenth largest in Europe, it is known mainly for a single wine—Tokaji.

Of the 200 000 ha of vineyards, as much as 85% are devoted to wine grapes; the rest are used for table grapes, juices, jams, and distillation. Currently the state owns 17% of vineyards, co-operatives 31%, and private owners 52%. More than half of all Hungarian wines are produced by two state wineries. Annual production is around 3 700 000 hL of wine and average consumption is 30 L per capita. Hungary exports 3 000 000 hL of wine annually, making it the sixth-largest exporter in the world.

Few Hungarian wines are known in North America except Tokaji, and perhaps the inexpensive red Szekszárdi. Wine buyers should not overlook Hungary's wines; not only do they offer a range of styles at affordable prices, they are also benefiting from modern techniques, improved technology, and the country's economic rebirth.

LAWS AND REGULATIONS

The National Institute for Wine Qualification was established in 1894. It issues state seals to wines that successfully pass rigorous laboratory tests as well as taste tests. Hungarian law recognizes three categories of wine—table wine, quality wine, and special quality wine.

Állami Pincegazdasag (AH-lah-mee pin-sa-GAHZ-da-sahg)—state cellars
asztali bor (AH-stah-lee bor)—table wine
aszú (AH-soo)—botrytis-shrivelled grapes
édes (AY-desh)—sweet
eszencia (ay-SEN-zee-ah)—an extremely rare wine made from the free drippings of aszú grapes, without pressing
fehér (FAY-hair)—white
félédes (FAY-lay-desh)—semisweet
gönci (GEN-shee)—136-L barrel originating from the village of Gönc
habzó (HAHB-zoh)—sparkling
kimert bor (KEE-mairt)—ordinary wine
különleges minöség (KEW-lun-lay-gesh MEE-nuh-sheg)—special quality wine
Magyar (MUG-yar)—Hungarian
minöségi bor (MEE-nuh-shay-ghee)—quality wine
putton (POOH-tun-yuh)—harvesting container holding 26 L or 20 kg–25 kg of aszú grapes and used in Tokaj to indicate the level of sweetness in Tokaji wines. (The plural is puttonyos.)
szamorodni (SAH-more-ud-nee)—as harvested
száraz (SAH-ruz)—dry
vörös (VUR-ush)—red

Wine Regions of Hungary

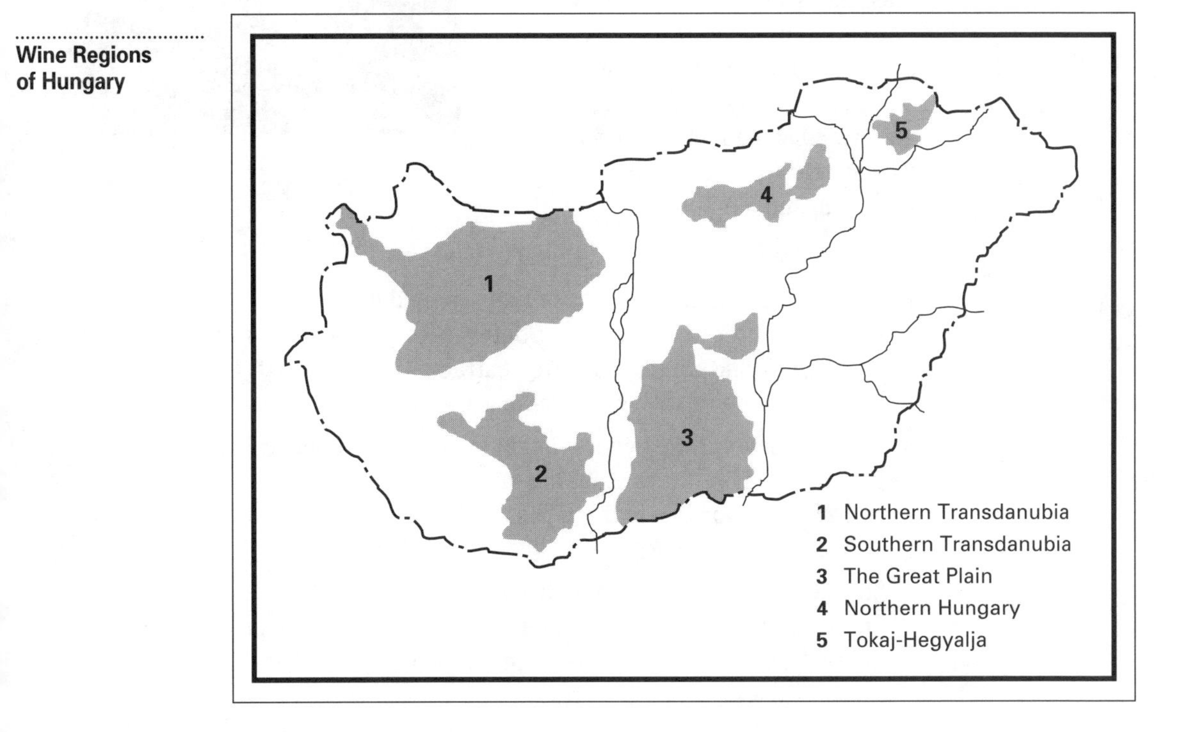

Table Wine

This is the lowest classification. Table wine has a minimum of 8% alc./vol., 16 g/L sugar-free extract for white wine, and 18 g/L sugar-free extract for red wine. The must has 14% sugar content by mass.

Quality Wine

Quality wine has a minimum of 10% alc./vol., 21 g/L sugar-free extract for white wine, and 22 g/L sugar-free extract for red wine. The must has 17% sugar by mass. Also, quality wine has qualities characteristic of the vineyard of origin, the grape variety, and the vintage.

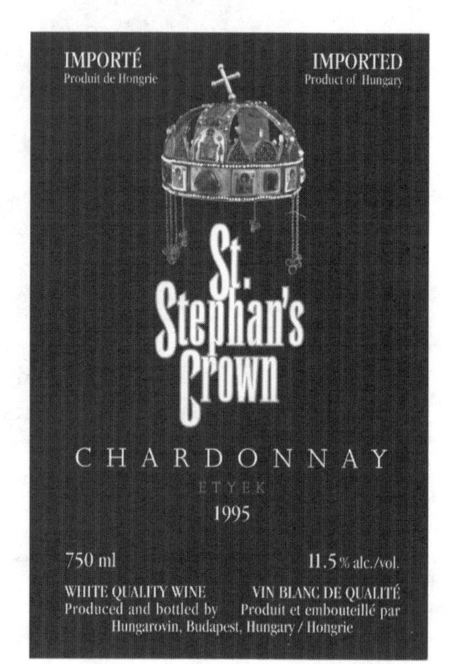

Special Quality Wine

Special quality wine is made from the must of fully ripened (or overripe) grapes, or those affected by noble rot. The wine has excellent flavour, as well as aromas that are characteristic of the grape variety, soil, vintage, and winemaking method. This wine carries the state's seal of approval, and a number.

There are four types of special quality wine:

- Late Harvest wine—made from fully ripened grapes harvested after normal harvest dates, with a minimum of 8% alc./vol. and 24 g/L sugar-free extract. Also, the must has a minimum of 19% sugar by mass.
- Selected Late Harvest wine—made from selected berries fully ripened on the vine, with a minimum of 8% alc./vol. and 25 g/L sugar-free extract. The must has a minimum of 20.5% sugar by mass.
- Shrivelled-berry wine—made from selected overripe grapes that are partly or completely shrivelled. The must has a minimum of 22% sugar by mass, and the wine has at least 26 g/L sugar-free extract.
- Aszú wine—produced from selected overripe berries affected by noble rot, with at least 16.5% alc./vol. and a minimum 30 g/L sugar-free extract. Aszú is a category used only for wine from the Tokaj-Hegyalja region.

SPARKLING WINE

Hungary's best producers of sparkling wine are the François, Hungaria, Pannonia, and Törley wineries. Sparkling wine is produced using either of the following methods:

- méthode Champenoise—The wine must be aged at least one year. If the sparkling wine is aged two or more years, it is labelled *matured in bottle*.
- charmat method—of varying qualities. The best wine is matured for six to nine months in tanks before being bottled and released.

GRAPE VARIETIES

A great majority of Hungary's wines are made from native grapes, but Chardonnay, Riesling, Sémillon, and Sauvignon are gaining in importance as the international market opens up to Hungary. The red wines made from Pinot Noir, Cabernet, and Merlot enjoy a certain popularity and these varieties are increasingly planted.

RED GRAPES OF HUNGARY

Kadarka
Kékfrankos (Blaufränkisch)
Kékoporto ("blue Portugieser")
Médoc Noir
Nagyburgundi ("great Burgundy")
Zweigelt

WINE PRODUCTION

Most Hungarian wines are named for their place of origin, followed by the grape varietal. In many cases, the letter "i" is added to the place name, meaning "from." Badacsonyi Szürkebarát, for example, is a wine from Badacsony, and Tokaji is a wine from Tokaj.

WHITE GRAPES OF HUNGARY

Cirfandli
Dinka
Ezerjó ("a thousand boons")
Furmint
Hárslevelü ("lime leaf")
Leányka ("little girl")
Juhfark ("sheep's tail")
Kéknyélü ("blue stalk")
Király ("royal")
Mezesfehér ("white honey")
Muskotály (Muscadelle or Muscat Ottonel)
Olaszrizling (Italian or Welschriesling)
Rizlingszilváni (Riesling/Sylvaner cross)
Szürkebarat ("grey friar"; a Pinot Gris)
Tramini (Gewürztraminer)
Veltelini (Veltliner)
Zöldszilváni (Green Sylvaner)

WINE REGIONS

Hungary has five wine regions, each of which contains numerous wine-producing districts. The Danube (or Duna) River separates the western regions (Northern and Southern Transdanubia) from the eastern regions (the Great Plain, Northern Hungary, and Tokaj-Hegyalja); Lake Balaton separates the regions of Northern and Southern Transdanubia. The vast expanse of flat land composing the Great Plain and the Puszta (steppe) occupies all the land east of the Danube, with the exception of the hilly northern border.

Soil varies greatly among the districts, but a good many areas consist of Pannonian clay or sand (referring to the flat land resting on the dry bed of the ancient Pannonian Sea), or chernozem soil. Chernozem soil is a dark coloured soil with a deep rich humus. It is found in temperate to cool climates, or rather low humidity, and composes a large part of Hungarian soil.

NORTHERN TRANSDANUBIA

Located in the northwest of the country, it is subdivided into three main wine districts—Sopron (across the border from Austria's Neusiedlersee regions); the northern shore of Lake Balaton between Keszthely and Csopak; and Mór. Parts of the area have been producing wine since Roman times. The following are the region's best-known wines:

- Badacsonyi Kéknyelü—from Kéknyelü grapes grown in sloping vineyards, on Pannonian sand and loam settled on volcanic materials, around the town of Badacsony. The wine is fragrant, firm, full bodied, well balanced, and powerful, with a deep golden colour.
- Badacsonyi Zöldszilváni—made from Zöldszilváni grapes originating from Austria. This wine has a discreet fragrance and is best drunk while young and fresh.
- Badacsonyi Rizlingszilváni—aromatic and soft with low acidity.
- Badacsonyi Olaszrizling—made from the most-planted vine in the district of Badacsony. The wine is fragrant with good acidity and a lovely aftertaste.
- Badacsonyi Szürkebarát—a rich, complex wine that has a touch of bitterness in the aftertaste, and high alcohol content. This golden wine has extraordinary extract. Some exceptional wines are marketed in numbered bottles.

In the great vintage, 1969, it won first prize in the World Wine Competition. This wine had 57 g/L residual sugar, 33 g/L sugar-free extract, and 13.2% alcohol.

- Balatonfüredi Rizling—one of the best Italico Reislings, from the town of Balatonfüred, near the town of Csopak. The lakeside vineyards on south-facing slopes are ideal for the grape. The subsoil is of slate covered by lime, sandstone, and Pannonian sand. High iron oxide in the soil gives it a red hue peculiar to the region. The wine is full bodied, with deep colour, and a spicy fragrance of reseda (sweet mignonette).
- Csopaki Rizling—made from Italico Riesling grapes grown around the town of Csopak. This wine has higher acidity and a more herbaceous bouquet than Balatonfüredi Rizling.
- Somlói Furmint—luscious and heady sweet wines made from shrivelled Furmint grapes grown on 500 ha of vineyards west of Lake Balaton, on the slopes of an extinct volcano. The soil is broken basalt and crumbled porous lava, so resembles the soil of Mount Vesuvius. Because of the area's temperate climate, dry wines are also made. The dry wines are similar to dry Tokaji Szamorodni—full bodied, with a delicate acid balance, and requiring two or three years' ageing to fully develop their character.
- Akali Muskotály—a dessert wine from the village of Balatonakali on the northern shore of Lake Balaton. Greenish in colour, the wine has an intense Muscat bouquet, a sweet taste, and low acidity.
- Móri Ezerjó—made from Ezerjó grapes from the Mór wine area, which lies amidst mountains between Sopron and Budapest. Grown on Mór's volcanic soil (covered with loess and Pannonian sand), the grapes produce a lively greenish-white wine with firm acidity, balanced by 1.5% residual sugar. Elsewhere, Ezerjó wines are only mediocre.
- Soproni Veltelini—made from Veltliner grapes, which distinguish themselves in this cool area in the foothills of the Alps. The wine area lies around the town of Sopron, and has above-average rainfall. Lively when young, this wine is well balanced with good acidity and a spicy bouquet.
- Soproni Kékfrankos—made from Kékfrankos, the most widely planted grape in the area. This deep purple, full-bodied wine is high in tannin and acidity and benefits greatly from ageing.

SOUTHERN TRANSDANUBIA

South of Lake Balaton and east of the Danube, this region is known for its good red wines. The districts are as follows:

- Dél Balaton (South Balaton)—the newest wine district. The rich and fertile soil is mostly sand and loess of high nutritive power. The climate is tempered by Lake Balaton, Central Europe's largest lake. The modern Balatonboglar State Farm and research station on over 1000 ha is planted with new grape varieties such as Sémillon, Sauvignon Blanc, Chardonnay, and Merlot, as well as Leányka, Muskotály, Tramini, Olaszrizling, Kékfrankos, and Nagyburgundi. The result is wines with good balanced acidity, predominantly fruity aroma, and lower alcohol. Most of the wines are on the sweet side, and that includes the reds. These modern wines already have earned a fine reputation, thus helping to rejuvenate the Hungarian wine industry.
- Mecsek—vineyards surround the town of Pécs on the southern slopes of the Mecsek Hills. Here the subsoil is comprised of volcanic granite and slate, with loess and Pannonian sand on top. The regional specialty is white wine (often with a touch of sweetness). The area grows Furmint, Olaszrizling, Chardonnay, and Cirfandli. Pécs Cirfandli has a unique spicy aroma.
- Villány-Siklós—Hungary's southernmost wine district is famous for earthy, plummy, dark, full-bodied Cabernet Sauvignons and Merlots. The soil is limestone, covered by loess and loam rich in calcium. These vineyards once produced exceptional Kadarka wines, but have been replanted with Kékfrankos and Kékoporto grapes. The Kékfrankos is bottled under the Nagyburgundi label, and the Kékoporto is the best produced in Hungary today. Zweigelt and Pinot Noir are also grown with success. The Siklós district produces most of the area's whites from Olaszrizling, Chardonnay, Tramini, and Hárslevelü.
- Szekszárd—one of the oldest wine-growing districts of Hungary, with some vineyards dating back to Roman times, it is located between the Szekszárd Hills and the Danube. The Kadarka grape made the region famous. Today, the best Kadarka wines are hard to find, even in the cellars of the local vintners. Merlot, Kékoporto, Kékfrankos, and Cabernet are the new grapes.

THE GREAT PLAIN

This region (known as Nagualföld in Hungarian, as distinct from Kissafföld, or "little plain," in the west) is the country's largest wine producer, covering most of the eastern half of the country, except for the mountainous area of the north. It is considered the least important region for quality wines.

The region is just south of Budapest, Hungary's capital, and extends from the Danube, south to the border town of Kunbaja, and in the east to Jaszberény and Szeged. Temperatures are extreme—reaching 60°C at soil level in summer and at the limit of frost tolerance for grape vines in winter. Summers are regularly very dry. The soil is mainly chalky sand with alluvial, alkali, and chernozem soils.

Most of the region's wine is white. The red wine is concentrated in the southernmost part of the region, around Vaskút and Szeged. The most renowned wines of the Great Plain are the following:

- Jászberényi Rizling—made from Olaszrizling around the village of Jaszberény, in the sandy soil of the basin of the River Zagyva. The wine is quite fruity and has a characteristic fragrance of sweet mignonette. It is soft, dry, and distinctly green when young, but develops a bright golden hue with age.
- Kecskemeti Cirfandli—produced east of the town of Kecskemét (in the central part of the region) on sandy soil from Cirfandli grapes (known as Zierfandler in Austria). The heady wine is greenish-yellow in colour with a delicate, spicy aroma.
- Halasi Veltelini—made from Veltliner grapes grown on sandy soil. The wine, which is soft and well balanced, is derived from the name of the town of Kiskunhalas, located in the southern area of the Great Plain.
- Kunbajai Olaszrizling—produced in the sandy, sunny south of the region, from Olaszrizling grapes. The area has 2050 hours of sun yearly, so is very productive. The wine is dry with almond overtones.
- Kunbajai Hárslevelü—named for the ancient grape variety Hárslevelü. The wine is medium sweet and full bodied, with a particularly oily, floral bouquet.
- Kunbajai Rizlingszilváni—made from Müller-Thurgau, the Riesling Sylvaner hybrid. The wine is fresh, with a pleasant bouquet, a light body, and a balanced acidity.

- Kunbajai Leányka—made from Leányka grapes, which originated in Transylvania. The wine has a greenish hue and is soft, with a candied nose and a medium-sweet taste.
- Bajai Hárslevelü—named for the town of Baja, on the Danube. The grapes grown in the sandy soil of the plain yield a full-bodied, golden wine.
- Hajósi Cabernet—produced from Cabernet Franc grapes in Hajós, in the southwestern part of the region. The wine has firm tannin, full body, and a dark ruby colour. It ages well.
- Vaskúti Nagyburgundi—a rich, red wine produced near the town of Vaskút.
- Kiskörösi Kadarka—made from Kadarka, which migrated from Albania. Around the town of Kiskörös, Kadarka produces especially robust and aromatic wines with good tannin and a lovely ruby colour.
- Kunbajai Kékfrankos—made from Kékfrankos, the wine is fruity, with medium tannin and a deep red colour.

Northern Hungary

This region is located between the Danube and Tisza rivers north of the Mátra and Bükk mountains. The town of Eger has long been the hub of Hungarian wine trading abroad; by the early seventeenth century it was already famous for "Bull's Blood," a wine that is still produced here. The region includes the following wine districts:

- Eger—in the foothills of the Bükk Mountains, this is the source of Egri Bikavér ("Bull's Blood"). Soils are black volcanic clay, low in lime, that rest on rhyolite tuff. The northern climate is fairly dry, with a late spring and cool nights. While Bull's Blood traditionally used primarily Kadarka grapes, now Kékfrankos is the main grape, to which Cabernet Franc, Cabernet Sauvignon, Kékoporto, and Merlot are added.
- Mátraalja—a vast district west of Eger that mainly produces white wines. The Mátra mountains were once an island rising from the Pannonian Sea that covered Hungary. The rich soil is composed of rhyolite tuff, Pannonian clay, and yellow loess. The mountains (climbing to 1000 m) provide a sunny exposure and shield the vineyards from the harsh northern climate. The best-known grapes are from the village of Debrö. Debröi Hárslevelü has a spicy aroma and good acid, sugar, and alcohol levels; it is one of the country's biggest exports.

Other wines from Olaszrizling, Rizlingszilváni, Tramini, Szürkebarát, Leányka, Zöldveltelini (gold Sylvaner), and particularly Muskotály are produced in a multitude of villages. Muskotály have the typical Muscat nose and a rich aftertaste. The Gyöngyös estate produces fresh crisp Sauvignon and Chardonnay. Nagyred has become known for a crisp, fruity rosé.

- Bükkalja—east of Eger to the town of Miskolc, this district mainly produces white wines.

TOKAJ-HEGYALJA

In the northeast corner of the country, this wine region encompasses twenty-eight villages, and includes the town of Tokaj. The 526 ha of vineyards are at elevations of 150 m–300 m above sea level, on the south-facing slopes of the Eperjes-Tokaji mountains (or Zemplén foothills). The soil is volcanic and loess. The best wine villages are Mád, Tallya, Tarca, and Tolcsva.

Tokaj gained its reputation for many reasons: soil, climate, viticulture, the grape varieties themselves, the cavernous cellars cut into loess, the unique barrels, and traditional methods of vinification. Together these produce Hungary's unique wine known as Tokaji.

The three main grapes planted for Tokaji are Furmint (70%), Hárslevelü (20%), and Sargamuskotály or Ottonel (5%). Furmint and Hárslevelü grapes have a unique ability to retain a high percentage of acid while they concentrate the sugars. The prunings are precise and rigorous. Harvesting traditionally starts on October 28, and lasts through to late November. (Some less daring farmers start in mid-September.) These late harvests allow for noble rot to develop in this area of warm autumns and dewy nights. Grapes affected by noble rot are called *aszú*. Because noble rot does not affect the grapes every season, the vintages vary.

The region's wines are matured in cellars cut out of rock or loess. Here the temperatures range between 8°C–12°C. The walls of the cellars are lined with a dark mould, Racodium cellare. The wine is matured in long barrels (gönci) that each hold 136 L. These barrels are never completely filled in the early stage of maturing, thus providing the wine free contact with the dank air in the cellar. This exposure helps the wine develop its characteristic bouquet, and in return, feeds the mould with acidic vapours.

The categories of Tokaji wines include the following:

- Tokaji Szamorodni—szamorodni ("as it was grown" or "self made") wine is made from an entire harvest, without specially selecting aszú grapes. The grapes are first de-stemmed and crushed, and then left to macerate for between twelve and twenty-four hours; the few aszú berries release their sugar and aromatic substances. The must is pressed and fermented like an ordinary wine, which may be sweet or dry. Száraz Szamorodni is a rich wine with good acidity and relatively high alcoholic content (13%–15%). Edes Szamorodni is a delicate, well-balanced wine, with a minimum natural sugar level of 10 g/L–25 g/L.
- Tokaji Aszú—all aszú wines are made from late-picked, overripe Furmint and Hárslevelü grapes that are naturally affected by noble rot. The original 20% sugar content increases to 40%–60% levels without loss of acidity. The shrivelled aszú grapes are individually picked and, after about seven days, crushed to a pulpy dough. Tokaji must (or new wine) is poured over this. The mixture is stirred several times during thirty-six hours of soaking, in order to extract the maximum sweetness and flavour. Then it is pressed, poured into a gönci, and matured for several years in local cellars hollowed out from the hillsides.

  The eventual level of sweetness depends on the proportion of aszú grapes added. Thus, Tokaji Aszú wines are labelled according to the number of puttonyos added to each gönci—e.g., *Tokaji Aszú 4 Puttonyos*.

  The categories are:

  - 2 puttonyos—permitted by law but never marketed.
  - 3 puttonyos—aged in wood at least five years. This wine is medium-sweet and has a minimum natural sugar level of 60 g/L and 30 g/L sugar-free extract.
  - 4 puttonyos—aged in wood at least six years. This wine is sweet and has a minimum natural sugar level of 90 g/L and 35 g/L sugar-free extract.
  - 5 puttonyos—undergoes a longer fermentation and is aged at least seven years. This wine is very rich and sweet and has a minimum natural sugar level of 120 g/L and 40 g/L sugar-free extract.

    - 6 puttonyos—aged in wood at least eight years. This wine is rare and very rich. It has a minimum natural sugar level of 150 g/L, and 45 g/L sugar-free extract. It equates to a German beerenauslese.
- Aszú Eszencia—quoting the Hungarian Wine Law, "Tokaji Aszú Eszencia is a superb aszú wine of an exceptional vintage obtained from a careful delimited area, the quality of which cannot be measured any more by the aszú number of puttonyos." The grapes are individually selected, botrytized grapes that are de-stalked. The wine is made from the free run of the aszú grapes. It is the equivalent of 7 or 8 puttonyos, with the addition of a small amount of eszencia. The wine spends ten to twenty years in wood, and has a minimum natural sugar level of 180 g/L (but often reaches as high as 240 g/L) and a minimum of 50 g/L sugar-free extract. It is magnificent, rare, and expensive.
- Eszencia—an extremely rare wine made from the free drippings of aszú grapes, without pressing. The average production is around 150 L a year.

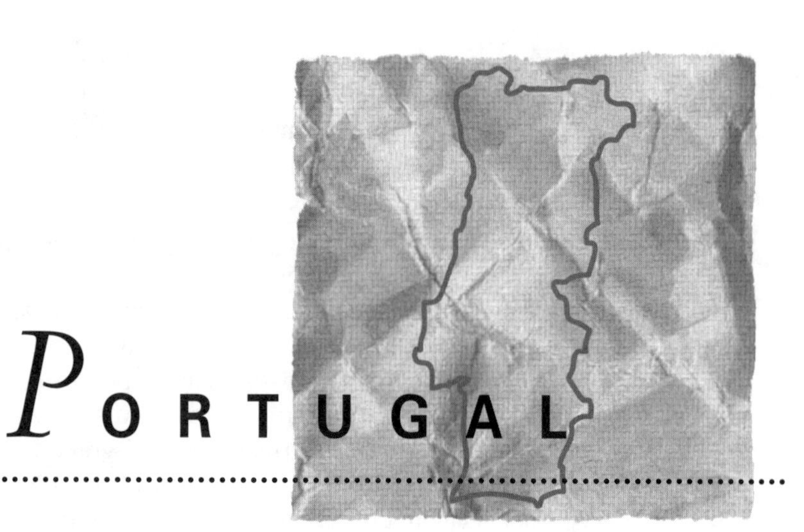

PORTUGAL

Portugal's 401 000 ha of vineyards produce upward of 4 600 000 hL of wine annually; consumption is 55 L per capita. The country has a long history of exporting wine, and Port wines have long dominated this trade. Approximately one-quarter of the rural population is employed in the wine trade.

adega (ah-DAY-ga)—cellar or winery
aguardente (ug-gwu*CH*-DEN-tee)—brandy, a spirit distilled from wine
bagaceira (ba-ga-SAY-ra)—marc, a spirit distilled from skins and seeds of grapes after pressing
branco (BRUNK-oo)—white
doce (DOE-see)—sweet
espumante (esh-poo-MUN-chee)—sparkling
garrafeira (ga-*CH*a-FAY-ra)—red wine of a superior vintage that has been aged for a minimum of two years in vats and one year in the bottle. The alcohol level must be 0.5% above the legal minimum.
quinta (KEEN-tah)—farm, estate, or vineyard
seco (SAY-koo)—dry
solar (so-LU*CH*)—wine estate
tinto (TEEN-too)—red
vinho generoso (VEEN–yoo zhay-nay-ROE-zoo)—dessert wine rich in alcohol
vinho maduro (ma-DOO-roo)—mature table wine

GEOGRAPHY

Portugal forms a rectangle on the western, Atlantic coast of the Iberian Peninsula. Including the Madeira Islands and the Azores in the Atlantic Ocean, Portugal covers 92 082 km^2.

Strangely, the climate seems more Mediterranean than Atlantic, with hot dry summers blistering the interior—the ocean moderates only a thin strip of coastal Portugal. Winters are severe and quite wet in the extreme north, with an average of 2500 mm of rain in the provinces of Minho and Beira. The south is much more arid, receiving barely 500 mm of precipitation annually.

Two great rivers, the Tagus and the Douro (Duero in Spain), divide the land into three distinct wine zones. South of the Tagus River, the Algarve and Alentejo regions are hot and arid with mild winters. The central region between the Tagus and Douro rivers is more temperate, producing fine wines in Bairrada and Dão. North of the Douro, the Tras-os-Montes e Alto Douro is a mountainous area with extreme weather; Douro Litoral and Minho are damp fertile regions with green and prolific vineyards.

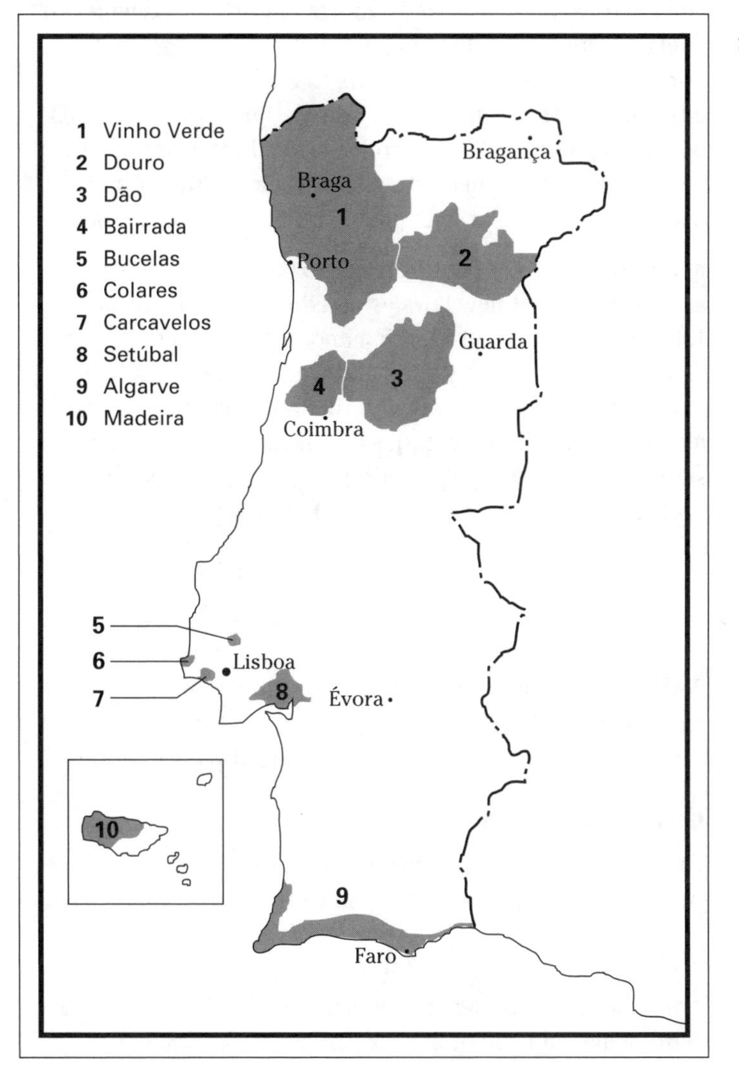

Wine Regions of Portugal

HISTORY

It is quite possible that some Greek adventurer came to the shores of Portugal and planted wine grapes. The ancient Greek technique of growing vines up into trees has long been practised in the region of Vinho Verde. Whatever the case, local tribes traded wine with the Carthaginians as far back as the third century B.C. The wine trade increased under the Romans, who fought the local Lusitanians in the second and first century B.C., and finally conquered them under the Roman emperor Augustus (63 B.C.–A.D. 14). Because Rome was too far away for the Romans to import the precious liquid, local wine production prospered.

Over time, and with stimulation from the nobility and the Church, vineyards expanded and wine became a valuable commodity. The Treaty of Windsor (1386) allied England and Portugal and increased exports of Portuguese wines to England. However, after King Philip II of Spain annexed Portugal in 1580, wine shipments to England were interrupted for seventy years. The Methuen Treaty of 1703 boosted exports to England once again. In the eighteenth century, Madeira wines established their character and became very popular in England and the American colonies.

In the twentieth century, Portugal experienced upheaval and isolation, but new laws established in the 1930s and the 1960s regulated the production and exportation of its wines. Since 1970, the country's red and white table wines have undergone a major revival with the introduction of more modern vinification techniques such as the use of stainless steel vats and temperature-controlled fermentation. In 1986, Portugal became a member of the European Union.

LAWS AND REGULATIONS

Portugal's wine law and system of demarcating wine regions date back to the eighteenth century. Since 1991 they conform to the EU framework. There are three categories.

Vinho Regional

This is the lowest level of classification: wine made from grapes grown from a specific area. It is similar to France's vin de pays.

IPR

Indicação de Proveniencia Regulamentada (IPR) is the next tier of wine, equivalent in rank to France's VDQS.

DOC

Denominação de Origem Controlada (DOC) is the equivalent of AOC in France, and is Portugal's highest rank of quality wine. Prior to Portugal joining the EU, these zones were known as Regiões Demarcadas (demarcated regions). Requirements for DOC cover the grape varieties used, cultivation practices, methods of vinification, minimum natural alcohol level, production per hectare, chemical analysis, and taste tests. Only Vitis vinifera grapes may be used.

Portugal has thirteen DOC zones and thirty-one IPR zones.

The IPRs are Chaves, Valpaços, Planalto-Mirandês, Varosa, Encostas da Nave, Lafões, Castelo Rodrigo, Pinhel, Cova da Beira, Encostas d'Aire, Tomar, Alcobaça, Portalegre, Chamusca, Santarém, Óbidos, Almeirim, Cartaxo, Alenquer, Torres, Coruche, Arruda, Borba, Palmela, Redondo, Evora, Arrábida, Reguengos, Granja-Amareleja, Vidigueira, and Moura.

DOC ZONES

From roughly north to south:

Vinho Verde
Douro/Porto
Dão
Bairrada
Bucelas
Colares
Carcavelos
Setúbal
Lagos
Portimão
Lagoa
Tavira
Madeira

WINE REGIONS

The following presents Portugal's wine regions, roughly from north to south. The unique region of Porto and its wine will be treated in the next chapter.

VINHO VERDE

The region of Vinho Verde is located in the historic province of Minho, in the northwest corner of the country between the Douro River and the Minho River at the Spanish border. The 35 000 ha of vineyards occupy 15% of the region's agricultural area and produce 2 300 000 hL of wine, representing nearly 20% of total national production. There are more than 100 000 producers, but as much as 60% of them make less than 1 hL each. (Only 5% produce more than 10 hL wine annually.)

Vinho Verde wines were first recorded in Monção in 1261; the region was delimited in 1908 and demarcated in 1929. Vinho Verde is supervised by the Comissão de Viticultura da Região dos Vinhos Verdes, which supplies the selo (stamp) endorsing authenticity.

Vinho Verde is produced in the districts of Viana do Castelo, Oporto, and Braga as well as the municipalities of Arouca, Castelo de Paiva, Val de Cambra, Mondim de Basto, Ribiera de Pena, Cinfães, and Resende.

GEOGRAPHY

The climate is fairly humid and cool, from proximity to the Atlantic, and the region's many rivers. Fog is frequent. The average annual precipitation is 1500 mm–2000 mm. The soils are mostly granitic and low in lime and phosphoric acid, but sufficient in potassium.

GRAPE VARIETIES

Vines are generally trained up trees in what is called the hanging system, or on traditional pergolas. These ancient systems are slowly being replaced with cross-shaped cruzetas, garlands, and fencing. The permitted grape varieties are Azal Tinto, Borraçal, Espadaneiro, Vinhão, Brancelho, Pedral, and Rabo-de-Ovelha for the red grapes; Alvarinho, Loureiro, Trajadura, Azal, Avesso, Batoca, and Pederña for the white.

WINE PRODUCTION

The name Vinho Verde ("green wine") refers to the crisp acidity of the wine rather than to its colour.

Most (90%) of the region's wines are red, and all but 10% of these are consumed locally. The wines are dark, with a low alcohol content ranging from 8% –11.5%, and a grapey aroma.

White wines are made solely from white grapes. The best quality wine is made from Alvarinho and Loureiro grapes, and has an alcohol content of 12%. The average white Vinho Verde is around 8.5%. It is tart and refreshing because of the light carbonation resulting from malolactic fermentation. Vinhos Verdes do not age well.

The region is also known for aguardente and aguardente bagaceira—brandies with alcohol content ranging from 45%–60%.

Douro

The Douro region lies along the upper Douro River valley, from the village of Barqueiros in the west to the Spanish border. It covers 240 000 ha, but only a tenth of it is planted. The area is sheltered from the damp Atlantic winds by the Montemuro and Marão mountains.

The climate is extreme, with cold, frosty winters, damaging hailstorms, and scorching summers; annual rainfall is around 6000 mm. The stony soils are predominantly schist, with some granite, limestone, and gneiss.

GRAPE VARIETIES

The main red varieties grown are Bastardo (known as Trousseau in France's Jura), Mourisco Tinto, Tinta Amarela, Tinta Roriz, Tinta Barroca, Tinta Francisca, Tinta Ção, Touriga Francesa, and Touriga Nacional. The main white grapes are Donzelinho Esgana-Ção ("dog strangler"), Folgazão, Gouveio (Verdelho), Malvasia Fina, Rabigato, and Viosinho.

WINE PRODUCTION

Local table wines have long been eclipsed by Port but now are making a comeback. Portugal's best wine (Barca Velha) is produced here. The reds are rich in colour and flavour; they have a complex finish and develop character when aged. Red wines can be bottled only after eighteen months. The whites are fresh and aromatic. They can be bottled after nine months. Both must contain a minimum of 11% alcohol.

Dão

Dão is located in the central area of Portugal, surrounded by the Estrela, Camamulo, Buçaco, Nave, Lousã, and Açor mountains. The rivers Dão and Mondego and their tributaries divided the region into three subregions:

- the north-central region—with granitic soil, at 200 m–600 m elevations.
- the south-central region—with schistous soil and average elevations of 400 m.
- the periphery—includes all the flat and fertile lowlands.

Dão was protected first in 1390, demarcated in 1908, and delimited in 1912. The demarcated region covers sixteen municipalities and three districts (Viseu, Guarda, and Coimbra), with 20 000 ha planted with vines. Of the 35 000 growers, most do not produce more than 50 hL each. Ten co-operatives are responsible for 55% of the total production.

The climate is relatively mild. The long, hot, dry summers help the grapes reach perfect maturity. Autumn is cool, but winter can be wet and mild. The region has an annual precipitation slightly above 1000 mm. Most of the best vineyards lie between 400 m–500 m. Around 800 m, the vineyards are terraced or planted among the pine trees on steep slopes. Vines are trained low and attached to wires or stakes.

GRAPE VARIETIES

The red grape varieties (70% of the total) include Touriga Nacional, Tinta Pinheira, Tinta Carvalha, Bago-de-Louro, Alvarelhão, Bastardo, and Alfrocheiro Preto. The remaining 30% are white varieties including Arinto, Dona-Branca, Barcelo, Fernão Pires, Sercial, and many others.

WINE PRODUCTION

Dão used to be the source of the best Portuguese table wines; now, however, other regions with more modern viticultural and winemaking practices have challenged them.

Most (90%) of the region's wines are red, and it is the reds that are best: soft, full bodied, and with intense colour. They age well, acquiring a pleasant bouquet.

These reds spend eighteen months in the cellar before being bottled. Whites must wait ten months before being bottled. They are soft, fresh, and light, with a pale straw colour. Both reds and whites have an average alcohol content between 11% and 12.5%. They are noteworthy for their high glycerine content (up to 12 g/L).

Bairrada

In spite of an early start in winemaking—in the tenth century—and praise for the region's wines recorded in the sixteenth, seventeenth, and eighteenth centuries, Bairrada was demarcated only recently, in 1979. One reason for this was the Marquis of Pombal's order in 1760 that all of Bairrada's vines be ripped up in order to protect the trade in Port to the north. Now the region has 18 600 ha of vineyards located between Dão and the Atlantic Ocean. There are three distinct areas—the lowlands, the gentle slopes, and the mountain range.

The climate runs to hot, dry summers and long, bitterly cold winters. Precipitation is 1100 mm. The soil is generally poor in minerals, and varies from sand to clay, with some areas of pure sandy soil.

GRAPE VARIETIES

Red grapes grown in the region include Baga (Poceirinha), Castelão (Moreto), Tinta Pinheira, Alfrocheiro Preto, Bastardo, Preto-de-Mortága, Trincadeira, Jaen, and Agua-Santa. Whites include Bical, Maria-Gomes (Fernão Pires), Rabo-de-Ovehla, Arinto, Sercial, Chardonnay, and Sercialinho. The vines are planted either free-standing or in trellis rows.

WINE PRODUCTION

As the quality of Bairrada wines continues to improve, they present the greatest challenge to the wines of Dão. In order for Bairrada wines to receive the seal of origin, maximum production must not exceed 55 hL/ha. The wines must be vinified in the region of Bairrada, must have a minimum of 11% alcohol, and must age in wood a minimum of eighteen months (for reds) or ten months (for whites).

The red wines are generally fermented on the skins. They are full bodied and deep in colour; they have good tannin, dry extract up to 25 g/L, and improve considerably with age (minimum five years). The white wines are made in two styles: to produce either wines that are young and light with a fruity taste, or those that mature well to give a golden straw hue and a strong, particular aroma.

The region also produces rosés and sparkling wines (from Chardonnay grapes, using the méthode Champenoise). Aguardentes and bagaceira are made, too. All wines must be submitted for an analysis and taste tests.

BUCELAS

Begun by the Romans, this wine region lies 25 km north of Lisbon, in the valley of the Trançāo River around the town of Bucelas. Its wine was known as Charneco in England during Shakespeare's time. The climate is one of mild summers and cold winters. The area is tiny, with limestone-clay and sandy soils. The wines are all white, dry, and acidic, made from Arinto and Esgana-Çāo, and are normally aged for a long period in wood.

COLARES

Located on the Atlantic coast, just northeast of Lisbon, the area was delimited in 1908. The maritime climate is mild and humid, so that the region is frequently covered with fog.

Colares was saved from the ravages of phylloxera by the region's sandy soils. The sand rests on clay and granite. In order to plant the Ramisco vine in the clay subsoil, a trench must be dug (sometimes as deep as 10 m). As the vine grows, sand is pushed back into the trench. This situation makes it impossible for the insect to reach the roots.

The best wines are red, and must be made with 80% Ramisco grapes. In addition to Ramisco grapes, the following red grapes are permitted: Molar, João de Santarém, Tintureiro, Moscatel (Muscat), Preto, Castelão Tinto, Parreira-Matias, Parreira-da-Velha, Tinta Miúda, and Santarém. The wines are deep ruby and quite astringent in their youth, turning brownish and much smoother with age.

The permitted white grapes are Malvasia, Bual, Arinto, and Jampal.

Carcavelos

This region, the third smallest, is near Lisbon. It was defined in 1908. Completely engulfed by the sprawling capital city, the area is left with only one vineyard, the Quinta dos Pesos. The amber wine is fortified to 18%–22% alcohol and sweetened with sugar (10 g/L–15 g/L). Ageing normally takes four years in wood, the rest in the bottle. Regrettably, this region is on its way to extinction.

The climate is maritime (influenced by the Atlantic Ocean), with an annual average rainfall of 590 mm. The soil on the estuary of the Tagus consists of limestone and sandy clay.

The grape varieties permitted are Galego-Dourado, Buais, and Arinto for the whites; Trincadeira, Espadaneiro, and Negra-Mole for the reds.

Setúbal

This region, located between the rivers Tagus and Sado on the Setúbal Peninsula, encompasses the districts of Setúbal and Palmela. It was demarcated in 1907. The area to the southwest is mountainous and the northern side is marked by hills of 100 m–500 m.

Precipitation is limited to 450 mm despite the maritime climate. The soil includes sandy clay, sand, and limestone.

Grape varieties permitted include the red Periquita ("parrot"), Espadeiro, Monvedro, Moreto, Bastardo (or Bastardinho), and Moscatel-Roxo, as well as the white Arinto, Manteúdo, Fernão-Pires, Buais, Tamarez, Moscatel-de-Setúbal, and Moscatel-do-Douro.

Setúbal is best known for its complex, topaz-coloured Moscatel de Setúbal. There are two styles produced:

- an aromatic, dark, aperitif wine released when five years old.
- a golden, complex wine released when at least twenty years old.

At least two-thirds of the grapes used for these wines must be Moscatel; the rest may include local white grapes. During the fermentation, the wine is fortified with aguardente to retain the level of sweetness desired. Skins of Moscatel are then added and left to steep until the following winter, thus transmitting their flavour to the wine. The finished wine generally has an alcohol content ranging from 18%–20%, and a sugar level of around 20 g/L.

The firm of José Maria da Fonseca is credited with having done the most to improve the quality and promotion of these famous wines. Fonseca is also the producer of the popular red wine Periquita, on the property of the same name. In the middle of the nineteenth century, the Fonseca family grafted Cova da Periquita with Castelão Frances to create the Periquita grape. Pasmados and Camarate are other Fonseca wines, made with a small percentage of Cabernet Sauvignon.

Algarve

This region runs along the southern coast of Portugal, from the Spanish border in the east, up the Atlantic coast to the Seixe River. The Mediterranean influences the weather here. It is hot and dry, with scorching temperatures and a low average rainfall. This region gets an average of 3188 hours of sunshine annually.

The vines are planted deep in sand or gritty soils and produce wines high in alcohol but low in tannin and acidity. The vast majority of Algarve wine is red with a minimum alcohol content of 12.5%. Production is around 60 hL/ha and the average annual production is 75 000 hL. Only 3% of this is white. The normally soft and aromatic wines must be aged a minimum of eight months for the reds and six months for the whites.

First demarcated as a single growing region, the Algarve's 499 123 ha are now subdivided into four DOCs.

From east to west, the DOCs are as follows:

- Lagos—the red Negra-Mole and Periquita are cultivated as well as the white Boal Branco (Bual). The wines must contain at least 70% of these grapes.
- Portimão—Negra-Mole and Periquita make up the red plantings; Crato-Branco is the sole white. The wines must contain at least 70% of these grapes.
- Lagoa—the red varieties grown here include Negra-Mole, Monvedro, and Periquita. The main white variety is Crato-Branco. The wines must contain at least 75% of these grapes.
- Tavira—the varieties planted are the same as for Portimão. The wines must contain at least 75% of these grapes.

Alfonso III is the name of an aperitif wine produced locally from Crato-Branco grapes. Aged in wooden casks, it has an alcohol content of 15%. The wine is dry and citrus-yellow, with characteristics that make it quite similar in style to Sherry.

Madeira

Located 1000 km from mainland Portugal and 750 km off the coast of North Africa in the Atlantic Ocean, Madeira is actually a group of islands, with the islands of Madeira and Porto Santo forming the wine region.

The Island of Madeira

The Arabs called it the "Turquoise Island" for its luxuriant forests. Its present name (Portuguese for "wood") also recalls these forests. When João Gonçalves Zarco and Tristão Vas Teixeira claimed the island of Madeira in 1418, they started a fire to make a clearing, and to discourage the Arabs from returning. This fire raged for seven years and destroyed every last tree on the island. However, it did enrich the volcanic soil, and soon Malvasia vine stocks from Napoli di Malvasia were planted. Later Roxa, Bual, Malvasia, and Verdelho were planted with equal success.

Madeira wine as we know it today evolved from the days of the seafarers. It is said to have started when some wines carried in the hull of a ship were not sold abroad, and returned to the island. These wines were found to be greatly improved after their voyage. Apparently, the hot, damp conditions on the ship had softened and stewed the wines, creating a richer, more mellow flavour. Smart merchant mariners charged more for the wine according to the number of sea crossings. Madeira wine became famous throughout Europe. In Shakespeare's *Henri IV*, Falstaff was prepared to sell his soul to the devil for a leg of capon and a cup of Madeira.

The island of Madeira has a climate that is consistently damp and mild, with little variation from summer to winter. Its soil is composed of volcanic ash and basalt. In contrast, the island of Porto Santo is hot and arid, with sandy soil. The best districts for growing vines are Campanário, Ponta do Pargo, Câmara de Lôbos, and Estreito.

GRAPE VARIETIES

The main grape varieties planted are Malvasia, Sercial, Bual, Verdelho, Malvasia Rosso (a red Malvasia), Tinta-de-Madeira, and Terrantez. A few other grapes (Cândida, Tinta Negra-Mole, Negra Mole, Alicante, and Bastardo) are also planted here.

WINE PRODUCTION

Today, Madeira wines are specially aged to produce their unique burnt and mellow character. Most are aged using a system called *estufagem*. The wines are kept in a heated room, or passed through pipes circulating hot water that must not exceed 50°C. This cooking lasts for about three months (one hundred days). Wines aged naturally are called *vinhos de canteiro*, but this is an extremely slow and costly process lasting anywhere from twenty to a hundred years. Madeira can be aged longer than any other wine. All Madeiras are fortified with aguardente, giving them an average alcohol content ranging from 18% to 21%. Wines known as Surdos, from Porto Santo, are sweet and also fortified with aguardente.

The types of Madeira wines are as follows:

- Sercial—the driest, with a yellowish hue. It develops a nutty flavour with age.
- Verdelho—a bit less dry than Sercial, yet more aromatic, with a golden colour.
- Bual—medium-sweet, burnt topaz in colour, rich, aromatic, and full bodied.
- Malvasia—very sweet, dark in colour, rich, and luscious. With age, it tends to dry out somewhat, producing a very madeirized nose.
- Terrantez—semisweet and very, very rare these days.

Madeira also produces table wines, but they are low in quality, and are not included in the Madeira DOC. Thus, they can not be called Madeira nor given the seal of quality.

MADEIRA WINE

The following terms on Madeira wine labels indicate the level of quality, from oldest to youngest.

colheita—wine of a single vintage, made from a traditional grape variety, with not less than twenty-two years in wood.

garrafeira or frasqueira—vintage wine made from traditional grape varieties, aged with a minimum of twenty years in wood and two in the bottle.

superior—wine showing exceptional qualities, made from a traditional grape variety.

reserva velha—wine that shows select quality and that has been aged for a minimum of ten years in wood.

reserva—wine of good quality aged more than five years

seleccionado—wine showing good quality and aged for a minimum of three years.

solera—wine that has been blended and periodically refreshed with younger wines of the same type until it is bottled. If the solera wine is vintage dated, the year refers to when the ageing process began.

canteiro—wine fortified right after fermentation and then aged in casks for a minimum of two years.

rainwater—a light, medium-dry wine with a yellow to gold colour. The name derives from when this wine would be diluted with rainwater during its shipment across the Atlantic. Today the rules about this style are relatively vague, so shippers can personalize their product.

By continually adapting to the times, the Portuguese region known as Porto (or Port), and its distinctive wine, have survived wars, revolutions, misfortunes, and changes in drinking habits. Port has modernized itself without lowering its standards, by preserving traditions, and incorporating state-of-the-art techniques.

GEOGRAPHY

Defined as a wine-producing territory in 1756, Port was the world's first wine region to be demarcated. Now it has 240 000 ha in the northwest corner of Portugal. It includes the valleys of the Douro and its tributaries, from Barqueiros, to Barca d'Alva on the Spanish frontier.

The valleys are sheltered from Atlantic winds by the Marão Mountains. The climate is severe, with bitterly cold winters and scorching summers. During the summer months, the sun bakes the rocky slopes to raise the temperature to 60°C. In winter, torrential rains average 600 mm (up to 1000 mm) of precipitation. Hail storms and frost are frequent. From November to February, there are fogs, especially in lower areas near water.

Vineyards are planted on rocky and treacherous mountain slopes along the riverbanks. For centuries, these slopes have been terraced with stone retaining walls, to prevent erosion and facilitate cultivation. The soils are mainly schist with some granite outcrops. A few areas contain gneiss, limestone, and sedimentary deposits.

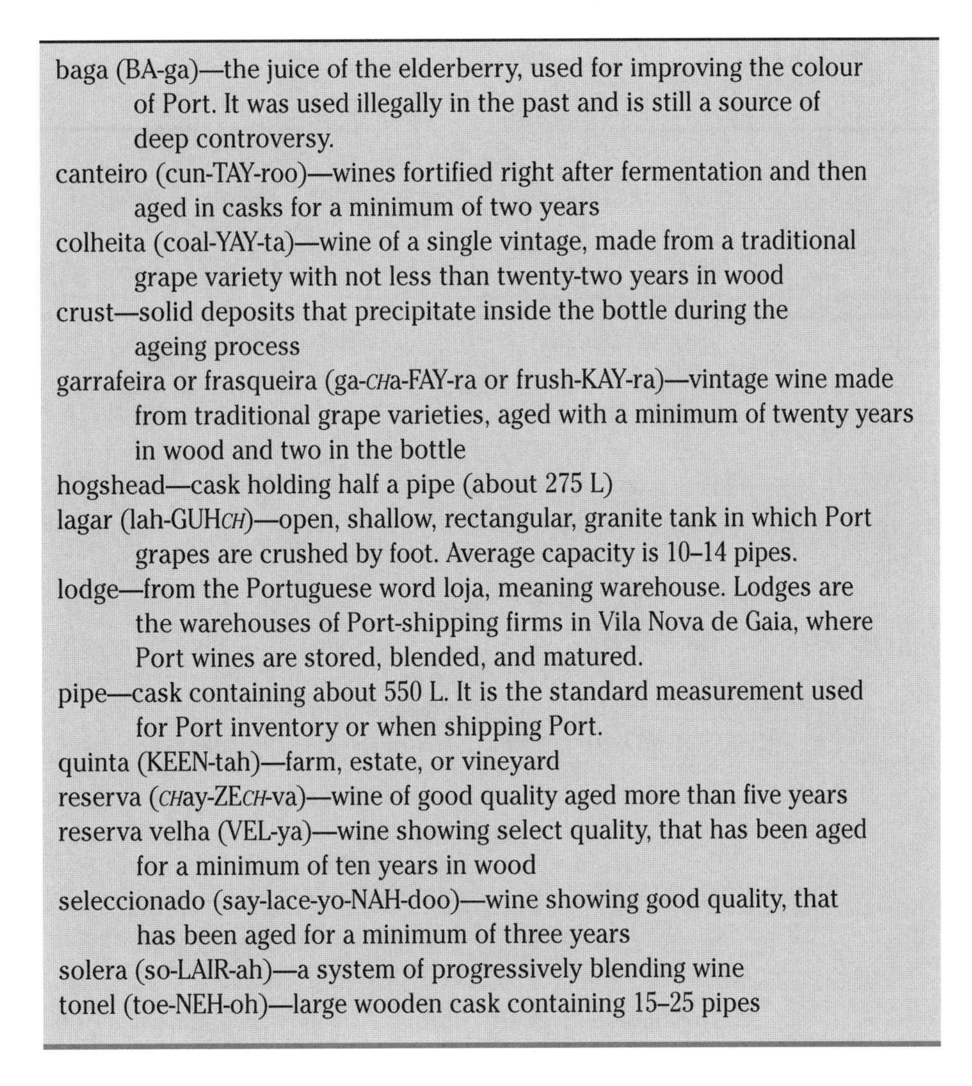

baga (BA-ga)—the juice of the elderberry, used for improving the colour of Port. It was used illegally in the past and is still a source of deep controversy.

canteiro (cun-TAY-roo)—wines fortified right after fermentation and then aged in casks for a minimum of two years

colheita (coal-YAY-ta)—wine of a single vintage, made from a traditional grape variety with not less than twenty-two years in wood

crust—solid deposits that precipitate inside the bottle during the ageing process

garrafeira or frasqueira (ga-*CH*a-FAY-ra or frush-KAY-ra)—vintage wine made from traditional grape varieties, aged with a minimum of twenty years in wood and two in the bottle

hogshead—cask holding half a pipe (about 275 L)

lagar (lah-GUH*CH*)—open, shallow, rectangular, granite tank in which Port grapes are crushed by foot. Average capacity is 10–14 pipes.

lodge—from the Portuguese word loja, meaning warehouse. Lodges are the warehouses of Port-shipping firms in Vila Nova de Gaia, where Port wines are stored, blended, and matured.

pipe—cask containing about 550 L. It is the standard measurement used for Port inventory or when shipping Port.

quinta (KEEN-tah)—farm, estate, or vineyard

reserva (*CH*ay-ZE*CH*-va)—wine of good quality aged more than five years

reserva velha (VEL-ya)—wine showing select quality, that has been aged for a minimum of ten years in wood

seleccionado (say-lace-yo-NAH-doo)—wine showing good quality, that has been aged for a minimum of three years

solera (so-LAIR-ah)—a system of progressively blending wine

tonel (toe-NEH-oh)—large wooden cask containing 15–25 pipes

The Wine Region of Port

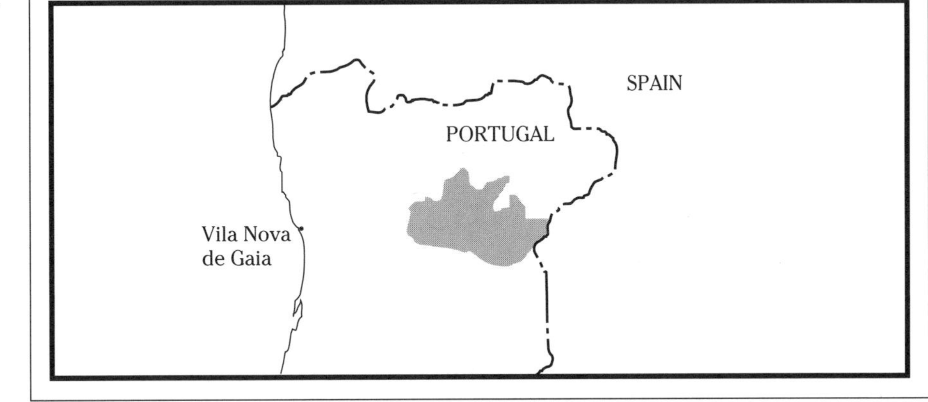

HISTORY

On May 17, 1386, the Treaty of Windsor was ratified between England and Portugal, establishing Portugal as England's first ally. This agreement still stands, making Portugal and England the world's oldest allies. The agreement formed the basis of an ongoing and profitable association.

In 1654, both countries agreed to trade within their colonies, and gave English merchants distinct privileges in Portugal. By 1678, around 400 pipes of wine were exported to England. By 1693, shipments had grown to 13 011 pipes. This growth in wine exports was particularly helped by trade wars between England and France, when French claret from Bordeaux first fell out of favour, and then became unavailable in England.

The Methuen Treaty of 1703 solidified the trade relationship that had developed: it allowed England to sell its wool duty-free in Portugal, and it allowed Portugal to sell its wine in England with a lower duty than paid by France.

Soon British and Portuguese wine shippers formed associations to regulate the trade. The Portuguese formed the Companhia General da Agricultura dos Vinhos do Alto Douro, with the Marquis of Pombal at the helm. In 1756, he changed the course of the industry by implementing tough regulations, new demarcations, and protective tariffs.

The nineteenth century brought many significant changes to the region. During the brief capture of Oporto in 1809 by France, all wine stocks and vineyards were destroyed. Civil war broke out in 1832, leaving Oporto and the Port region unstable until 1850. Then in 1851, powdery mildew infected the vines and in 1868, phylloxera arrived to plague the industry for more than twenty years. But positive changes came to the region as well. For example, Baron Joseph James Forrester banned the adding to Port wines of baga made from elderberry juice. Today, Baron Forrester is considered the link between the old and the modern Port wine industry.

Since the turn of the twentieth century, Prohibition in the United States, two world wars, and other upheavals have whittled away at the export trade, but Port persists. The region was once again defined in 1907, and in 1921.

LAWS AND REGULATIONS

IVP, The Instituto do Vinho do Porto (Port Wine Institute) officially guarantees the authenticity of Port. It issues certificates of origin and guarantee seals to all wines that qualify as Port. These wines must be bottled in Portugal. Any wine not produced there is not a true Port.

GRAPE VARIETIES

Many grapes are authorized for the production of Port. The best red varieties are Touriga-Nacional, Touriga-Francesa, Tinta-Francisca, Tinta-Roriz, Tinta-Ção, Donzelinho, Mourisco, and Bastardo. The best whites include Gouveio or Verdelho, Malvasia-Fina, Esgana-Ção, Folgasão, Rabigato, and Viosinho.

Other varieties are acceptable. They include reds (Cornifesto, Malvasia-Preta, Mourisco-de-Semente, Periquita, Rufeto, Samarinho, Souzão, Tinta-Amarela, Tinta-da-Barca, Tinta-Barroca, Tinta-Carvalha, and Touriga-Brasileira), and whites (Arinto, Bual, Sercial, Codega, Malvesia-Corada, and Moscatel-Galego). Still other varieties are being phased out by regulations governing the replanting of vineyards.

Crushing and Fermentation

There are two methods for crushing and fermenting grapes that are used in the creation of Port. Some shippers use both methods. Of the two, the modern method is faster and less labour intensive. As well, it allows better control over the temperature of the fermentation. Today, 75% of Port is produced using the modern method.

Traditional Method

According to the traditional method, grapes are placed in open stone lagares (tanks) and trodden by barefooted men and women, for four to six hours. The slow crushing of the grapes extracts the maximum pigmentation from the skins, and avoids breaking pips and stalks, so no excess tannins and acids are released into the must. The natural warmth of legs slowly raises the temperature of the must, which encourages the yeast to start breaking down sugars. Once the fermentation begins, the cap of floating skins and pips is pushed down with long paddles, at regular intervals, to extract the maximum amount of colour. Fermentation takes anywhere from forty-eight hours in warm weather to eighty-four hours in cold weather. During this time, sugar levels are constantly monitored with a saccharometer. When the wine reaches the desired level of residual sugar, it is poured into a vat simultaneously with brandy, at a rate of 100 L brandy to 450 L wine. This early fortification effectively stops the fermentation, and so stops further loss of natural sugars.

Modern Method

Grapes are weighed and dumped into a centrifugal crusher/de-stemmer. Tannin content can be controlled by removing an appropriate percentage of stalks. The must is pumped into a large, concrete tank, connected to another tank above. The tanks are sealed and the must begins to ferment. During the fermentation, mechanical paddles in the tanks continuously stir to speed up fermentation and extract maximum colour. Carbonic gases released during fermentation create pressure on the floating cap of skins and pips, pushing the must upward through a pipe into the tank above. When the must reaches a certain level in the upper tank, a valve opens to pour the must back down again. This process is repeated over and over during the fermentation. Sugar levels and temperature are monitored throughout the process.

When the wine reaches the desired level of residual sugar, it is drawn off into a larger vat. At this time, brandy is added, at the traditional rate of 100 L brandy to 450 L wine.

The fortified wine is then ready to be shipped downstream to the Port shippers' lodges in the coastal town of Vila Nova de Gaia.

Selection of Wines

When the wines arrive at the lodges in Vila Nova de Gaia, they are tasted and classified into lots according to style, sweetness, and colour. The best casks are left to mature on their own for up to eighteen months. If the results are good, they may be selected to become vintage Port. The lesser wines are destined to be blended.

While in the lodges, the wines are assessed many times and, according to their evolution, may be reassigned to different categories. Market demand and the amount held in stock can also influence the direction these wines will take during the selection process.

Types of Port

All Port wines are fortified to an alcohol level ranging from 19%–22%. They can be divided loosely into two categories: wood-aged Ports and bottle-aged Ports.

Wood Ports

Since all wood Ports are matured in oak, they tend not to throw a sediment in the bottle. They are generally ready to drink when bottled and do not benefit from cellaring.

- White Port—light, white wine with a pleasant nose. It can be sweet or dry. It is made from a blend of white grapes and aged in casks.
- Ruby Port (Tinto Aloirado)—made from a blend of sweet, young wines aged in casks. It has a bright, ruby colour and a straightforward, fruity nose.
- Tawny Port (Aloirado)—aged somewhat longer in casks, it develops a burnt-topaz colour. Tawny is smoother in texture and more elegant in taste than ruby Port. When aged ten to forty years in oak, it achieves complex, charming, and subtle qualities. Less expensive tawnies are made simply by blending red and white Ports to create the desired colour.

Bottle Ports

- Vintage Port—the highest quality, and the longest-lived. It is produced only in outstanding vintages that are "declared" by Port shippers during the second spring following the harvest. Vintage Port is always a blend of the very best casks from a mix of vineyards. It is bottled after more than two but less than three years of ageing. Deep purple and extremely tannic in its youth, it needs long ageing in the bottle to develop its sophisticated bouquet, its silky texture, and its intricate, lingering finish. It can improve in bottles for fifty years, and may last well over a hundred years. Vintage Port must be decanted to separate the crusty sediment from the clear wine; decanting also helps to aerate the wine and develop its bouquet. The character of vintage Port varies with every vintage and according to the house style. The word *Vintage* and the year of harvest must be printed on the label.
- Single Quinta Vintage Port—a vintage Port that has been produced from grapes taken from a single vineyard. The name of the quinta (estate) is printed on the label.
- Port Wine with the Date of Harvest—this is produced from a single good-quality harvest that has not been "declared." It is aged in oak for a minimum of seven years, and must display the vintage date and the year of bottling on the label. Less expensive than vintage Port, it offers good value for early consumption.
- Crusted or Vintage-Character Port—blended from several vintages, bottled after an average of four years in the barrel, and treated as vintage Port. It must be aged in the bottle to develop character, and also requires decanting.
- Late Bottled Vintage Port (LBV)—the wine of a single harvest, aged in oak casks for no less than four and no more than seven years. It is lighter in body, ready to be consumed immediately, and good value. LBV must show the vintage and the year of bottling on the label.

PORT TONGS

Port tongs are long pincers made of wrought iron, used to heat and break the neck of an old Port bottle that may have a crumbling cork. Red hot tongs are placed around the neck of the bottle just below the base of the cork. A cloth dipped in cold water is then applied to the neck of the bottle. With a little flick of the wrist, the top of the bottle snaps off cleanly, together with the entire cork.

Vintage Ports are the calling card and the pride of every firm, but ruby and tawny Ports are the bread and butter. The reason for this is that vintage Ports tie up large amounts of capital and move slowly; in contrast, wood Ports are sold quickly for a faster return. France is the largest market for wood Ports, and England the best market for vintage Port.

Young vintage Port is one of the best investments in a cellar. It can reach very high prices once mature. A good selection of Ports on a wine list is a clear indication of a dining establishment's standards.

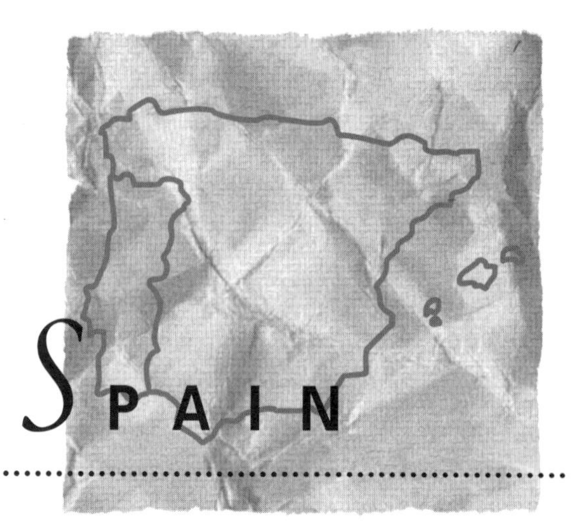

SPAIN

Spain is the third-largest wine producer in the world, following Italy and France. It has the largest surface area under vine in Europe with 1 500 000 ha, but average yields are low, around 25 hL/ha compared to 40–50 hL/ha in France and 80–110 hL/ha in Germany. Total production is approximately 19 000 000 hL, and exports have reached 7 700 000 hL. Spaniards consume an average of 38 L of wine per capita.

abocado (uv-oh-KAH-though)—semisweet wine
afrutado (uff-roo-TAH-though)—fruity
albero (ull-VAIR-oh)—chalky, as in soil
amontillado (ah-moan-tee-YAH-though)—a classification of Sherry
anejo (a-NAY-hoe)—old
anejo por (a-NAY-hoe pore)—aged by
año (UN-yoh)—year, as in "3° año" meaning "bottled in the third year after the harvest"
árrope (AHR-roe-pay)—very sweet, brown syrup produced by boiling down grape juice
barrica (bah-REE-kah)—225-L wooden barrel
blanco (BLUNK-oh)—white
bodega (bo-DAY-ga)—wine cellar or winery
butt (BOOT)—standard cask of 500 L made of American oak, used for maturing Sherry. The larger butt (called *bocoy*) contains 600 L.
Cava (KAH-vah)—meaning (since 1970) any quality sparkling wine made by the méthode Champenoise.
consecha (cone-SAY-cha)—vintage year

crianza (cree-UN-thuh)—young wine with little or no cask ageing
doble pasta (DOE-vlay PUS-tah)—wine rich in colour and body, made from one part must and two parts grape skins
fino (FEE-no)—a classification of Sherry
flor (floor)—the film-forming yeast found on Sherry; also, a classification of Sherry
gran reserva (grun ray-SAIR-vah)—quality wine produced only in good vintages and aged for at least two years in oak casks, plus three years in the bottle for red. Whites and rosés must age for a minimum of four years with at least six months in oak casks.
granvas (GRUN-vuss)—sparkling wine made using the Charmat method
joven (HOE-ven)—young, that is, wine destined for early consumption
mistela (mee-STAY-lah)—grape juice prevented from fermenting by the addition of brandy
oloroso (oh-lo-ROH-so)—a classification of Sherry
palo cortado (PA-loh kor-TAH-though)—a classification of Sherry
rancio (RUN-thyoh)—old white wine, madeirized, and sometimes fortified
reserva (ray-SAIR-vah)—good quality wine aged for at least three years in the bodega. For reds, there must be a minimum of one year in oak casks; for whites and rosés, a minimum of six months in oak casks.
rosado (roe-SAH-though)—rosé
solera (so-LAIR-ah)—a system of progressively blending wine
tinto (TEEN-toe)—red
vino de color (VEE-no day co-LORE)—a thicker árrope
vino de crianza (cree-UN-thuh)—"wine of breeding." This wine has been aged according to the rules of the Consejo Regulador. Red wines must be aged for a minimum of two years with at least six months in oak casks; white and rosé wines must be kept for a minimum of one year with six months in oak.
vino espumoso (es-spoo-MO-so)—sparkling wine made using the transfer method
vino espumoso, metodo traditional—made like a Cava, but outside of the areas approved for Cava production
vino gasificado (gus-ee-fee-KAH-though)—carbonated wine, usually made by dissolving blocks of solid carbon dioxide in a pressure tank
vino generoso (hay-nay-ROH-so)—fortified wine
vino joven (VEE-no HO-ven)—young wine made for immediate drinking, and bottled immediately after fining
vino maestro (ma-ACE-troh)—grape juice fortified at 7% with brandy and fermented to 16% alcohol
vino tierno (TYAIR-no)—sweet wine made from grapes that have been dehydrated in sunshine, fermented, and then fortified with brandy

GEOGRAPHY

Spain is the second-largest country (after France) in western Europe, with an area of 504 750 km^2. Together with Portugal, it covers the entire Iberian Peninsula. The main rivers are the Tajo, Guardiana, Guadalquivir, Ebro, and Duero (known as Douro in Portugal). Spain is a quite mountainous country with the Pyrenees to the north on the French border, the Cordillera Cantábrica (Cantabrian mountain range) and the Guadarrama, Morena, and Nevada mountains surrounding the central plateau of Spain. This Meseta (high plateau) has an altitude around 650 m.

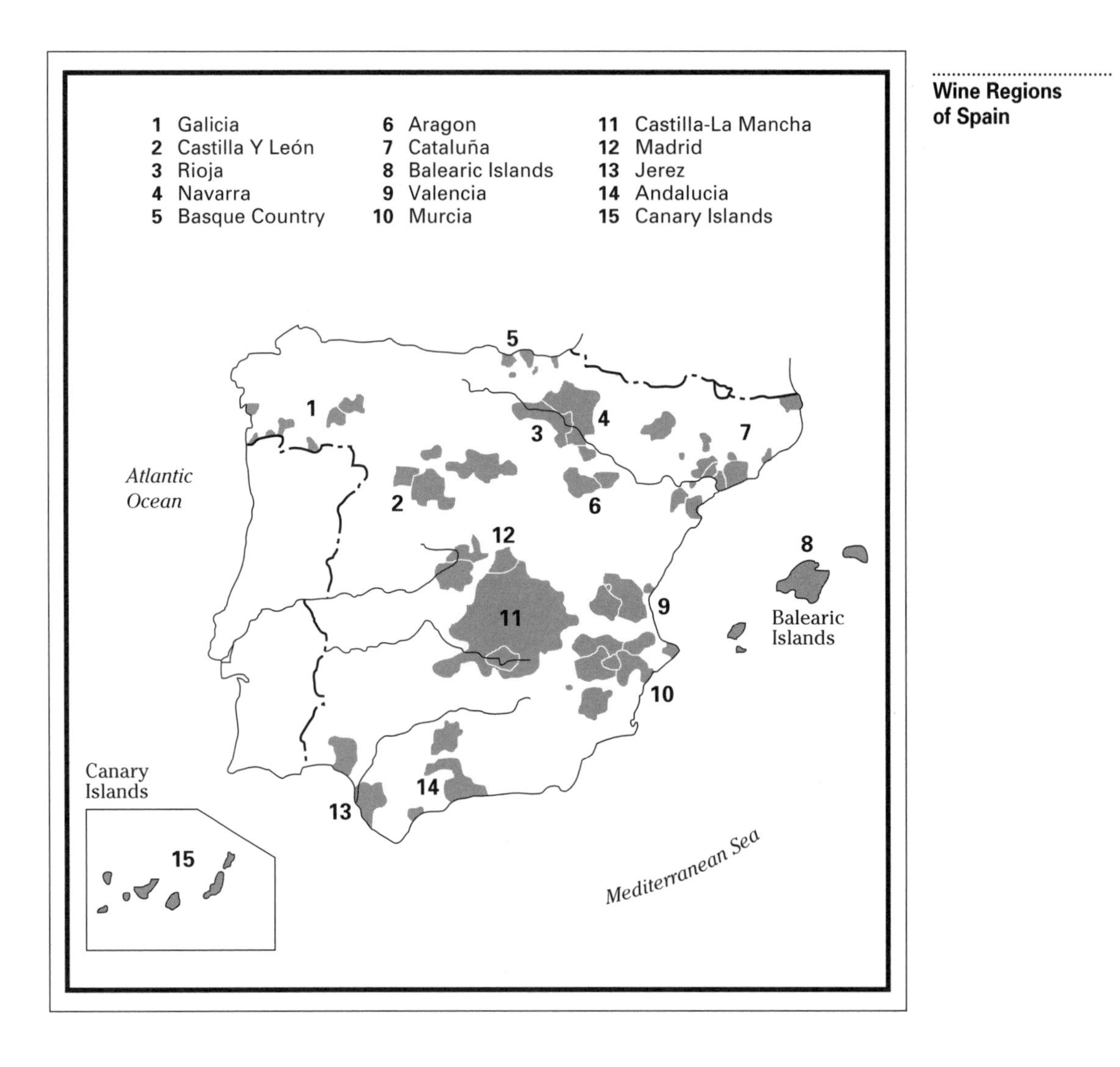

Wine Regions of Spain

Climatically, there is a wide array of conditions, from the long, wet seasons of northern Spain, to the dry and arid central plain, with severe winters and scorching summers, to the more temperate winters of Andalucia. It is not surprising there are so many types and styles of wine produced.

Spain's Constitution of 1978 divides the country into seventeen autonomias (autonomous regional governments) to form a union much like the United States. The autonomias are further divided to give a total of fifty provinces plus two overseas territories (the Balearic Islands and the Canary Islands). Madrid is the capital city and the name of the autonomia surrounding it. The official language of Spain is Castellano, but other autonomias have their own language, e.g., Catalan in Cataluña, Gallego in Galicia, and Euskara in País Vasco (Basque country). Some DOs spill over onto more than one autonomy or province. For example, Rioja extends into Navarra and into País Vasco.

HISTORY

The oldest evidence of winemaking in Spain goes back at least 3000 years. Phoenician traders established the first fortified settlement in what they called Gadir (Cádiz). Later, they moved inland to build the town of Xera (now Jerez). The hot climate was favourable to vines, and trade in wine prospered as a colonial industry. Greek, Carthaginian, and Roman forces invaded and exploited the generous land. However, when flourishing Spanish vineyards began to threaten the domestic wine industry, the Romans retaliated by uprooting vines west of the Alps. This was the first of many setbacks to the rich, sweet wines of Spain.

Under the Moors (A.D. 711–1492) drinking wine and involvement in its commerce were prohibited to the followers of the Koran; this ended with the reconquest of Spain by the Catholic monarchy, completed in 1492. Wine production was reborn, led by local monasteries.

Spanish influence in winemaking extended when the conquistadores travelled to the Americas. It was a drink that kept better than water during the long ocean voyages and it prevented scurvy and other ailments. Cortés (1485–1547) went so far as to order the colonists to plant ten vines yearly per Native American living on a property.

In the second half of the nineteenth century, Spain experienced a boom in wine sales after phylloxera began its progressive destruction of all the vineyards of northern Europe. Many French winemakers fled to Spain at that time because it was still untouched by the devastating insect. With them they brought new skills and new vines to the regions of Rioja, Ribera del Duero, and Navarra. However, powdery mildew posed its own problems in the 1850s, and around 1876, phylloxera finally reached Spain. Pestilence, civil strife, and the two world wars of the twentieth century took their toll on the vineyards and on the industry as well.

In the 1950s, the wine trade rebounded: quality and quantity both increased. By 1975, Spanish wines were known to be excellent buys and demand shot up again in the North American market. Finally, joining the EU in 1986 brought Spain an even greater market.

LAWS AND REGULATIONS

The Consejo Regulador is the regulatory body appointed by Spain's Ministries of Agriculture and Commerce, as well as by shippers and winemakers, to regulate the wine trade. The regional councils that make up the Consejo Regulador maintain laboratories in order to analyse wines, to inspect vineyards, and to regulate viticultural and vinicultural practices.

EU wine regulations took effect in 1991. As a member of the EU, Spain has two broad categories of wine—table wines (vinos de mesa) and quality wines (Vinos de Calidad Producidos en Regiones Determinadas or VCPRD)—and, like many other members, has some categories within these classifications.

The categories of wine are as follows:

Table Wines

- Vino de Mesa (VdM)—ordinary wine, blended from unclassified vineyards.
- Vino de la Tierra (VdlT)—ordinary wine of superior quality, from an unclassified but defined area.

Quality Wines (VCPRD)

- Denominación de Origen (DO)—wine similar in classification to France's AOC and Italy's DOC.
- Denominación de Origen Calificada (DOC)—added in 1988, a higher category, with more stringent regulations.

INDO (the National Institute for Denominations of Origin) licenses and registers twenty-eight Vino de la Tierra regions, forty-six DOs, and one DOC (Rioja). The DOs are guaranteed by the Consejo Regulador for a demarcated area. Each area has its own emblem or stamp printed on the label or bottle.

There are some additional designations, including:

- Vino Comarcal—this classification is for local wine combining characteristics of both Vino de Mesa and Vino de la Tierra. There are twenty-eight Vino Comarcal areas registered.
- Denominación Especificada, Denominación Especificada Provisional (DE/DEp) —these are classifications awarded to a Vino de la Tierra that is on its way to becoming a DO.
- Denominación de Origen Provisional (DOp)—this is a wine of quality seeking promotion, but in a region where too few bodegas show enough quality to grant a DO to the entire region.

Of Spain's total production of 18 954 000 000 hL, about 10 500 000 hL are from the forty-seven DO/DOC, with vineyards covering 691 822 ha.

The wine production varies considerably throughout Spain's autonomias. Cavas are treated separately at the end of this chapter, and the distinct region of Sherry is treated separately in the next chapter.

WINE REGIONS

GALICIA

Located on the extreme northwest of the Iberian peninsula, north of Portugal, this region on the Atlantic coast differs climatically from the rest of the country. Cool winds and rain from the Atlantic and off the Bay of Biscay temper the climate significantly. The Miño River (known as the Minho in Portugal) flows through here. Rich granite soils yield luxurious green vegetation similar to that of the northern Minho vineyards of Portugal. Here too, the vines are often strung high on wires in the traditional pergola method to avoid the cool dampness of the soil. Most producers have tiny vineyards, often less than a hectare each, and the wines tend to be made and distributed by co-operatives.

GRAPE VARIETIES

So far, 136 varieties of grapes have been catalogued in Galicia. Not surprisingly, Galicia's east coast is one of the best areas of Spain for white wines. The Albariño grape shines in the production of crisp, refreshing white wines with a trickle of carbon dioxide, similar to Portugal's Vinho Verde. Typically, they undergo malolactic fermentation. The warmer, east side of Galicia (around Monterrey and Valdeorras) is better suited for red grapes.

The Albariño grape is believed to be the Riesling brought by German monks making pilgrimage to Santiago de Compostella.

WINE PRODUCTION

As Galicia undergoes the modernization taking place elsewhere in Spain, its woody, oxydized wines are being replaced by lively, fruity wines. The fruity, acidic, local wines are often blended with the bland, fat whites of La Mancha to yield a good balanced wine. There are four DOs:

- Ribeiro—awarded the DO in 1957 and amended in 1976. Located in the province of Orense, around the River Miño, the area extends to the Portuguese border. The vineyards are terraced into the alluvial soil of the valleys, over a granite bedrock, at altitudes of 90 m–300 m. The most widespread cultivars are the white Treixadura and the red Caiño. The best white wines are made from Albariño, reds from Garnacha and Mencia, and rosado wines from Garnacha and Loureira (Loureiro in Portugal). The majority of Ribeiro wines are made for immediate drinking.
- Valdeorras—awarded the DO in 1957 and amended in 1977. This area is in the east of the province of Orense, where Galicia meets Castilla y León across the River Sil. In the late 1980s, the region began updating equipment and methods to be more competitive with other regions. The vineyards are planted on fertile alluvial soil with outcrops of limestone, at altitudes of 230 m–300 m. This warmest part of Galicia produces white wines mainly made from Godello grapes, and reds from Mencia grapes. All are for immediate drinking.
- Rías Baixas—("lowland rivers") was awarded the DO in 1988. It is located in the fjordlike coves and inlets that skirt the northwest coast. It is the coolest and wettest wine area of Spain. The DO is divided into three subzones in the province of Pontevedra: Val do Salnés on the coast around Cambados, O Rosal on the coast along the Portuguese border, and Condado do Tea further inland along the River Miño.

The vineyards, on alluvial topsoil and granite, average 450 m altitude. The main production is white wines made from Albariño grapes, blended at 70% with Treixadura and/or Loureira Blanca in the two latter subzones. A small amount of reds are mainly made from Brancellao and Caiño Tinto.

- Monterrei—the latest DO, awarded in 1996. This area bordered by Portugal lies around the town of Verin and along the River Tàmega. Protected by the Sierra de Larouco, the vines are trained low "a la castellana" to produce red wine, mainly made from Alicante Negro, Garnacha, Mencia, Tinto Fino, Tinta de Toro, and Monstelo. Whites are mainly made from Godello, Dona Branca, Treixadura, and Palomino.

Castilla y León

This is an austere land located in the northwest of Spain. The city of León is at its centre. Its scorching summers, harsh winters, and low rainfall make the cultivation of grapes a back-breaking pursuit. Vineyards are declining, except in some districts (such as Pesquera, Janus, and Vega-Sicilia), where wines of great quality are produced. Modernization of wineries is slow, but continued improvements have resulted in better wines. The region may yet be "discovered."

Grape Varieties

The region has a multitude of grape varieties. They will be dealt with in the wine production listings below.

Wine Production

The area's five DOs are as follows:

- Rueda—awarded the DO in 1980 and amended in 1992. This area is located mainly in the province of Valladolid, but also in Avila and Segovia. The vineyards lie between 600 m–720 m on the central Spanish Meseta. The soils vary from alluvial to sandstone, with some iron and limestone deposits. Rueda produces only white wines, mainly made from the Verdejo grape; some Verdejo sparkling, fortified, and flor wines from the Palomino grape are also made.

- Ribera del Duero—awarded the DO in 1982. Although mainly in the province of Burgos, the area spills over into the provinces of Soria, Valladolid, and Segovia along the River Duero. The vineyards lie between 760 m–900 m, among the highest in Spain. The chalky soil is planted with red Tinta del País, Garnacha Tinta, Cabernet Sauvignon, Malbec, Merlot, and some white Albillo. Only the red wines are classified as DO.
- Toro—awarded the DO in 1987. The vineyards are mainly located in the province of Valladolid and around the town of Toro, at an average altitude of 700 m. The various soils (alluvial deposit in the south and limestone and sand in the north) are mainly planted with Tinta de Toro, a variation of Tinta del País. There is also some Garnacha. White wines are made from Malvasia and Verdejo grapes.
- Bierzo—awarded the DO in 1989 and amended in 1991. Located in the northeast part of the region and bordering Valdeorras, the vineyards average 550 m and are planted on alluvial and slate soils. The main red grape is the Mencia (believed to be related to Cabernet Franc), followed by Garnacha, and Tintorera. The main white grape is Palomino, with increased planting of Dona Branca and Godello.
- Cigales—awarded the DO in 1991. Located north of Valladolid on the central Meseta, at an altitude of 730 m, the rocky limestone soil produces one of the best Spanish rosado wines, mainly made from Tempranillo. Other red grapes include Tinta del País, Garnacha Tinta, and Garnacha Roja. The white varietals are Verdejo, Albillo, and Viura. There are no DO white wines, but some DO rosés are made from a blend of red and white grapes.

Rioja

Rioja was the first wine district to be demarcated, in 1926, was awarded the DO in 1947, amended in 1976, and is the only one to be elevated to DOC status, in 1991.

It lies south of Navarra and the Basque country, northeast of Castilla y León, and 120 km from the Atlantic coast. The capital is Logroño.

Despite the proximity to the ocean, Rioja is shielded in part from the cold sea winds and rain by the Cordillera Cantábrica. Early springs are warmed by a southern wind called *solano*. Summers are hot with cool nights, autumns are mild with showers, and winter is relatively short.

The wine region itself stretches for 120 km on both sides of the River Ebro and has a maximum width of 40 km. It is divided into three major zones:

- Rioja Alavesa—the northernmost vineyards in the province of Alava (Araba in Basque). They run from the steep hills of the Cordillera Cantábrica to the River Ebro. The soil is practically all ferruginous clay with layers of limestone.
- Rioja Alta—the highlands in the northwest and southwest of the River Ebro, with vineyards found at an altitude of 720 m. Its soil is a mixture of clay and iron, or clay and limestone, with richer alluvial deposits along the river.

Both Rioja Alavesa and Rioja Alta have a cool climate, ideal for red wines of high quality and fine aroma. Spring showers from the Atlantic provide sufficient moisture, and the summers are long and hot. Temperatures drop below 0°C for a short time in winter, but generally the snow remains above the vineyards, on the mountain peaks.

- Rioja Baja—located in the lowlands in the southeast, with an enclave in Navarra. The climate is drier and warmer than in Rioja Alta and Rioja Alavesa, with definite Mediterranean influences coming up the River Ebro valley. Springs are early and short, followed by long, very hot, and semi-arid summers. Autumns are short and wet, and winters are mild with no frost or snow. The vineyards average 300 m in aititude. The soils are mainly alluvial silt with clay and ferrous compounds. Compared with the other two districts, this district produces coarser wines with deep colour and a higher alcoholic level.

Rioja

Theories about its name vary: some say it derives from Rio (river) Oja; others believe it is named for its sometimes red (roja in Spanish) soil; some Basque scholars suggest the name comes from the Basque word Ería-ogia ("land of bread," alluding to the expansive cornfields that were once there) and corrupted to Errioja; and finally, the name may derive from the pre-Roman grape-growing tribe called Ruccones, later Rugiones. Whatever the case, the name has a long history, and (as Rioxa) it is found in documents dating from 1092. Wine is known to have been made in this area for nearly 2000 years.

GRAPE VARIETIES

As many as forty grape varieties can be found in the Rioja. The Consejo Regulador officially permits only seven—four reds and three whites.

Red Grapes

These make up 76% of the total planting, with the early ripening Tempranillo as the principal (40%). The other reds are Garnacha Tinta (which provides warmth and ripeness), Graciano (for style and character), and Mazuelo (for colour and depth). Rioja wines are mostly a blend of these grapes, typically Tempranillo 70%, Garnacha 18%, Mazuelo 12%, and Graciano 12%.

White Grapes

These are dominated by Viura (also known as Maccabéo). It is high in acid, adds brilliance, and resists oxidation. Malvasía de Rioja gives a deep yellow colour but is low in acidity and sugar. Garnacha Blanca is disease-resistant, and vigorous, but produces rather dull wines with high alcohol and low acidity.

The name Viura derives from a Basque word, possibly zuri (white), ori (yellow), or urín (juice).

White Rioja are often blended from these three varietals.

WINE PRODUCTION

The wines are categorized as follows:

- Rioja—classified by vintage and by how long they have been aged in oak barricas. Wine is bottled for sale in the first year.
- Red Crianza—must spend one year minimum in barricas and cannot be released for sale until the third year.
- White Crianza—must spend a minimum of six months in barricas and cannot be released until the second year.
- Red Reserva—aged at least one year in barricas plus one year in the bottle, and cannot be released for sale until the fourth year.
- White Reserva—aged a minimum of six months in barricas and may be released in the third year.
- Gran Riserva—red wine made only from the finest vintages, and aged for a minimum of two years in barricas and three years in the bottle. It may be released in its sixth year.

Navarra

This large region produces an average of 35 000 000 L of wine. It is to the west of Aragón, rolling from the Pyrenean Mountains to the River Ebro. The grape-growing regions extend south of Pamplona, the provincial capital.

GEOGRAPHY

Climatic conditions and soil formation naturally divide Navarra into five subzones: Valdizarbe in the north, Tierra Estella in the northwest, Baja Montana in the northeast, Ribera Alta in the centre, and Ribera Baja in the south. Their diversity results in a range of wine characteristics and personalities. The vineyards lie at altitudes from 560 m in the north to 250 m in the south. The best wines seem to come from Tierra Estella and Valdizarbe.

The soil in the north consists of rich, loamy topsoil, over gravel and a chalky bedrock. The more arid Ribera Baja has drier, sandier soil.

GRAPE VARIETIES

Navarra's most widely planted grape is Garnacha (73%). It produces one of the best rosados of Spain and some pleasant reds as well. The other recommended red varieties are Mazuelo (Carignan), Tempranillo, Cabernet Sauvignon, and Graciano. Prefered white grapes are Viura, Moscatel de Grano Menudo (small grained Muscat), Malvasia, Chardonnay, and Garnacha Blanca.

WINE PRODUCTION

Navarra was awarded the DO in 1967. Although long overshadowed by the powerful neighbouring Rioja, Navarra is making a swift comeback with modern vinification methods. A modern research station run by the autonomia government called EVENA (Estación Viticultura y Enología de Navarra) provides growers with recommendations and assistance for improving both vineyards and winemaking.

Basque Country

The autonomia of País Vasco (Euskadi) has two DOs:

- Chacolí de Getaria (Getariako Txakolina)—around the three towns of Zarautz, Getaria, and Aia in the province of Gipuzkoa.
- Chacolí de Bizcaia (Bizkaiko Txacolina)—in the provinces of Gipuzkoa and Bizcaia.

Both areas are small and mainly produce a dry, crisp, white wine from the native grape Ondarrabi Zuri. The equally crisp reds are made from Ondarrabi Beltza. The wines are at their best young, when they retain a certain petillant character that results from malolactic fermentation.

Aragón

There are 82 000 ha of vineyards producing around 75 000 000 L of wine annually. The region lies south of the French border, at the foot of the Pyrenean Mountains. The climate is harsh: freezing winters and hot summers. It is regularly swept by strong winds and has a low level of rainfall.

Wine Production

The four DOs of Aragón are as follows:

- Cariñena—first regulated in 1696, demarcated in 1932, awarded the DO in 1960, and amended in 1990. The vineyards range in altitude from 400 m–760 m. Soils vary from limestone, to slate and alluvial deposit. This ancient area had a reputation for rancio wines. Today the main wines are a fresh, white wine made from 90% Viura, rosado made from Garnacha, and red from Garnacha, Tempranillo, Mazuelo, Cabernet Sauvignon and the local grape Cariñena. Other varieties include the reds Monastrell and Juan Ibáñez, and the whites Moscatel Romano, Garnacha Blanca, and Parellada.
- Campo de Borja—awarded the DO in 1980 and amended in 1989. Located in the province of Zaragoza and eastward along the south bank of the River Ebro, the vineyards range in altitude between 300 m–650 m. The topsoil is sandy, over a stony limestone base.

The area produces white wine from Viura grapes and rosado from Garnacha. Red wine is mainly made from Tempranillo with some Cabernet Sauvignon, Garnacha Tinta, and Mazuelo.

- Somontano—awarded the DO in 1985. The area is located in the province of Huesca in a semi-arid climate, at the foot of the Pyrenean Mountains. The vineyards average 450 m in altitude. The soils are reddish with sandstone and clay, and alluvial deposits near the rivers. All have a good proportion of limestone. The native grapes of Somontano are the white Alcañon and the red Moristrel (not to be confused with Monastrell). The principal varieties are Viura, Garnacha Tinta, and Garnacha Blanca. New plantings include Tempranillo, Pinot Noir, Cabernet Sauvignon, Merlot, Chenin Blanc, Chardonnay, Riesling, and Gewürztraminer.
- Calatayud—awarded the DO in 1990. Located in the province of Zaragoza, the vineyards stretch out on the southern side of the Sierra de la Virgen at altitudes ranging from 500 m–900 m. The climate is continental with semi-arid areas. The soil consists of sandy loam with some limestone over marl in the north, and slate and gypsum in the south. The main red grapes are Garnacha, Tempranillo, Mazuelo, and Monastrell. White varieties include Viura, Malvasia, Garnacha Blanca, and the local Juan Ibàñez.

Cataluña

The land rises like a staircase from the Mediterranean coast to the Pyrenean peaks at the Franco-Spanish border. Cataluña is also known as Catalunya and Catalonia; it is adjacent to Languedoc-Roussillon in France. Barcelona is the capital. This area is the source of the widest variety of wines in all Spain: it produces table wines, sweet Solera Tarragona, brandies and liqueurs, rancio wines (Vi Rancí), Vermouth, and almost all (92%) of Spain's sparkling wines.

GEOGRAPHY

The region has mild winters and long, hot summers tempered by Mediterranean breezes. Because of the climate it has consistently good vintages.

HISTORY

During the phylloxera epidemic in France, Cataluña was planted extensively, and supplied France with wines of average quality. That all changed in the 1960s, when stainless steel tanks and temperature-controlled fermentation were introduced, and when the area returned to the practice of using new barrels in the Bordeaux tradition. Since then, Catalan wines have gained worldwide acclaim. Miguel Torres' wines were among the first: his Black Label Gran Coronas 1970 (Mas La Plana) was judged better than Latour 1970 at the Gault-Millau Wine Olympiad in Paris in 1979.

GRAPE VARIETIES

The DO regulations authorize 121 varieties for planting in Penedés alone.

WINE PRODUCTION

Cataluña produces an average of 250 000 000 L of wine within its eight DOs:

- Alella—granted the DO in 1956, amended in 1976, and 1989. The area is located north of Barcelona, and runs from the foothills of the Cordillera Catalana to the sea. The low coastal vineyards have a sandy topsoil (low in calcium and minerals) over granite bedrock. The Vallés area is higher at 300 m and has a similar sandy topsoil, but over a limestone bedrock. The authorized white grapes are Pansá Blanca (Xarel-lo) and Garnacha Blanca. The reds are Ull de Llebre (Tempranillo), Garnacha Tinta, and Garnacha Peluda.
- Ampurdan-Costa Brava—awarded the DO in 1975. Located in the province of Gerona at the foothills of the Pyrenees, where the mountains adjoin the sea, the vineyards extend from 200 m down to sea level at the coast. The fertile, brownish topsoil rests on limestone-based bedrock. Rosado wines made from Garnacha or blended with Cariñena represent 60% of the production. Red wines are made from these same grapes, and whites are made from Maccabéo and Xarel-lo.
- Priorato—awarded the DO in 1975. This small area (known locally as Priorat) is located in the Montsant Mountains and in the province of Tarragona.

The steep terraced vineyards climb from 200 m–1200 m, and are planted over a unique soil known as Llicorella. Of volcanic origin, it consists of alternating bands of reddish quartzite and black slate. It gives the wine a distinctive character. The red grapes are Garnacha Tinta and Cariñena. For white wines, the main grapes are Garnacha Blanca, Maccabéo, and Pedro Ximénez.

- Tarragona—awarded the DO in 1976. This is the largest DO of Cataluña and is located in the province of Tarragona, between the sea and the province of Lleida (Lerida). The DO is divided into two subzones: Tarragona Campo rises from the sea to about 200 m and has loamy topsoil mixed with limestone, over alluvial subsoils; Falset in the south rises from the Ebro valley up to 450 m and has a loamy soil mixed with limestone, over granite outcrops. The authorized red grapes are Garnacha Tinta, Cariñena, and Ull de Llebre. White varieties are Maccabéo, Parellada, Xarel-lo, and Garnacha Blanca, with some experimental plantings of Cabernet Sauvignon, Merlot, and Chardonnay.
- Penedès—awarded the DO in 1976. The area is located south of the city of Barcelona, partly in the province of Tarragona. Vineyards rise from the coast to nearly 800 m in the interior. Sandy topsoil over limestone in the lowlands is replaced by a mixture of clay and chalk in the highlands. The ten authorized varieties are Garnacha, Mazuelo, Monastrell, Ull de Llebre, Samsó, and Cabernet Sauvignon, for the reds; Maccabéo, Xarel-lo, Montonec (Parellada), and Subirat Parent for the whites.
- Terra Alta—awarded the DO in 1985. It is the southernmost DO of Cataluña and, like Priorato, is a mountainous area. Located around the town of Gandesa, the vineyards are situated on slopes consisting of deep topsoil over limestone and clay. The authorized white grapes are Garnacha Blanca and Maccabéo. Red grapes are Cariñena and Garnacha. Rancio wines are also made.
- Costers del Segre—awarded the DO in 1988. Located in the province of Lleida, the DO is divided into four subzones: Artesa in the north, Riucorb and Les Garrigues in the south, and Raimat in the west. In spite of this division, the soil is consistent: sandy soils with a little clay, over a limestone bedrock. Authorized grape varieties are the red Garnacha, Ull de Llebre, Cabernet Sauvignon, Merlot, Monastrell, Trepat, and Mazuelo. Whites include Maccabéo, Parellada, Xarel-lo, Chardonnay, and Garnacha Blanca.

- Conca de Barberà—awarded the DO in 1989. Conca ("basin") de Barberà lies 80 km inland from Barcelona. The vineyards are located between 200 m–400 m in the river valleys. The topsoil contains chalk and alluvial deposits, over a limestone bedrock. Authorized red grapes are Garnacha, Trepat, and Ull de Llebre. Whites are Maccabéo and Parellada.

Balearic Islands

The sun-drenched green and fertile Mediterranean islands of the Baleares had 27 000 ha under vines before phylloxera struck. Today, there are only 2589 ha, with the majority of the planting in Mallorca (Majorca). The problem, from the grower's perpective, is that it is more lucrative to develop resorts and villas than it is to cultivate grapes.

The only DO is Binissalem, granted in 1991. It is located on the plain northeast of Palma de Mallorca, at an altitude between 250 m–300 m. In summer the temperature reaches 35°C, and in winter never dips below 0°C. Autumn rainfall averages 550 mm. The soil tends to be light and poor in nutrients, but with some limestone over clay, well-suited for vines. The main red grapes are Manto Negro (93%), Callet, Tempranillo, and Monastrell. The white grapes are Moll (also known as Prensal Blanc), Parellada, and Maccabéo.

Valencia

Valencia is the third largest city of Spain and the most important wine port. It has three DOs: Valencia, Utiel-Requena, and Alicante. Technically, wines or grapes from a DO zone cannot be mixed with those from another DO zone without losing the appellation name. However, Valencianos have a clause (in the small print) of their local rules permitting them to ship grapes or must from one DO zone to another "when in difficulty." So, it happens that a wine with the DO Valencia can contain grapes harvested in Alicante or in Utiel-Requena. This process has been used to upgrade the wines, rather than to increase the bulk.

Valencia is a unique region in Spain because it has always directed its wine trade toward the export market. As well, it has always been on the cutting edge.

Valencia's Exports
Valencia's wines are present in all European supermarkets, but the area may be best known for its oranges. It also produces rice—in fact, it has the largest area of rice paddies outside southeast Asia.

Quality control was a local priority even before the DO regulations were established. As well, the region was installing stainless steel tanks when the rest of Spain was using epoxy-lined concrete tanks, and Valencianos had computerized cold fermentation when others were still questioning the use of stainless steel tanks.

The region was once famous for its red Doble Pasta. This was used locally (and exported) to increase colour and extract in thinner wines. Today, red Doble Pasta has been replaced by grape concentrate, which is itself a major business in Valencia.

The best wines from this DO are rosados and red wines made from Bobal, Tempranillo, and Garnacha. The white grapes account only for 16% of the vineyards; they include Maccabéo, Merseguera, and Planta Nova. Cabernet Sauvignon and Chardonnay are starting to show up in newer plantings.

More than half of the production is exported to Switzerland.

WINE PRODUCTION

- Valencia—awarded the DO in 1957, amended in 1991. The vineyards are divided into four subzones: Clariano in the extreme south of the province, Moscatel de Valencia southwest of the city, Alto Turia northwest of the city, and Valentino in the northwest bordering the province of Cuenca (Castilla-La Mancha).

  The climate is Mediterranean on the coast and continental up in the hills. Temperatures can rise or fall 25 Centigrade degrees in a day. Highs of 35°C are registered in summer and lows of –5°C in winter. The soils vary from alluvial deposits in the lowland, slowly changing to clay in the middle elevations, to sandy soil over a limestone base in the highlands of Alto Turia.

  Authorized red grapes are Garnacha, Monastrell, Tempranillo, Tintorera, and Forcayat. The preferred white grapes are Maccabéo, Malvasia, Merseguera, Moscatel de Alejandría, Pedro Ximénez, Planta Fina de Pedralba, plus some Planta Nova and Tortosi.

- Alicante—awarded the DO in 1957, amended in 1991. It is divided into two subzones: La Marina, around the coastal town of Denia, northeast of Alicante; and Alicante, encompassing the city and west to Yecla. The vineyards have a Mediterranean climate and rise from the sea to the high hills (400 m). The topsoil is alluvial, over a limestone bedrock. Authorized red grape varieties are Monastrell, Garnacha Tinta, Garnacha Tintoreta, Bobal, and Tempranillo. Whites are Merseguera, Maccabéo, Moscatel Romano, and Planta Fina. Although Alicante still makes its famous Doble Pasta, its reputation rests on fresh wines modestly priced (for the export market), fortified wines, and sweet Muscat wines.
- Utiel-Requena—awarded the DO in 1957, amended in 1991. This is the farthest inland and highest DO of the province of Valencia. The summers are quite hot (rising to 40°C) and the winters are frosty (falling to –15°C). The average temperature is just 14°C. Rain falls heavily in the spring and autumn to give an annual average of 450 mm. Ranging from 600 m–900 m, the vineyards are located on slopes facing southeast. The soil is a mixture of clay and marl, on a sandstone base with limestone outcrops.

Murcia

Murcia is located in southeastern Spain, bordered by Andalucia, Valencia, and Castilla-La Mancha.

Wine Production

Although the wines of Murcia have a reputation of being rough, alcoholic, and coarse, some wineries are starting to change this image in the three DOs of Jumilla, Yecla, and Bullas:

- Jumilla—awarded the DO in 1975, amended in 1977 and 1986. The vineyards average 700 m in altitude, and are located around the town of Jumilla in the provinces of Murcia and Albacete.The climate is semi-arid with temperatures rising above 40°C. The reddish-brown soil is sandy, over limestone. Jumilla produces robust, red wines from Monastrell (90% of the vineyards), Garnacha, and Cencibel. The white wines tend to be heady and have low acidity; they are made mainly from Merseguera Airén and Pedro Ximénez. Some joven-style wines are locally appreciated. With the help of foreign investors, the region is rapidly modernizing.

- Yecla—awarded the DO in 1975. It is a small enclave next to Jumilla, on a plateau ringed by mountains. The climate is continental with hot summers and cold winters. The soil consists mainly of limestone, over a clay subsoil. White wines are made from Verdil and Merseguera. Red wines from red Monastrell and Garnacha may contain some Cabernet Sauvignon and Tempranillo.
- Bullas—awarded the DO in 1994. This area covers most of the western half of the province of Murcia. Vines are planted in the valleys, on terraced slopes, and are often seen among olive and almond trees. The summers can be dry and very hot with temperatures staying above 37°C for many days. The most important grapes are Monastrell and Tempranillo for red wines, with Maccabéo and Airén for whites.

CASTILLA-LA MANCHA

Located on the vast central plateau of Spain at altitudes of 500 m–800 m, this region supplies nearly 50% of the nation's wine production. The climate is arid, with long, hot summers averaging 200 days of full sunshine. The grapes build up a large amount of sugar to produce low-acid, coarse, alcoholic wines. There is an overabundance of white wines and a scarcity of reds. The yield is an incredibly low 16–18 hL/ha. The area is the home of 485 wine co-operatives, many of which distil wine for brandy, or as spirit for addition to liqueurs.

WINE PRODUCTION

There are four DOs:

- Mentrida—awarded the DO in 1960, amended in 1976 and 1991. The area is located in the province of Toledo and spills over into the region of Madrid. The climate is continental. The vines are planted in sandy and clay-based soils, mixed with a little limestone in places. The area produces only red wines, for immediate drinking, made from Garnacha, Tinto de Madrid, and Cencibel.
- La Mancha—awarded the DO in 1966, amended in 1976. This is Spain's largest DO zone, spreading over four provinces. It meets the DO Vinos de Madrid in the north, surrounds the DO Valdepeñas in the south, and occupies most of the southern part of the central Meseta.

The climate is full continental with very hot, dry summers and bitterly cold winters. The flat land consists of red-brown sandy soil with patches of limestone. The majority of the grapes are white: Airén, Pardilla, Verdoncho, and Maccabéo. The sporadic red grapes are Cencibel, Moravia, Garnacha, and Cabernet Sauvignon.

- Valdepeñas—awarded the DO in 1964, amended in 1976. This enclave in La Mancha has a long tradition of winemaking, from before the time of the Romans. Valdepeñas ("valley of stones") has a rocky soil consisting of broken limestone mixed with alluvial clay, over a chalky subsoil. The climate is the same as La Mancha but with occasional, damaging, massive rainstorms. The principal grapes are the red Cencibel and the ubiquitous white Airén.
- Alamansa—awarded the DO in 1966, amended in 1975. It is located on the western edge of the region, in the province of Albacete, overlapping in the province of Valencia. The climate is continental. The vineyards are planted in lowlands consisting of rich topsoil over some limestone base. The production is mainly red wines made from Monastrell, Cencibel (Tempranillo), and Garnacha. The few whites are made from Merseguera.

Madrid

Madrid, from the Moorish word Majerit, began its central role when Philip II moved his court from Valladolid to the small village of Madrid in 1561. Now the city offers the local wine producers a receptive market, which includes two of the world's most expensive hotels and eighteen restaurants carrying Michelin stars. This incentive has encouraged local winemakers to modernize, and to upgrade quality.

- Vinos de Madrid—awarded the DO in 1990. The vineyards are located to the west, south, and east of the city. They are divided into three subzones: San Martín de Valdeiglesias (to the west of the city, it is the wettest, with rich topsoil over granite), Navalcarnero (in the south, with light sandy soil over clay), and Arganda (in the southeast, it is the driest, with clay and marl over granite).

  The main red grapes are Garnacha (28%), Tempranillo, and Tinto Fino. The main white varieties are Malvar, Albillo, and Airén.

Extremadura

The southwestern part of Spain, bordering Portugal, is a scantily populated area of high sierras where dark-skinned wild pigs and goats roam freely in deep forests. The region has no DO and produces few wines. These are mostly in a Sherry- or rancio-style, or distilled and used for blending. There are a few exceptions, such as Tierra de Barros (DOp Badajoz).

The white table wines are typical of hot viticultural regions, with low acid, high alcohol, heavy body, and short finish. They are produced mainly from the Cayetana Blanca, Zalema, Alarije, Bomita Airén, and Marfil grapes. The reds—heady, dark, and high in alcohol—come from Garnacha Tinta, Graciano, Tempranillo, Morisca, and Palomino Nero.

Andalucia

This second largest autonomia (also called Andalusia) is in the extreme south of Spain. Its eight provinces cover 9 000 000 ha. It is the cradle of Spain's wine industry. The long, hot summers and mild winters favour the production of fortified, cooked, heavy wines that are high in alcohol. These can be dry or sweet.

WINE PRODUCTION

The four DOs in the region are diverse both in size and production. Of the four, perhaps the best known is Sherry (Jerez), which is discussed in the next chapter.

- Condado de Huelva—awarded the DO in 1964, amended in 1979. Condado ("county" or "earldom") de Huelva is situated at the westernmost end of Andalucia, east of the Portuguese border and west of Seville (Sevilla). For a long time, these wines, grown on chalky soil, were overshadowed by the wines of Jerez and Montilla, and were sent there for blending with Sherry. When the Sherry market began to dwindle, the region's producers had to rethink their wines, as well as their strategy for marketing them.

The vineyards are barely 20 m above sea level, and planted in a sandy soil over a limestone bedrock. Although the wines (Condado Palido and Condado Viejo) are made using the solera system, they cannot rival the best of Jerez. However, there is an interesting light wine, cool-fermented from white Zalema grapes. Listán (Palomino), Garrido Fino, and Moscatel are also grown.

- Montilla-Moriles—classified in 1933, and awarded the DO in 1945, amended in 1985. The vineyards lie 45 km south of Córdoba, on the Meseta slopes leaning toward the southern coast at an altitude between 300 m–700 m. They survive the hottest and driest climate Spain can offer. There are two main types of soil in the area: albarizas (rich in chalk and similar to the best soils of Jerez) make up the central "Superior" vineyards; and arenas (sandy soils) make up the perimeter of the zone designated as "Ruedos."

  Two types of wine are produced in this DO, one in a modern style, the other more traditionally.

Modern Style

These wines are called Joven Afrutado. They are fruity, dry, white wines made from early picked Baladi, Airén, Torrontés, and Pedro Ximénez grapes.

Traditional Style

These wines are made in the style of Sherry, but most are not traditionally fortified. Instead, they are fermented and matured in huge clay jars called *tinajas*. (More often now, one sees stainless steel tanks with temperature-control jackets replacing these picturesque amphorae of the past.)

Traditional-style wines come in two versions, categorized according to colour and alcoholic strength. Those which have achieved 13% alcohol, but no more than 15%, are aged for one year in oak casks, and are known as Crianza. Wines exceeding 15% are known as Generoso. They go through the solera system, where they develop a distinctive character. They are labelled according to their level of sweetness: Dry, Medium, Pale Cream, and Cream. They may also bear the name Fino, or Amontillado ("having developed a flor"), or Oloroso.

The authorized grapes are Pedro Ximénez (95% of vineyards), Moscatel, Baladi, and Airén.

- Málaga—first demarcated and classified in 1937, regulated DO in 1976. Málaga wines were traded around the Mediterranean by Greek, Phoenician, and Roman sailors, and sold for high prices in every port of call. They were most popular between the seventeenth and nineteenth centuries, so much so that Málaga became a trendy place for wealthy tourists. But the hunger for real estate for hotels, apartments, and villas superseded the thirst for this ambrosial "Mountain Wine." As a result, the DO area has shrunk to several small patches of vineyards.

  Málaga is divided into two subzones: a small area around Estepona on the coast, and the main area stretching along the coast from the city of Málaga, north along the border with Granada and west along the border with Córdoba. This main area is further divided into three: Molina in the northwest, Axarquía along the coastline, and Mountain in the extreme north around the town of Cuevas de San Marcos.

  The soil varies in each subzone, but is dominated by gravelly topsoil over ferruginous clay. Some areas contain chalk, mica, and quartz.

  Pedro Ximénez and Moscatel are the only grapes grown in Málaga. The regulations recognize different types of Málaga wines according to grape, colour, and sweetness. For example, Seco (dry), Abocado (semisweet), or Semi-Dulce (medium sweet) may be used to distinguish the various types. The wines are fermented dry, fortified, and enriched with a variety of sweet wines such as árrope, vino de color, or mistela. With all these options in blending, the winemaker can vary the sweetness and establish a personal style.

  Traditionally, these wines are fermented in large cement vats, cold-stabilized to remove tartrates, and aged in a solera system.

  There are two other particular wines: Málaga Dulce is made from grapes exposed to the sun after the harvest, to shrivel and concentrate the juice. The wine is fortified during fermentation to about 18% alcohol. Málaga Dulce Lágrima is similar, but made only from the free-run must (without pressing).

CANARY ISLANDS

The seven Canary Islands lie in the Atlantic, 115 km west of the coast of Morocco and 1150 km from Spain. The islands got their name from the Latin *Insulae Canium* (Dog Islands), but in 1402 were renamed, by the Spanish, *Las Islas Canarias* (from old Spanish *can*, meaning "dog"). The climate is subtropical, humid, with frequent drizzle, foggy days, and an average temperature of 16°C. Phylloxera never reached the Canary Islands, so all vines are planted on their natural roots. The most well-known wine is the Malvasia of Lanzarote, also known as Canary-Sack.

WINE PRODUCTION

The islands DOs include:

- Tacoronte-Acentejo—awarded the DO in 1992. Located on the island of Tenerife, the terraced vineyards lie at altitudes between 200 m–800 m on the fertile north- and west-facing volcanic slopes of the mountain. The main red grapes are Listán Negro and Negramoll, plus some Gual and Negra Común. The main white grapes are Malvasía, Moscatel, and Listán Blanco, plus some Verdello and Vijariego.
- Ycoden-Daute-Isora—awarded the DO in 1994. Also on the island of Tenerife, the vineyards climb up to 900 m on the northwest-facing slopes of the extinct volcano which gave birth to the island. Reds, whites, and rosados (as well as sweet and fortified wines) are made here from nineteen authorized varieties. The main white grapes are Vermejuela, Gual, Moscatel, Pedro Ximénez, Verdello; the main reds are Listán Blanco and Negramoll.
- La Palma—awarded the DO in 1994. The vineyards cover most of the island and are planted on the black volcanic ash of the windy mountain slopes, at altitudes of 300 m–1200 m. Six red grapes are authorized, of which Negramoll is the principal variety. Gual, Malvasía, and Verdello are the main varieties of the fourteen white grapes authorized.
- Lanzarote—awarded the DO in 1994. The DO covers most of the island of Lanzarote. The fertile topsoil is mainly composed of black volcanic ash. Strong winds in some areas compel growers to plant the vines in scooped-out hollows, sometimes 3 m deep.

Lanzarote is the island which prompted the name Canaries: invading Romans sighted large packs of wild dogs there.

Lanzarote's vineyards are planted with ancient local white grapes such as Burrablanca, Breval, and Diego, plus Listán Blanco, Malvasia, Moscatel, and Pedro Ximénez. The authorized reds are Listán Negro and Negramoll.

CAVA WINES

Cava means "cellar."

Cava is a QSWPSR (Quality Sparkling Wine Produced in a Specific Region) made exclusively using the méthode Champenoise. The DO was established in 1986. The cork is branded with a star. Production of Cava is approximately 150 000 000 L, one-third of which is exported.

The first bottle of Cava was made in 1872 by José Raventos, head of the family firm of Cordoníu.

Cava wines are not restricted to any one area. However, most Cava wines (95%) are produced in Cataluña, and more than 90% of those in San Sadurní de Noya, in Penedès, in the province of Barcelona. But, under EU law, the wines cannot be called Cava del Penedès.

To be considered DO Cava, the wine must:

- rest on the lees for at least nine months before disgorgement (most good Cavas get one to two years ageing on the lees).
- achieve, at 20°C, 4 atmospheres of pressure.
- attain an alcoholic strength of 10.8% to 12.8%.

In 1970, the Spanish government agreed to discontinue the use of the name Champagne (or Champaña), and to replace it with the word Cava. As a member of the EU, Spain was permitted to use the term *Metodo Champenoise* only until 1994. Since then, the term *Metodo Tradicional* has been permitted (but not required) on labels.

Cataluña tends to use mainly Maccabéo, Parellada, and Xarel-lo, typically in a proportion of 50%, 20%, and 30%. Cavas produced outside Cataluña are usually made from Viura (Maccabéo) grapes, occasionally blended with a little Parellada or Malvasia Riojana. Chardonnay and Pinot Noir are used only on an experimental basis. (Two Cava houses currently make a sparkling wine entirely from Chardonnay.) Small amounts of Garnacha and Monastrell are used to produce pink Cavas.

Spain's Other Sparkling Wines

Spain produces a multitude of other sparkling wines:

- Vino Espumoso, Metodo Traditional—made like a Cava but outside of the areas approved for Cava production.
- Vino Espumoso—made using the transfer method, although the label may indicate it was bottle-fermented. It is aged for at least two months. The cork is marked with a rectangle.
- Granvas—made using the charmat method, in a tank (gran vaso means "large vessel"), and aged for a minimum of three weeks. These wines cannot be made in the same building as Cava. The cork is branded with a small oval.
- Vino Gasificado—carbonated, usually by dissolving blocks of solid carbon dioxide in a pressure tank. No ageing is required. The cork is marked with an equilateral triangle.

STYLE OF WINES

brut de brut brut nature brut reserva vintage	very dry	less than 6 g/L sugar
brut	dry	up to 15 g/L sugar
extra seco	fairly dry	12 to 20 g/L sugar
seco	semisweet	17 to 35 g/L sugar
semi seco	sweet	33 to 55 g/L sugar
dulce	very sweet	over 50 g/L sugar

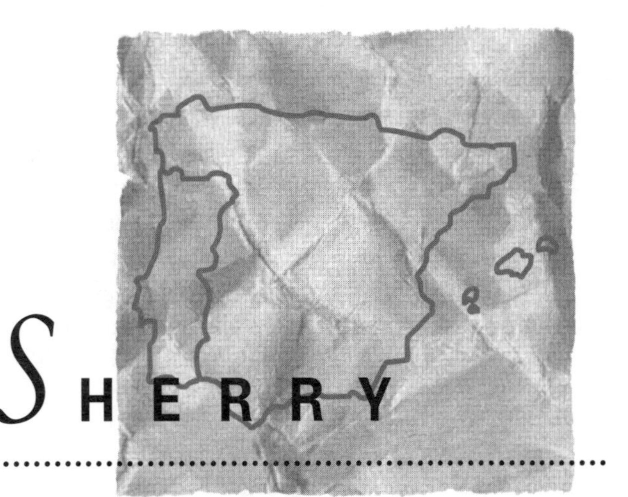

SHERRY

Sherry is the wine that is most misunderstood—not only by the average consumer but also by serving staff. The common image is of a sweet wine consumed on its own, after a meal. However, good Sherry is one of the most complex and intriguing wines in the world, and the resilient Andalusian people continue to provide it. Traditional Sherry production survives despite a weak market.

albariza (al-ba-REE-tha)—best type of soil in Jerez, very chalky
amontillado (ah-moan-tee-YAH-though)—a classification of Sherry, with flor, amber and sweet
amoroso (ah-mo-ROH-so)—a classification of Sherry, sweeter than oloroso
arena (ah-RAY-nah)—least preferred type of soil in Jerez; sandy
barro (BAR-roh)—type of soil in Jerez; dark, heavy
brown—a classification of Sherry, dark and very sweet
butt (BOOT)—standard cask of 500 L made of American oak, used for maturing Sherry. The larger butt (called *bocoy*) contains 600 L.
cream—a classification of Sherry made by sweetening a fino Sherry
fino (FEE-no)—a classification of Sherry, with flor, pale and dry
flor (floor)—the film-forming yeast found on Sherry; also, a classification of Sherry
manzanilla (mun-tha-NEE-yah)—a classification of Sherry produced only in Sanlúcar de Barrameda
oloroso(oh-lo-ROH-so)—a classification of Sherry, with little or no flor
palo cortado(PAH-lo kor-TAH-though)—a classification of Sherry, with no flor, rare
plastering—the addition of enseyado, gypsum, or calcium sulphate to crushed grapes used in making Sherry
solera (so-LAIR-ah)—a system of progressively blending wine

GEOGRAPHY

The Wine Region of Sherry

The town of Jerez is located near the town of Cádiz in the south of Spain, between the Rivers Guadalquivir and Guadalete. The demarcated zone embraces the towns of Jerez de la Frontera, Puerto de Santa Mariá, and Sanlúcar de Barrameda. The best vineyards lie within a 30-km range north and west of the town of Jerez.

More than 20 000 ha are planted, of which 15 000 ha are known as Jerez Superior because they are planted exclusively on albariza soil. A minimum of 4% of the stock of all Sherry producers must come from this zone.

The Mediterranean climate provides 295 days (Europe's highest) of intense sun, without a single drop of rain between the vine's flowering and the harvest. Annual precipitation is limited to 70 days of rain, with 40% of them in the months of October to December and the remaining 60% in the period from February to May. The average temperature is 17.5°C but temperatures can reach 40°C in the summer. Hot prevailing winds blow in from North Africa.

There are three classes of soil:

- White albariza soil contains up to 70% calcium carbonate (chalk) mixed with magnesium and clay. It yields the best wine, but is less productive than other areas. This soil originated 50 million years ago from sedimentation of algae and marine protozoa, alternating with layers of sand. The fossil deposits retain water well and store it for the long, dry summer. The layers of sand facilitate root growth as deep as 12 m. Even the light colour of the ground surface is useful, reflecting sunlight up to the vines and concentrating the sugars.
- Barro soil is darker and heavier, made of clay, mud containing iron oxide, and less than 30% chalk. This soil is usually found in low-lying valleys.
- Arena soil is light and yellowish, composed mainly of sand, with 10% chalk, silica, and alumina. It is much more fertile (giving double the yield of albariza) and is therefore avoided altogether.

HISTORY

The region that produces Sherry is known as Sherry, Jerez, or Xérès. The oldest indications of vine cultivation in the region go back 3000 years, to a time when the Phoenicians established a settlement called Xera ("fortress on a river").

During the second century B.C., Romans took control of the area and renamed it Ceres (for the goddess of the harvest) and later Ceritium. At the collapse of the Roman Empire, the Vandals invaded southern Spain and called it Vandalusia (now Andalucia, or Andalusia). In A.D. 414, the Visigoths moved in and stayed for 300 years until the Moors defeated them. At that time Ceres became known as Seris.

The region prospered under Moorish rule, and it was from the Moors that the secret of distillation was learned. (Distillation at that time was used for making perfumes, medicines, and so on.) Methods were passed on even as the Moors lost control of Seris to King Alfonso X, in 1264. At that time, the region was renamed Jerez, to become Jerez de la Frontera 100 years later under King Juan I.

During the time of the Catholic Queen Mary Tudor of England (1496–1533), exports of Sherry to England grew. The wine was then known as sack or sherrisack, from the Spanish sacar ("draw out"), which came to mean "export."

This relationship soured during a prolonged feud after England became Protestant. The confiscation of Gilbraltar by the English in 1704 dashed any hopes that trade between the two countries would rebound. Then, during the Napoleonic Wars (1803–1815), Jerez was destroyed by French armies.

In the nineteenth century, the wine trade began to bounce back. Exports expanded from 17 000 casks in 1840 to 70 000 casks by 1873, but powdery mildew and phylloxera set the industry back.

In 1933, the Consejo Regulador re-established laws (Reglamento de las Denominaciones de Origen Jerez-Xérès-Sherry y Manzanilla-Sanlúcar de Barrameda) to control the production and sale of Spanish wines. These were later modified and supplemented. Today, the Denominacion de Origen Jerez (DO) is regulated by revisions made in 1977.

Jerez was one of the first wines of Spain to be regulated and demarcated. In 1733, a vintner's guild (Gremio de la Vinateria) was established in Jerez. Its aim was to maintain the reputation, quality, and origin of the wines. In 1933, the Consejo Regulador re-established laws to control the production and sale of Spanish wine.

GRAPE VARIETIES

All the grapes used in the production of Sherry are white.

- Palomino Fino and Palomino de Jerez—the quality grapes cultivated in 90% of the vineyards, mostly in albariza soil. Palomino grapes are also known as Listán.
- Pedro Ximénez—grown in smaller amounts in barro and arena soils. It produces wines rich in sugar and is therefore used mainly for the sweetening wine Paxarete (commonly referred to as PX), to blend into sweet sherries. There are some dessert wines, such as those of the nearby Málaga and Montilla-Moriles regions, made from this grape alone.
- Moscatel (Muscat of Alexandria)—still grown, but since they tend to lack acidity, used almost exclusively as a sweetener.

WINE PRODUCTION

Sherries are fully fermented wines. The sweetness derives from the addition of vino dulce, which is made from over-ripe or sun-dried Moscatel and Pedro Ximénez grapes, dried on esparto grass mats to concentrate the sugars.

Making Sherry

Harvesting

Harvesting lasts three to four weeks, normally starting on the 8th of September. The maximum permitted yield is 80 hL/ha for Jerez Superior and 100 hL/ha for the other zones. Grapes are hand picked into wicker baskets, each holding 15 kg. These are emptied onto trucks and transported to the bodegas. Depending on their maturity, the grapes may undergo raisining—that is, be spread out in the sun on straw mats to dry, in order to further concentrate the sugars. Today, major wineries forego this stage and opt instead for a must with lower sugar and tannin but higher acidity.

Crushing, Plastering, Extracting Juice, and Pressing

At the bodegas, the grapes are crushed and plastered. Plastering has its origin in the times when the grapes were transported in open wagons along dusty roads, and became coated with the dust. As was learned later when transportation methods changed, the dust added to the flavour of the resulting wine, and encouraged formation of the flor.

Now the effect is created through plastering (enseyado in Spanish), which involves adding gypsum or calcium sulphate to the grapes, at a rate of 2 kg/45 hL.

Next, the crushed grapes are pumped into a dejuicer, which consists of a long, inclined, perforated cylinder with a screw that pushes the grapes up. The free-run juice drips through the perforations to produce 70% of the must. This is called the primo yema ("best").

The grapes then fall into a second identical dejuicer where some pressure is applied to the pulp. This releases the secundo yema ("second best"), which accounts for 15%of the must. Finally, the remaining pulp is conveyed to a press where a final squeeze extracts the prensa ("press").

The primo yema is reserved for the production of finos. Prensa is used only for distillation.

In contrast to the traditional method (pisadores wearing hob-nailed boots crush and press the grapes), the current method is much more efficient (only requiring three people to process forty tons of grapes per hour), and the yields are higher.

Fermenting

The traditional method of fermenting involves first pouring the must into new butts (casks) made of American oak.

These butts have a capacity of 500 L. A 10% air space is left in each one. Fast and furious fermentation follows, lasting only three to four days. The vino mosto ("new wine") immediately begins a slow malolactic fermentation, which lasts until the beginning of winter.

Nowadays, more than a third of Sherry is made using an alternative method. The must is pumped into large temperature-controlled decanting tanks (desfangado) where it sits for twelve to thirty-four hours. Sulphur dioxide is introduced, to prevent oxidation and to kill off wild yeasts. Alcoholic fermentation takes place in large temperature-controlled stainless steel tanks for four to seven days, at temperatures of 22°C to 24°C.

Flor

Flor is a yeast that grows in a layer on the surface of certain Sherries. It occurs spontaneously only in Jerez, Jura (France), and the Caucasus. The traditional Jerez method is based on this natural formation of flor.

For the longest time, the formation of strange white patches resembling flowers (hence the name) on the surface of wine was a mystery, considered miraculous.

Flor occurred unpredictably, in random casks, and to varying degrees. Eventually, scientists (from the Estación de Viticultura y Enologia in Jerez) uncovered the riddle. The formation of flor is a continuation of the fermentation process and is dominated by three types of Saccharomyces yeast: ellipsoideus, oviformis, and mangini.

The process begins only when the temperature is between 15°C and 20°C and the humidity is around 60%. Tannins cannot exceed 0.01% and sulphur dioxide must be no more than 0.02%. If the wine has been fortified to an alcohol level of less than 16.4%, the yeasts will rise to the surface of the wine. There the yeasts mutate to become aerobic yeasts (Saccharomyces beticus, rouxii, montuliensis, and/or cheresiensis). Of these, the best-performing and most sought-after yeast is Saccharomyces beticus.

Under ideal conditions, spotty, flowerlike patches grow on the surface of the wine. These first form a thin lacy veil, then a thick, white crust resembling cottage cheese. Some cells die and fall to the bottom of the cask, where they leave spores that grow anew every spring and autumn. No two casks develop in exactly the same way.

Flor has the following impact on the wine:

- the residual sugar is consumed completely.
- levels of alcohol, tartaric acid, volatile acids, and ethyl-acetate are reduced.
- acetaldehydes increase, giving fino Sherries their characteristic nose.

As the wines start to develop flor (which can happen near the end of fermentation or after) they are sorted and classified into two categories. Wines with a thick skin of flor and a strong aroma and taste will become fino or amontillado Sherries. Wines which have a thin coating and less of the flor bouquet will be destined for the production of olorosos or sweeter Sherries. The latter are immediately fortified to 18% alcohol to kill off any remaining flor and to protect them from acetobacter, which would turn the wine to vinegar. A second classification is carried out six months later to select wines for the solera.

The term "solera" derives from the Latin *solum* or the Spanish *suelo,* meaning

The Solera

The solera system is a method of blending wines to smooth out differences between vintages. The system consists of a minimum of three rows of oak casks stacked one upon the other. Each row, or level, contains Sherry of a single type and vintage.

An exceptional vintage is used to fill the ground-level or first layer of casks (called the solera). After the next harvest, the second layer of barrels (criadera primera), stacked atop the solera, is filled. The next year's harvest goes in to the third layer (criadera secunda), and so on. The top row is always filled with fresh new wine (añada).

North Americans may sense an inconsistency in naming the second layer as "primera." But in Europe, levels are counted from above ground level. The European "first floor" is the first floor above ground level; in North American style, this is the second floor.

After a minimum of three years in wood, Sherry that is to be bottled is drawn from the solera (the bottom row). These casks are then topped up with wine from the first criadera, which in turn is topped up with the wine from the second criadera, and so on.

The purpose of the solera system is to create a perpetual blend, eliminating all the variations in quality that exist from one vintage to another. Because the system mixes wines of various harvest, Sherries cannot be vintage dated in the usual way, nor labelled for their vineyard. Sherries are identified by the name of the solera in which they were raised (created) and any date given on a label indicates the year in which the solera was started. So, a 1903 fino Sherry is not a wine from the 1903 vintage, but a fino Sherry issued from a solera that was started in the year 1903.

Types of Sherries

Sherry must age no less than three years before being finally classified. The Denominación de Origen Jerez-Xérès-Sherry recognizes four types of Sherry—really, two broad categories, each with two further divisions. Flor Sherries are either fino or amontillado. Non-flor Sherries can be classified palo cortado or oloroso.

Flor Sherries

- Fino—a pale, straw-coloured wine which has grown flor, with a sharp and distinctive aroma, usually five to nine years old. It is dry and slightly acidic, with an alcoholic strength of 15.5%–17.5%. Fino is ready (and best) as soon as it is bottled (typically, between five and nine years old). It should always be consumed chilled. Once the bottle is open, it should be finished within a week.

The word Amontillado used in Jerez originates with the style of the wines of Montilla.

- Amontillado—a fino that has been allowed to age for a further period, up to eight years or more in the cask. It is typically sold at between ten and fifteen years old. Amontillado is amber in colour, sweeter, with a fuller body and nuances of hazelnuts in the bouquet; it gets darker with age. Its alcoholic content is usually 16%–18%, although some older amontillados reach 24%.

Non-Flor Sherries

- Palo Cortado—rare and often expensive. It cannot be made intentionally; it just happens. This unusual wine has a nutty nose characteristic of amontillado, yet a flavour reminiscent of oloroso. Alcohol content ranges between 18%–20%.
- Oloroso—means "fragrant." It is the wine that results when very little (or no) flor grows. It is typically sold at between ten and fifteen years old. Much darker and sweeter than amontillado, it has a strong aroma of walnuts. The alcohol content is usually between 18%–20%.

OTHER SHERRIES

- Manzanilla—a pale, delicate, fino wine with a sea-salt aroma and a spicy bitterness in the aftertaste. It has its own appellation since it can be produced only in Sanlúcar de Barrameda. It is typically sold at between five and nine years old. Much like fino Sherry, it should be drunk chilled and used quickly once the bottle has been opened. The alcohol content is 15.5%–17.5%.
- Amoroso—means "loving." This is a sweeter style of oloroso. It is produced by adding some Pedro Ximénez wine as well as a bit of vino de color (colouring wine).

  It is typically sold at between five and fifteen years old. Amoroso can improve with ageing in the bottle. The alcohol content is 18%–21%.
- Cream—a very sweet Sherry with thick body. It is made by sweetening a fino Sherry with some Dulce Apagado, a wine with very high residual sugar (made by arresting the fermentation with brandy). It is typically sold at between five and fifteen years old. The alcohol content is 18%–21%.
- Brown—a dark, very sweet Sherry, made from a blend of oloroso, Muscatel, and Pedro Ximénez vino de color. It is typically sold at between five and fifteen years old. The alcohol content is 18%–21%.

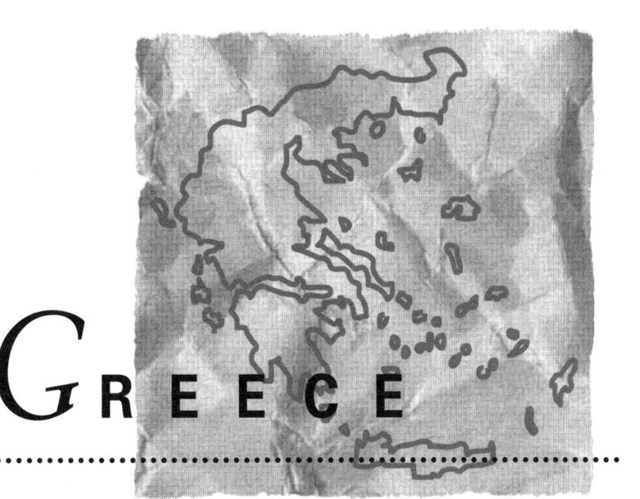

GREECE

Greece has the longest history of winemaking of any nation—dating back to Ancient Greece, when Dionysus was worshipped as the god of wine. Grapes and olives were the major cash crops of the day, and wine was traded on a large commercial scale. Now, in the modern state of Greece (founded in 1913), the wine trade is regaining momentum and finding a wide market among other members of the EU, and beyond. Although the names of its wines and grapes may be unfamiliar, recent advances in Greece's wine industry deserve the attention of every serious wine lover. Wine consumption in Greece is 30 L per capita.

GEOGRAPHY

Greece and its many islands are surrounded by the Mediterranean, Ionian, and Aegean seas. The country lies between 33°N and 41°N, so has consistently hot temperatures. The major factors determining the various macroclimates are altitude, and proximity to the sea. Some vineyards can be found clinging to sheer sea cliffs at altitudes of 650 m. Vineyards are generally located on poor and arid soils, consisting mainly of rock, chalk, loam, schist, sand, or clay. Some areas, such as the island of Santorini, are made up of large tracts of volcanic pumice, lava, and crystalline limestone. There are 186 000 ha of vineyards in Greece but only half are dedicated to wine grapes.

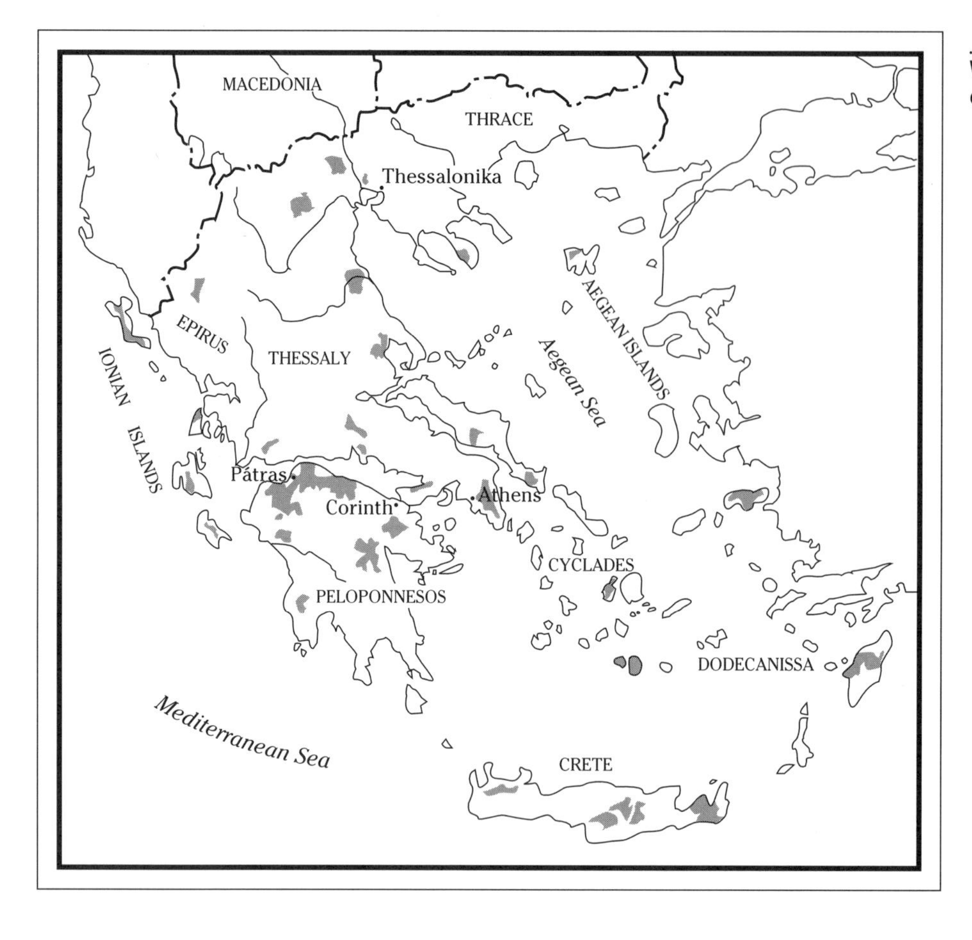

Wine Regions of Greece

HISTORY

Even earlier than the Neolithic Age (6000 B.C.–2500 B.C.), grapes were cultivated both on the mainland of Greece and on various islands in the Aegean Sea. The Minoan civilization (3000 B.C.–1000 B.C.) developed a high level of viticulture and understood well the meaning of microvinification, that is, the vinification of separate vineyard's grapes. Homer (c. 800 B.C.), Hesiod (c. 700 B.C.), Herodotus (c. 484 B.C.–425 B.C.), Xenophon (c. 430 B.C.–355 B.C.), and Aristotle (384 B.C.–322 B.C.) all describe wines of the day. Theophrastus (c. 372–287 B.C.) listed as many as ninety grape varieties and a hundred and thirty wines produced.

In the heyday of Ancient Greece, wine was the standard currency throughout the Mediterranean Basin. As Greek sailors established trading posts and colonies, they brought with them their vines, and propagated vineyards as far as Portugal and Magna Graecia (Sicily and southern Italy). Wars and rule by the Romans, however, brought an end to the expansion of Greek commerce. When the Roman Empire disintegrated in the fifth century A.D., the door opened to a series of new rulers of Greece—the Byzantines, Franks, Venetians, and Ottoman Turks. The Ottoman subjugation lasted for centuries and was particularly unfavourable for winemaking.

In the nineteenth and twentieth centuries, Greece underwent a series of political and social upheavals. The Greek War of Independence (starting in 1821), the Balkan Wars (1912–1913), two world wars, and civil war depleted the labour force and rocked the economy. As well, phylloxera struck in 1890 and took its toll until 1960. (Even today, Crete has serious problems with the louse.) Vineyards were devastated. All this was further aggravated by constant bleeding of the work force through emigration.

In the 1960s the wine industry got a new stimulus through a big investment of capital. Joining the European Community (now the EU) in 1981 further boosted the country's prospects for its wine trade. This opened up a huge European market and brought substantial agricultural subsidies to develop vineyards and modernize wineries. Companies started to expand vineyards and replant ancient sites. Young Greek winemakers studied oenology in Germany, Italy, Bordeaux, Montpellier, and Dijon. Californian-style wine boutiques popped up, and wine lovers internationally realized that Greece produced good wines as well as the famous Retsina.

LAWS AND REGULATIONS

The Wine Institute of Athens was created in 1937 to give support to winemakers. In 1952, a set of regulations was established to provide a system for creating superior wines. These wine regulations were upgraded and written into law in 1971. A decade later, when Greece joined the European Community (now the EU), it had to ensure that its wine laws and regulations complied with those of the EC. Wine laws and regulations are overseen by the Ministry of Agriculture.

Greek wines are divided into two categories: table wine and quality wine. Each of these has two subcategories.

Table Wine

Epitrapezeos Oino

This includes table wine, cava, and wine of appellation by tradition:

- Cava—a high quality, aged table wine. Cava White must have two years of cellar ageing with optional wood ageing. Cava Red must have a minimum of six months of ageing in new casks or twelve months in older casks. Minimum cellar maturing is three years.
- Appellation by Tradition—an exclusive and typical wine controlled by EU regulations. The obvious example is Retsina.

Topiko Oino

This is the equivalent of the French vin de pays.

Quality Wine

OPE (Controlled Appellation of Origin)

OPE is the equivalent of the French VLQPRD. These sweet wines bear a blue seal. Sámos Muscat and Mavrodaphne Pátras are the only two classified sweet wines.

OPAP (Appellation of Superior Quality)

OPAP is the equivalent of the French AOC and VQPRD, and is Greece's highest category of wine. There are twenty-seven appellations from proven regions of superior quality. Dry wine carries a pink seal printed in red.

Additional designations are as follows:

- Reserve White and Reserve Red—a minimum ageing of two years, with six months in cask.
- Grande Reserve White—a minimum ageing of one year in cask and two years in bottle.
- Grande Reserve Red—a minimum ageing of two years in cask and two years in bottle.

GRAPE VARIETIES

Greece now grows more than 300 grape varieties. Because most are simple or multi-hybrid mutations of older cultivars, modern Greece's viticulture is like a time capsule—it has perhaps the most valuable collection of ancient Vitis vinifera.

Here are some of the main varieties:

White Grapes

- Aïdani—mainly grown in the islands of the Aegean Sea and, in particular, the island of Santorini, where it develops an aroma of jasmine.
- Assyrtiko—an old variety most successful on Santorini, where some vines are 150 years old. This grape makes full-bodied wine in many styles, from crisp and dry, to lusciously sweet. It is one of Greece's best varieties.
- Athiri—often blended with Assyrtiko, produces wines that have a rich texture and strong citrus scents. The finest results come from Rhodes.
- Debina—loves high altitudes, particularly in the region of Epirus. It yields crisp and pale wines with pear and apple aromas. A light, fresh, semi-sparkling wine is a speciality unique to Epirus, especially around Zitsa.
- Lagorthi—a rare and ancient varietal, mainly found in Achaïa, that thrives up to 800 m. Its wines are high in acidity with a pronounced quince aroma.
- Malagousia—a clone of Malvasia Giala propagated in the 1960s for its peculiar mint and lime aromas.
- Malvasia—the name derives from the words moni emvasis ("single entrance") and relates to the town of Monemvasia in southern Peloponnesos. Because the grape is rich in sugar, it is often used to make sweet wine. The aroma is reminiscent of dry peaches or figs.
- Moscophilero—an outstanding semi-aromatic grape believed to be related to Traminer. It is an indigenous cultivar, a late ripener, low in sugar, and high in acidity. There are two clones: Asprofilero, with a pinkish skin; and the darker Mavrofilero. It is planted in the Peloponnesos, Arcadia, Mantinia, Messinia, and Achaïa.
- Muscat—the oldest white variety known. It is found mainly around Patrai and on the islands of Sámos and Cephalonia. The grape can produce wines high in sugar and with rich aromas. Muscat comes in several varieties, and clones can vary in shape and colour. Aspro Muscat (also known as Sámos Muscat, or in Beaumes-de-Venise as Muscat à Petit Grains) is considered the best.

- Muscat of Alexandria—a low-acid grape that is rich in sugar and produces wines with strong aroma. The grape is predominant on the island of Lemnos.
- Robola—believed to have been brought to Greece by Venetian merchants during the late thirteenth century, and thought to be a clone of the Ribolla Gialla from Friuli in Italy. It gives a very dry wine with citrus aromas, and is one of the most expensive grapes of Greece.
- Rhoditis—best when grown at altitudes of 350 m–850 m. Red and white clones of this grape are used. The red clone, the best of which is grown in Achaïa, is one of the more complex Greek varieties. Low-lying vineyards yield grapes that are flat in taste and end up being used in blends.
- Savatiano—a topaz-skinned grape, which loves heat. It makes up 15% of total wine grape-plantings in Greece. The grape gives a wine that is low in acidity and has strong body.
- Sideritis—a native of Achaïa. Its wines are high in acid and have a peppery nose. They are often blended with Rhoditis and distilled into brandy.
- Tsaoussi—grown on the island of Cephalonia, where its wines are refreshing and have a herbaceous nose. This grape makes more complex wines when blended with other grapes such as Robola.
- Vilana—particular to the island of Crete. It makes easy-drinking wines that are refreshing when young. They have a flowery aroma and a distinct green-apple tang.

Other white grapes used include Chardonnay, Grenache Blanc, Sauvignon Blanc, Sémillon, Viognier, and Ugni Blanc.

Red Grapes

- Aghiorghitiko—also known as St. George and Agiorgitiko, this (along with Xynomavro) is one of the best red Greek varieties. Wines made from this grape have a deep colour and soft tannin and can age well. Grapes grown above 350 m develop more complexity from the cooler environment and acquire a distinctive black currant aroma. The best examples come from Neméa.
- Amorghiano—a rare variety from Rhodes that yields a complex wine full of fruit and character.
- Avgoustiatis—found on the island of Zakinthos. The very dark and small berries can produce wines of great quality and depth when aged for no less than four years.
- Kotsifali—found in Crete, producing spicy, flavourful, and full-bodied wines that are soft in tannin.

- Krassato—a particular specialty of Thessaly. In the Raspani appellation, Krassato, Xynomavro, and Stavroto grapes are blended in equal proportions.
- Liatiko—an ancient and early ripening grape from the island of Crete. The name comes from *loulios*, Greek for the month of July, when it ripens. It is blended with the two best red grapes of Crete, Mandelari and Kotsifali, to create a special barrel-aged dessert wine.
- Limnio—also known as Kalambaki. A very ancient grape mentioned by Aristotle, it originates from the island of Lemnos. Its wines have a peculiar sage, thyme, and pepper nose that blends well with Cabernet Sauvignon.
- Mandelari—from the Cyclades and Crete. Its wines are high in tannin, deeply coloured, and aromatic. They are often toned down with wines that have lighter body and colour.
- Mavrodaphne—a versatile grape used to create table wines from dry to sweet, as well as fortified styles. Its name means "black laurel."
- Mavroudi—a prolific clone of the Aghiorghitiko grape. The small, dark berries produce soft wines low in tannin.
- Negoska—yields a soft wine, high in alcohol ,with blackberry aroma. These wines are good for blending with lighter wines.
- Thiniatiko—a very rare clone of the Mavrodaphne. It yields a soft, rich wine in Cephalonia, and has great potential.
- Volidza—saved from extinction by a few growers. This variety with its particularly large berries yields very aromatic, soft wines with a distinctive raspberry-mushroom nose and an earthy character. It has great potential.
- Xynomavro—one of the greatest red varieties, grown in the northern part of Greece. In cool environments, it yields full-bodied wines, with a nose of red berries, low tannins, and good acidity. When successfully vinified and matured, these wines resemble Barbaresco and Burgundy. Modern vinification techniques, and a better understanding of the variety, could lead to a better image for one of Greece's best red wines. Its name means "acid black."

  Sparkling rosé and blanc de noirs are also made from this grape, using the méthode Champenoise.

Other red grape varieties include Cabernet Sauvignon, Cabernet Franc, Grenache, Merlot, Syrah, Petit Syrah, Refosco, and Sangiovese Grosso.

WINE PRODUCTION

Greece produces upwards of 3 000 000 hL of wine annually. When the country joined the European Community, its wine grapes were used primarily to produce brandy, fruit spirits, and ouzo. The majority of Greek wines were made for Greek taste (oxidized, sweet, or laced with pine resin) and were difficult to market internationally. However, this has changed dramatically. Now Greece's wine is good; in some cases, exceptionally good.

RETSINA

This wine, so unique and particular to Greece, evolved from the ancient practice of sealing wine in earthen pitharia (large clay vats) or amphorae, with a wood, leather, or plaster seal coated with pine resin. The antiseptic property of the resin prevented the wine from going bad but gave it a strong resinous flavour. Today, Retsina is flavoured with resin from the Alep pine in regulated amounts of 1 kg/100 L. The resin is added and left until the first racking. International annual consumption of Retsina is approximately 6.3 million bottles.

WINE REGIONS

To simplify, the following list of wine regions gives the appellations, each appellation's predominant style of wine, and its predominant grape varieties.

MACEDONIA AND THRACE

Amynteon—dry red and sparkling rosé (Xynomavro)
Côtes de Meliton—dry white (Rhoditis, Assyrtiko, Athiri); dry red (Limnio, Cabernet Sauvignon, Cabernet Franc)
Goumenissa—dry red (Xynomavro, Negoska)
Naoussa—dry red (Xynomavro)

THESSALY

Ankhialos—dry white (Rhoditis, Savatiano)
Rapsani—dry red (Xynomavro, Krassato, Stavroto)

EPIRUS

Zitsa—sparkling white (Debina)

PELOPONNESOS

Mantinia—dry white (Moscophilero)
Mavrodaphne Pátras—sweet red (Mavrodaphne, Korinthiaki)
Muscat of Pátras—sweet white (Muscat Aspro)
Muscat of Rion Pátras—sweet white (Muscat Aspro)
Nemea—dry red and sweet red (Aghiorghitiko)
Pátras—dry white (Rhoditis)

IONIAN ISLANDS

Cephalonia Mavrodaphne—sweet red (Mavrodaphne, Korinthiaki)
Cephalonia Muscat—sweet white Muscat Aspro)
Cephalonia Robola: dry white (Robola)

CYCLADES

Paros—dry red (Malvasia, Mandelari)
Santorini—dry white (Assyrtiko, Aïdani, Athiri);
dry red (Assyrtiko, Aïdani)

AEGEAN ISLANDS

Lemnos—dry white (Muscat of Alexandria)
Lemnos Muscat—sweet white (Muscat of Alexandria)
Sámos—sweet white (Muscat Aspro)

DODECANISSA

Rhodes—dry white (Athiri); dry red (Mandelari)
Rhodes Muscat—dry white (Muscat Aspro, Muscat Trani)

CRETE

Archanes—dry red (Kotsifali, Mandelari)
Daphnes—dry red (Liatiko); sweet red (Liatiko)
Peza—dry white (Vilana); dry red (Kotsifali, Mandelari)
Sitia—dry red (Liatiko); sweet red (Liatiko)

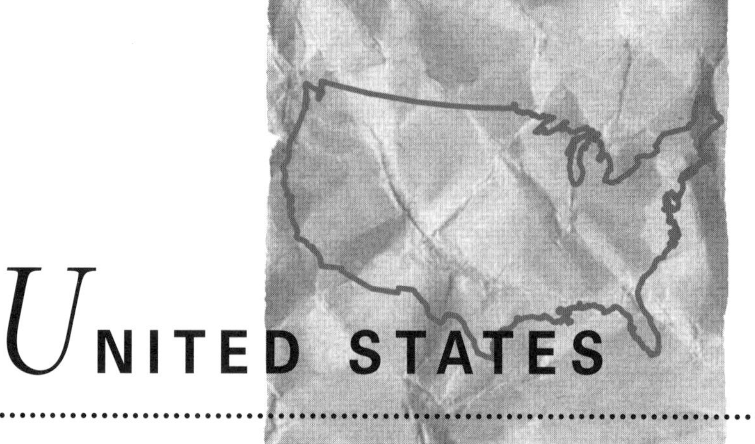

UNITED STATES

This vast country produces wine in forty-three of its fifty states. Most of the production (96%) takes place in the states of California, Oregon, New York, and Washington. In general, other states are just developing their wine industries and tend to concentrate on producing for local consumption. Today the United States ranks fourth in wine production (after Italy, France, and Spain), and has a yearly output of approximately 16 000 000 hL. Annual consumption is only 6 L per capita. Within the United States, there is a wide assortment of climates, soil formations, and local practices. These topics (where appropriate) are covered within the treatments of each of the key wine-producing states.

HISTORY

Wine production in what is now the United States began in the sixteenth century, with the arrival of the Europeans. They found wild Scuppernong grapes and native Vitis labrusca and Vitis rotundifolia. Between 1562 and 1564, French Huguenots made wine in a settlement near Jacksonville in Florida. The Jamestown colonists made wine in 1609, and Mayflower Pilgrims did likewise at Plymouth to commemorate their Thanksgiving, in 1623.

Because the native Vitis labrusca and Vitis rotundifolia produced wine with a peculiar (and generally disliked) foxy flavour, colonists tried to plant Vitis vinifera grapes imported from Europe. All attempts failed miserably. Harsh winters were blamed. Nobody suspected that insects (phylloxera in particular) and diseases were responsible for the devastation.

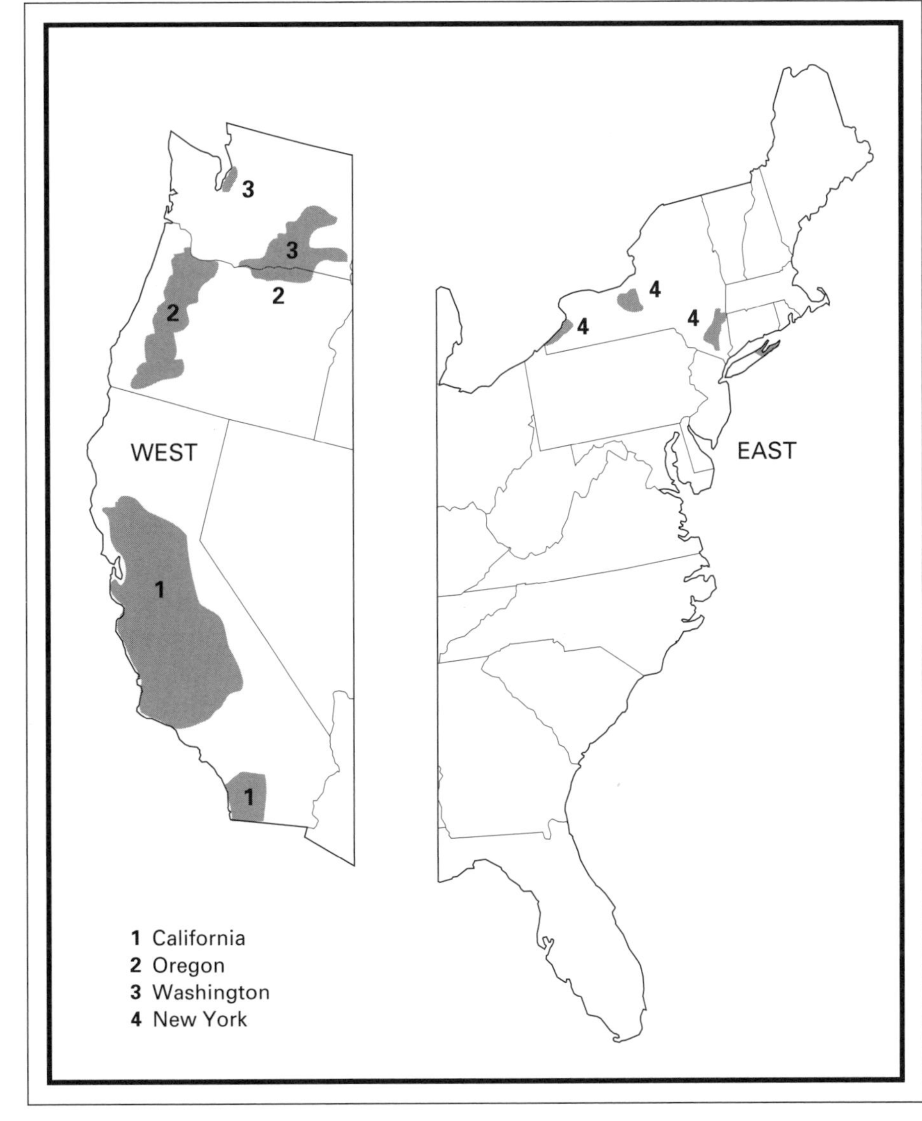

The Vinifera Failures

Throughout the eastern area of what was to become the United States, people and governments tried to transplant European viticulture. In 1619, Lord Delaware brought vines and vintners from France to establish wine grapes in Virginia. The Virginia Colonial Assembly passed a law in 1623 requiring each householder to plant ten vines, and later (1651–1693) offered prizes for wine. In Massachusetts, in 1632, Governor John Winthrop was granted an island in Boston Harbor to plant vines. Queen Christina of Sweden ordered Johan Printz to encourage settlers to plant wine grapes in New Sweden on the Delaware River in 1643. In 1662, Lord Baltimore tried, without success, to plant them in Maryland, and William Penn brought vines from France and Spain to plant near Philadelphia in 1683. The future president, Thomas Jefferson (1743–1826), failed also, at Monticello. The west coast had its successes, but these came later, by way of Mexico.

In the eighteenth and nineteenth centuries, some vineyards started producing wine for commercial purposes. European vine stocks were imported and succeeded well in California, but were less fortunate in the eastern states, where mildew and parasites decimated the plantings. Eventually, phylloxera found its way to California too, in the 1880s.

In the early twentieth century, the relatively young wine industry was growing. Many wineries were created by the steady flow of European immigrants. By 1913, American consumption had grown to 1 900 000 hL of wine annually.

But, during this same time, the Prohibition movement gained ground. From 1919 until its repeal in 1933, the Eighteenth Amendment to the Constitution prohibited the "manufacture, sale, or transportation of intoxicating liquors." California in particular suffered. Many vineyards were abandoned and left to grow wild; skills were forgotten. Some vineyards were maintained to provide grapes for the table, for home winemakers, for the production of grape juice, and for sacramental and medicinal wines. At the end of Prohibition, winemakers had to hurry production, so many of them produced cheap wines.

Around 1960, grape growing and winemaking began to regain its image as an honourable and profitable occupation in America, and in the 1970s, investment in the industry exploded. New vineyards and wineries began popping up all over the country like mushrooms after rain. Today the industry continues to flourish, Vitis vinifera varieties are successfully planted, some regions continue to use native varieties on a small scale, and native American rootstocks are used worldwide as a safeguard against phylloxera.

LAWS AND REGULATIONS

The Bureau of Alcohol, Tobacco, and Firearms (BATF) is part of the US Treasury Department and is the primary government agency for the regulation of the alcoholic beverage industry. The Bureau recognizes specific areas as American Viticultural Areas (AVAs). Each AVA is an area distinguished from surrounding areas by unique geographical features such as climate, soil, elevation, physical features, and (occasionally) history. The AVA regulations, although finalized in 1978, became mandatory in 1983. By approving an area as an AVA, the BATF does not endorse the quality of wine from the area; it simply identifies the area as distinct from others.

AVAs can overlap partially or entirely, as long as each individual area satisfies the criteria. As of 1996 there are 124 AVAs, 70 of which are in California.

Federal and state agencies enforce a variety of standards, including the following:

- wine labelled with the name of a vineyard must contain a minimum of 95% grapes grown in the named vineyard.
- if the label shows the name of an AVA, at least 85% of the grapes must originate from that AVA.
- wine labelled *California* must be made entirely from Californian grapes. Other states require a mimimum of 75% grapes for the wine labels to indicate the state.
- when the wine carries the name of a county, a minimum of 75% of the grapes must originate in the named county. Multi-county appellations—usually designated when the topography is similar across boundary lines—are permitted if percentages of grapes used from each variety are listed on the front label.
- for a wine to carry the name of a variety (e.g., Merlot, Chardonnay) on its label, it must contain a minimum of 75% of that variety.
- wine made from a blend of grapes, in the manner of the traditional Bordeaux blends (Cabernet Sauvignon, Cabernet Franc, Merlot, Malbec, and Petit Verdot for the reds; Sauvignon Blanc, Sémillon, and Muscadelle for the whites), is permitted to be called Meritage. (When blended successfully, such wine can be of very high quality.)
- wine with a vintage date must contain 95% of the stated vintage.
- chaptalization is forbidden in California.

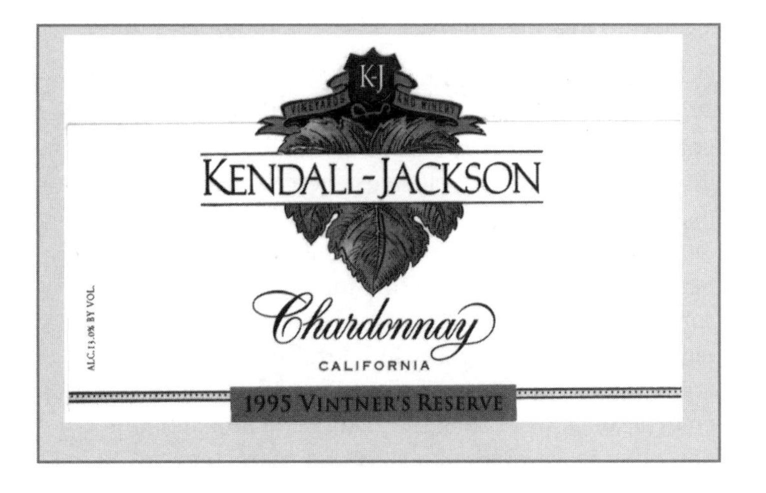

WINE REGIONS

The majority of American wines are identified by the varietal rather than by the geographical location alone, and most of the best wines are labelled as varietals. However, a large number of table wines are given generic names (e.g., Oregon Blush, Ruby Red, or Chablis). As well, wines may be labelled with fanciful names created by the winemaker, e.g., Le Cigare Volant ("the Flying Cigar").

Of the four key wine regions, California is the most important and biggest (85%–90%) producer, followed by New York, Washington, and Oregon. The following presents the three west coast producers from south to north, then New York State.

CALIFORNIA

California wines have, by far, the highest and most consistent quality of all American states. There are about 300 000 ha under vine, of which half is used for winemaking. About 800 winemakers produce 1836 million bottles of wines a year. The Gallo winery alone produces the equivalent of the yearly output of Chile.

GEOGRAPHY

In this sun-drenched land, grape growing is not a problem as long as there is sufficient water. During the summer months, there is virtually no rain. On the other hand, there are no extended periods of winter freeze; hail is rare and confined to northern vineyards; and early frosts are a minimal concern (only one crop in two decades might suffer significant loss due to frost).

Overall, climatic conditions in California are quite favourable to wine production, though tending toward the hot side. Unfortunately, California is still affected by the disease common in hot climates, Pierce's disease, as well as by phylloxera. In contrast to Europe, where grape-growers tend to look for the hottest sites, California growers search out the cooler locations so that they can produce better wines: coastal sites, northern and eastern exposures, hillside slopes, and higher altitudes. On warmer sites, grapes mature too fast and yield huge crops with high sugar content, but with little of the acidity and aroma that is needed to make wines of quality and character.

THE CALIFORNIAN CLIMATE

Based on the research of Professors Winkler and Amerine at the University of California, California's vineyards have been categorized into five climate regions based on the number of degree days for each area.

- **Region One**—cool regions with less than 2500 degree days: Napa, Hollister, Mission San Jose, Saratoga, Bonny Doon, Guerneville, Santa Rosa, Sonoma, Lompoc, Watsonville, Campbell, Aptos, Santa Cruz, Gonzales, Hayward, Peachland, and Santa Maria.
- **Region Two**—moderately cool regions with 2501–3000 degree days: Rutherford, St. Helena, Glen Ellen, Healdsburg, San Jose, Los Gatos, Santa Barbara, Gilroy, Sebastopol, San Luis Obispo, Soledad, San Jose, Grass Valley, Napa, Sonoma, and Placerville.
- **Region Three**—warm regions with 3001–3500 degree days: Calistoga, Hopland, Cloverdale, Livermore, Alpine, Ukiah, Paso Robles, Pinnacles, Cuyama, Santa Ana, King City, St. Helena, Healdsburg, Clear Lake Park, Jamestown, Camino, Mokelumne Hill, Potter Valley, Ramona, Mandeville Island, and Lodi.
- **Region Four**—moderately hot regions with 3501–4000 degree days: Davis, Manteca, Modesto, Ontario, Martinez, Escondido, Upland, Suisun, Colfax, Turlock, Linden, Vacaville, Sacramento, Clarksburg, Sonora, San Miguel, Fontana, Pomona, Stockton, and Auburn.
- **Region Five**—hot regions with more than 4000 degree days: Madera, Fresno, Delano, Visalia, Bakersfield, Chico, Red Bluff, Redding, Ojai, Oakdale, Brentwood, Antioch, Woodland, and Reedley.

The most favourable sites have the Pacific coast to moderate the effects of the sun. From early spring to late autumn, cool ocean waters and warm air create banks of fog that linger over coastal areas. In the morning, as the sun warms the inland areas, warm air rises to create a low-pressure system that draws in cooler air and fog from the sea.

This cyclical effect is literally an air-conditioning system, moderating the hot rays of the sun and improving the acid levels in the grapes. In the Sonoma region, the morning fog cools the vineyards as regularly as clockwork.

California's wide range of soils (chalk, clay, limestone, loam, volcanic) add character to the wine.

HISTORY

Baja means "low," Alta means "high."

Vitis vinifera was first brought from Mexico to Baja California and Alta California by Spanish priests of the Franciscan Order. They built a chain of missions, each with its own little vineyard to supply wine for Mass. Padre Junipero Serra is credited with planting the vineyard of Mission San Diego in 1769. The first winery was established in 1775 at Mission San Gabriel, where the famous Trinity wine was to flourish for over a century. From Mission San Gabriel, settlers later left to establish the Pueblo of Los Angeles. The grape they planted is now known in California as Mission, as Pais in Chile, and as Criolla in Argentina.

The first professional wine-grape grower in California was a native of Bordeaux with a most appropriate name: Jean Louis Vignes. His 95-ha vineyard in the heart of present-day Los Angeles lay just west of the Los Angeles River. He was the first grower to import and to experiment with new European varieties, and to age his wine before its release. His famous Aliso wine took its name from a lofty sycamore towering over the entrance to his vineyard. In later years he was known as Don Louis del Aliso. In 1855, he sold his property to his nephew Jean-Louis Sanssevain who set out to produce "champagne" to sell to the thousands of people enriched by the Gold Rush of 1849, and still arriving.

Many French grape growers and winemakers became attracted to California after hearing of its rich soil and balmy climate: Bontemps, Prevost, Peltier, Thée, Charles Lefranc, and Paul Masson all settled here during the nineteenth century. They brought with them Pinot Noir, Chardonnay, Cabernet Sauvignon, Malbec, and Sémillon vines.

In 1868, the University of California was founded at Berkeley. Soon after, the University's Eugene Hilgard became the driving force behind the effort to classify California's grape-growing regions according to their average temperatures.

He provided grape growers with analytical reports on the nature of vineyard soils and helped them to select the most suitable sites for specific grapes. The university continues its viticultural research at Davis, a small city at the north end of the Central Valley.

Many colourful characters took part in the growth of the California wine industry.

- The first Franciscan parishioner to become a wine producer was Governor Pedro Fages, who, in 1783, planted a vineyard near his residence in Monterey.
- Dona Marcelina Felix Dominguez planted 1800 vines at Montecito near Santa Barbara before her death in 1865. Better known as La Vieja de la Parra Grande ("The Grand Old Lady of the Grapevine"), she is reputed to have lived to the age of 105 and to have cultivated one vine that was to yield an incredible 4064 kg grapes in a good year.
- Two German musicians, Kholher (playing flute) and Frohling (fiddle) established the Anaheim Winery. Their wine was sold as far away as New York City, but in 1885 a bacterial infection known as Anaheim disease spread over the region, pushing grape growing further north. This disease later became known as Pierce's disease.
- Another towering figure in the industry was a Monterey-born General Mariano Guadalupe Vallejo, who at a mere twenty-six years old, became commander-in-chief of the Mexican army in California. He eventually settled in Sonoma and became a gentleman farmer. In 1846, a group of so-called "Yankis" kidnapped Vallejo, the then-governor of California, locked him up, and raised the Bear Flag to declare California a republic. Vallejo accepted the declaration. During the Gold Rush, Vallejo saw opportunities in providing wine and brandy to miners and the nouveau riche. He established new vineyards in Sonoma and was the first large-scale, non-missionary, wine producer of the region.
- Agoston Haraszthy is considered by many the father of the modern California wine industry. Haraszthy left Europe in 1840, arrived in Wisconsin, and trekked to California during the Gold Rush to try his hand at many ventures. In 1852, he purchased 182 ha near San Francisco, called it Los Flores, and planted vines from his native Hungary. Two of these vines are reputed to be the original Zinfandel and Muscat of Alexandria varieties now so ubiquitous in the state. Later, he moved to Sonoma and established Buena Vista Vineyards, in the foothills of the Mayacamas Mountains, not far from Governor Vallejo's vineyard. In 1858, the Colonel asserted that he had 165 successful varieties of European vines at Buena Vista. In 1861, he was elected president of the State Agricultural Society. The same year, he imported and planted more than 100 000 European vines. When he lost his bid to become state governor, Haraszthy departed for Nicaragua, where he started a sugar-cane plantation and built a distillery. The vines he brought to California multiplied and flourished until the advent of phylloxera, although energetic early grafting minimized the damage.

As the leader and generally the trendsetter in the American wine industry, California's history in recent years is much the same as the nation's. Relatively unique to California is the impact of irrigation. Whereas many inland valleys of the state once produced wheat, irrigation has enabled them to produce other crops, including wine grapes.

GRAPE VARIETIES

California has over 133 550 ha of wine grapes—more land than for any other fruit. The top white variety is Chardonnay (24 270 ha) and the top reds are Cabernet Sauvignon (13 990 ha) and Zinfandel (13 817 ha).

Of the fifty varieties grown in California, some names are peculiar to the state. They include the following:

- Emerald Riesling—a hybrid (cross of Riesling and Muscadelle) created by the University of California at Davis. It produces white wine with a fresh and tangy taste, but which does not age well and tends to oxidize and acquire a darker colour at any age.
- Grey Riesling—not a true Riesling, but the Chauché Gris of France. It yields a soft and mild wine with a sweet, spicy flavour.
- White Riesling/Johannisberg Riesling—the true Riesling of the Mosel and Rhine regions of Germany. This variety is distinct from other Californian wines labelled Riesling, many of which are made from Sylvaner or Franken Riesling.
- Sauvignon Blanc—the genuine Sauvignon from the Loire Valley, sometimes referred to as Sauvignon Blanc Fumé or Blanc Fumé. It is not the Sauvignon Vert, which is used in blending and in the production of wines of below-average quality.
- Red Pinot/Pinot St. George—a variety from the south of France, believed to be a clone of the Gamay, but no relation to Pinot Noir.
- White Pinot—on a label, usually indicates that the wine is produced from Chenin Blanc, not Pinot Blanc.
- Gamay Beaujolais—not a true Gamay, but a clone or subvariety of Pinot Noir.
- Petite Syrah—not related to Syrah, but to the Durif grape grown in the southern Rhône region of France.
- Ruby Cabernet—a hybrid (cross of Cabernet Sauvignon and Carignane) produced by the University of California at Davis in 1968.

WINE PRODUCTION

California is home to more than eight hundred wineries in five wine-growing areas: the North Coast, the Central Coast, the South Coast, the Central Valley, and the Sierra Foothills.

Californian AVAs

Within the growing areas, the following are the AVAs listed by county. Some AVAs spread over more than one county.

North Coast AVA

From Humbolt County to San Francisco, this area produces the largest quantity of quality wines.

- Humboldt County—Willow Creek
- Lake County—Benmore Valley, Clear Lake, Guenoc Valley, North Coast
- Marin County—North Coast
- Mendocino County—Anderson Valley, Cole Ranch, McDowell Valley, Mendocino, Potter Valley, North Coast
- Napa County—Atlas Peak, Howell Mountain, Los Carneros, Mount Veeder, Napa Valley, Oakville, Rutherford, Spring Mountain District, Stag's Leap District, Wild Horse Valley, North Coast
- Solano County—Solano County-Green Valley, Suisun Valley, Wild Horse Valley, North Coast
- Sonoma County—Alexander Valley, Chalk Hill, Dry Creek Valley, Knights Valley, Los Carneros, North Coast, Northern Sonoma, Russian River Valley, Sonoma Coast, Sonoma County-Green Valley, Sonoma Mountain, Sonoma Valley
- Trinity County—Willow Creek

Central Coast AVA

This area runs from San Francisco to Santa Barbara County.

- Alameda County—Livermore Valley, Central Coast
- Contra Costa County
- Monterey County—Arroyo Seco, Carmel Valley, Chalone, Monterey, San Lucas, Santa Lucia Highlands, Central Coast
- San Benito County—Chalone, Cienega Valley, Lime Kiln Valley, Mount Harlan, Pacheco Pass, Paicines, San Benito, Central Coast
- San Luis Obispo County—Arroyo Grand Valley, Edna Valley, Paso Robles, Santa Maria Valley, York Mountain, Central Coast
- San Mateo County—Santa Cruz Mountains
- Santa Barbara County—Santa Maria Valley, Santa Ynez Valley, Central Coast
- Santa Clara County—San Ysidro District, Santa Clara Valley, Santa Cruz Mountains, Central Coast
- Santa Cruz County—Ben Lomond Mountain, Santa Cruz Mountains, Central Coast

South Coast AVA

This area stretches from Ventura County to San Diego.

- Los Angeles County—South Coast
- San Diego County—San Pasqual Valley
- Riverside County—Temecula

Central Valley AVA

This area covers the valley between Sacramento and San Benito County.

- Fresno County—Madera
- Madera County—Madera
- Sacramento County—Lodi
- San Joaquin County—Lodi
- Yolo County—Clarksburg, Dunnigan Hills, Meritt Island

Sierra Foothills AVA

This is the western side of the Sierra Nevada Mountains.

- Amador County—California-Shenandoah Valley, Fiddletown, Sierra Foothills
- Calaveras County—Sierra Foothills
- El Dorado County—California-Shenandoah Valley, El Dorado, Sierra Foothills
- Mariposa County—Sierra Foothills
- Nevada County—Sierra Foothills
- Placer County—Sierra Foothills
- Tuolumne County—Sierra Foothills
- Yuba County—North Yuba, Sierra Foothills

Among the great number of AVAs in California, the following include the most important.

North Coast

Mendocino County

There are six main grape-growing areas, totalling 4860 ha of vineyard plantings. Warm summer afternoons allow the grapes to ripen slowly, and cool nights help the grapes retain their acidity (thus their fruitiness). These narrow coastal valleys and ridges are covered by rocky, well-drained soils. The cool Anderson Valley grows Gewürztraminer, Pinot Noir, and Chardonnay. Cole Ranch, a tiny area nestled between Ukiah and Boonville, produces excellent Cabernet Sauvignon and Johannisberg Riesling. East of the town of Hopland, the McDowell Valley yields good Syrah, Cabernet Sauvignon, and Zinfandel.

The larger Ukiah Valley is better known for its Chardonnay, Sauvignon Blanc, and Zinfandel. Redwood Valley and Potter Valley specialize in Chardonnay and Riesling.

Lake County

This region around Clear Lake registers the greatest differences between daytime and nighttime temperatures on the North Coast. The Mayacamas Mountains shield the county from coastal fog and excessive precipitation during the autumn, making this area ideal for producing grapes with good acidity and fruit. Most of the valley's vineyards are above 400 m. They cover an area of 1215 ha and have a wide range of soils.

Napa

This is probably the best-known name worldwide for Californian wines. Napa Valley ("valley of plenty") has 13 770 ha of vineyards that yield only 4% of California's entire wine output. The low average yield of 9.8 t/ha points clearly to quality rather than quantity. The area is entirely dedicated to grape growing and has 200 wineries dotting either side of the well-touristed Highway 29. With ten different soil types and five general climate zones, the area produces an unlimited range of wines. Note that a Californian state law requires that all wines labelled with a subarea of Napa Valley must also be labelled *Napa Valley*.

Most of California's grape varieties are produced here, but the most celebrated wines are made from either Cabernet Sauvignon or Chardonnay. Sauvignon, Pinot Noir, Merlot, and Chenin Blanc also can yield very good results for top winemakers. Some Bordeaux blends (Meritage) have achieved marvellous results in a dark and very fruity Californian-wine style.

Mount Veeder's 400 ha of vineyards in the Mayacamas Range are located at altitudes of 120 m–800 m. They produce intense Cabernet Sauvignon and Chardonnay (40 hL/ha). The same can be said about the Stag's Leap District, Diamond Mountain, and Howell Mountain, which also yield the most intense Zinfandel. Atlas Peak is unique in its use of the Sangiovese grape variety. Rutherford and Oakville are known for their herbaceous and tannic Cabernet Sauvignons.

Sonoma County

Because it is here that Agoston Haraszthy planted his imported European vines, Sonoma can be considered the birthplace of the modern California wine industry.

Sonoma is two-and-a-half times the size of Napa. But with about 14 000 ha of vineyards, it only just equals Napa's vine coverage, because Sonoma also grows other fruit, and flowers.

Annually, the 125 wineries produce all the major varieties. Each of the twelve AVAs has a unique situation in terms of soil and climate, although this variety does not always show in the wines. Chalk Hill produces interesting Chardonnay and Sauvignon on volcanic ash. The cool Green Valley is virtually dedicated to Chardonnay and Pinot Noir. Sonoma Valley yields good Cabernet Sauvignon and Merlot as well as some deeply flavoured Zinfandel grown on the upper slopes that border the valley. Sonoma Mountain produces rich Cabernet Sauvignon, Zinfandel, and Sauvignon Blanc, from only 145 ha of vineyards located above the morning fog at 440 m. Russian River Valley excels in producing flinty Riesling, and good Chardonnay and Pinot Noir. The Alexander Valley's rich soils yield reliable botrytis-affected Gewürztraminer and Riesling, crisp Sauvignon Blanc, and chunky Chardonnay. Dry Creek Valley succeeds best in catching the latest trend for Rhône-style wines, as well as in showing off certain Italian varieties; the Zinfandels are deeply coloured and spicy, while traditional Bordeaux grapes perform well in the gravelly soils.

Carneros

The Carneros area gets its name from a Spanish word meaning "ram"; it was once a ranch where sheep grazed on the scanty grass growing on the hills' thin soils. This area straddles both southern Napa and Sonoma counties, along the north shore of San Pablo Bay. The cool and dry climate is ideal for growing delicate Chardonnay and strawberry-scented Pinot Noir. Merlot is just beginning to be planted in substantial amounts and produces some good quality wines.

Central Coast

Livermore Valley

The valley is 25 km long and 16 km wide, and runs east to west in Alameda County. It was wine from this valley that gained California the gold medal at the International Exposition in Paris in 1889. As a result of that win, many European winemakers immigrated here. The valley is extremely favourable to grape growing. Proximity to San Francisco Bay creates an evening fog that lasts through to the morning; the sun burns off the fog each day at around noon; and finally cool marine air rushes east in the late evening to start the cycle again. Deep gravelly soil is favourable to Bordeaux varieties. Cabernet Sauvignon, Sauvignon Blanc, and Sémillon achieve magnificent results.

Santa Cruz Mountains

This AVA was created in 1982. Located 80 km south of San Francisco, this coastal region has about 80 ha of vineyards along the ridge of the low (1000 m) mountain range. The east side of the range is dedicated to Cabernet Sauvignon and the cooler west side (facing the Pacific Ocean) is planted mainly with Pinot Noir, Chardonnay, Riesling, and Zinfandel. The area produces wines of consistently high quality.

Monterey County

This coastal region has more than 13 400 ha of vineyards dominated by plantings of white grape varieties such as Chardonnay, Chenin Blanc, and Johannisberg Riesling. Smaller appellations such as Chalone, Arroyo Seco, and Carmel Valley succeed well with Pinot Noir, Merlot, Chardonnay, and Riesling. The arid climate makes most vineyards totally dependent on irrigation. This must be done cautiously— Cabernet Sauvignon, for example, tends to produce vegetal and grassy wines when the grapes have received too much water. When the irrigation is reduced (or stopped) by the end of June, this grape yields wine with better aromatics and more character. Monterey's sandy soils discourage the spread of phylloxera.

San Luis Obispo County

This area of 3868 ha of vineyards is divided into five AVAs. The two largest are Paso Robles and Edna Valley. Paso Robles is particularly known for its dark, spicy Zinfandel and soft, fruity Cabernet Sauvignon and Sauvignon Blanc; Edna Valley yields rich, opulent Chardonnay. Arroyo Grand Valley produces high quality méthode Champenoise wine, and York Mountain makes intensely concentrated Zinfandel and dark Cabernet Sauvignon. The Santa Maria Valley, which spills across into Santa Barbara County, produces some decent red wines but is better equipped for making whites.

Santa Barbara County

As happened elsewhere in California, the thriving early Mission vineyards of Santa Barbara were left to go wild after Prohibition. But in 1962, a Canadian named Pierre Lafond rekindled the local wine business. From his 4.5-ha plot, the vineyards have expanded to 4000 ha run by twenty-five wineries. The region is noted for its large production of rich Chardonnay, unctuous late-picked Johannisberg Riesling, distinctive Pinot Noir, and some pleasant Chenin Blanc and Sauvignon Blanc.

South Coast

Los Angeles County had its first commercial winery in 1824. However, Pierce's disease, ceaseless urban expansion, and the development of the cooler Napa and Sonoma wine regions diminished this area. Now the 8000 ha of vineyards produce mainly table grapes, grapes for the production of sweet wine and fortified wine, and grapes to be used in the distillation of brandy. The Temecula AVA produces Chardonnay on sandy soil. San Pasqual Valley AVA, northeast of San Diego, is hardly 40 ha.

Central Valley

San Joaquin Valley, 320 km long and 80 km wide, is the northern half of the vast Central Valley. Prior to the introduction of irrigation, this flat, hot, fertile region produced mainly wheat, but now produces vegetables, fruit, and wine grapes. From a total of almost 71 000 ha of vineyards, this area produces almost 80% of the total wine output of the state. Colombard and Chenin Blanc plantings (32 500 ha) imported from France dominate the vineyards. Zinfandel, Grenache, Barbera, Palomino, Tinta Madeira, and Mataro (known locally as Mondeuse) produce a huge amount of dessert wine, as well as some reliable everyday wines.

Sierra Foothills

In the late nineteenth century, this area accounted for over a hundred wineries, but phylloxera and then Prohibition devastated this remote region of California. During the wine boom of the late 1960s, dilapidated vineyards were once again cultivated. Better transportation options and an appreciation for the cooler areas brought about the rebirth of the Sierra Foothills' viticulture; growth continues today. (Real estate is more affordable here than in the rest of the North Coast.) The summers are dry and sunny, with cool nights due to the altitude of the vineyards (450 m–910 m). During the winter, most vines are tucked under the snow, minimizing the damage from frost. The result is robust and lushly-scented Zinfandel, Cabernet Sauvignon, and Barbera. Chardonnay, Sauvignon Blanc, Muscat, and Chenin Blanc also yield remarkable wines. If Nebbiolo is ever to be grown successfully in California, this will likely be the place.

OREGON

Situated north of California and south of Washington, this coastal state attracts international attention especially for its interesting Pinot Noirs. Considering the short history of vinifera grape cultivation in this state, Oregon's achievements are impressive. Following the pioneering efforts in 1963 of Hillcrest (near the town of Roseburg), the early 1970s brought an increase in investment, new wineries, and new plantings. Now 122 wineries cultivate 2400 ha of vineyards.

GEOGRAPHY

Most of the areas planted are located between the Pacific Ocean and the west side of the Cascade Mountains. The climate is cool, damp, and somewhat marginal. The best territories have a climate similar to that of Burgundy—which explains the keen interest of Burgundian houses such as Domaine Drouhin, and even Laurent Perrier from Champagne. The harvest occurs late in the season to get maximum ripeness and maturity.

WINE PRODUCTION

The cool climate is best suited to the production of fine Pinot Noirs, as well as some Chardonnays and Rieslings. Many of these wines have a refreshing crispness, fine fruit, and good complexity.

- Pinot Noir—the celebrity grape in Oregon. The wines vary from light coloured to deep toned, and exhibit a full range of aromas from black cherry to spice to floral. Unlike many other non-Burgundian Pinot Noir wines, these have a good level of heartiness. When left unfiltered, they can be very deep, with lots of extract and an intense, smoky bouquet unaffected by high alcohol. There are many styles, but the majority are attractive wines similar to the good reds of Beaune in France.
- Chardonnay—has great potential but is frequently given too much barrel ageing. The best wines undergo some malolactic fermentation to lower the acidity without losing freshness. These wines display fine aromas of ripe apple, spice, and vanilla when not masked by excessive oaking.

- Pinot Gris—most likely to become the star of the wine industry in Oregon. It was planted in tiny quantities, but many winemakers now recognize its potential, and are busy expanding their Pinot Gris vineyards. Good examples achieve deep fruit flavours with sufficient acidity to complement the nutty and smoky character.
- Cabernet Sauvignon and Merlot—prefer a warmer climate. As a result, the local grapes tend to lack character, though the best can be found near Oregon's border with California.

A few good dry Riesling wines and Müller-Thurgau wines can be found, along with rare amounts of good Gewürztraminers.

Oregon's five AVAs are as follows:

- Willamette Valley—south from Portland to Eugene. This area accounts for three-quarters of Oregon's wine production.
- Umpqua Valley—extending south of the Willamette Valley around the town of Roseburg.
- Rogue Valley—south from Grants Pass and Medford to the border with California.
- Columbia Valley and Walla Walla Valley—extend from the state of Washington.

WASHINGTON

North of Oregon and bordering Canada, this state is the fastest-growing grape region in the United States. It already ranks third after New York State, producing 3.6 million cases of wine. The majority of the total 13 900 ha of vineyards are in the Columbia River Basin. However, only a portion of the state's grapes are used for local wine production. Many are sold to California wineries, or exported across the Canadian border to British Columbia. Fully two-thirds of the state's cultivation is given over to Concord grapes for the production of grape juice.

GEOGRAPHY

Washington's grape-growing regions are found inland. The best sites are located on south-facing slopes near the Columbia and Yakima rivers, which moderate the extremes of temperature.

The climate of these regions is strictly continental, with hot summers and cold winters. Precipitation levels are relatively low, so irrigation during the growing season is necessary. The regions' cool valleys have relatively few spring frosts, and grape crops typically have an adequate balance of sugar and acidity.

Vitis vinifera vines must be hardened off to survive extreme winter temperatures; the technique is widely practised and the degree of winterkill has been minimized.

GRAPE VARIETIES

With the exception of Zinfandel, Washington state grows the same varieties as California, plus a red variety of German origin called Lemberger. This is usually produced in a Beaujolais style, with a great deal of fruitiness and freshness. Lemberger wines should be drunk young.

WINE PRODUCTION

Of the white varieties, Chardonnay and Riesling tend to steal the show. They develop good acidity and generally ripen fully. Chardonnays (when barrel fermented and given extended lees contact and full malolactic fermentation) are very likeable, although they can lack the full body of the Burgundies they attempt to emulate. Muscat and Chenin Blanc wines are produced in the sweet style with very good results. They retain enough acidity, natural sweetness, and aromatic depth to make them refreshing and pleasurable. Sauvignon Blanc seems to have trouble bringing out its grassy character; this weakness is aggravated by the many winemakers who choose to barrel age the wine.

Of the red varieties, Cabernet Sauvignon and Merlot are the stars. Considering that these vines are still quite young and that most winemakers are still learning to work with them, attempts to develop the best character in these wines have been most encouraging. They are all still a bit thin, herbaceous, and lacking some of the concentration that should come once the vines are older. The best Cabernets to date are produced in the eastern part of the state. They are deeply coloured, with plenty of extract and the classic plum, black currant, and sometimes slightly grassy aroma. Merlot is generally lower in acidity and tannin than Cabernet Sauvignon. The fussy Pinot Noir has yet to yield a wine of any significance.

Washington has three AVAs :

- Columbia Valley—the major producing region, with 2500 ha and more than half of all vinifera plantings. This arid region gets only 200 mm annually of rainfall, so requires irrigation. The hot summer provides the right conditions for grapes to fully ripen. This AVA extends into Oregon.
- Yakima Valley—located where the Yakima River joins the Columbia River, west of the town of Tri-Cities, the area consists of 1650 ha of vineyards. This AVA extends into Oregon.
- Walla Walla Valley—a smaller area of only 45 ha located between Tri-Cities and Walla Walla, south of the Snake River.

NEW YORK

This northeastern state is the second-largest wine producer in the United States. There are about 15 400 ha of vineyards from which 94 wineries annually produce an average of 1 140 000 hL of wine. The state's vines produce not only wine (45%), but also grape juice (50%), and table grapes (5%).

HISTORY

In the recent history of New York's wine industry, two periods stand out. In the 1950s, two European immigrants, Charles Fournier and Dr. Konstantin Frank, proved that viniferas could be grown successfully despite New York's harsh climate. Also, in 1976, the Farm Winery Act was passed: it lowered licence fees and allowed a farm winery's annual production to be increased from 1900 hL to 5678 hL. This act made it much more economically feasible to own and operate wineries, and so triggered a dramatic expansion—almost overnight, the number quadrupled.

GRAPE VARIETIES

New York is well known as a producer of the native Concord grape, which accounts for half of the total production. Fortunately, this grape is used mainly for juice, not wine. Increasingly, viniferas and hybrids are replacing it.

The history of the Lake Erie AVA is directly connected to the history of the temperance movement that culminated in Prohibition. In 1818, a local Baptist deacon named Elijah Fay planted the first vineyards in western New York with wild grapes he had brought from New England, then replaced them with Isabella and Catawba grapes. In 1859, his son opened the first winery. In 1835, the prohibitionist movement was advocating total abstinence. Chautauqua County, where Fay lived, became the centre of it all.

Prohibitionists Charles and Thomas Welch exhorted farmers to grow grapes for table consumption or grape juice. Thus, the Concord grape proliferated and caused the Lake Erie district to become a major region for juice and table grapes, instead of wine.

WINE PRODUCTION

The state of New York has six AVAs: Lake Erie, Finger Lakes, Cayuga Lake, Hudson River Region, North Fork of Long Island, and the Hamptons, Long Island.

Lake Erie

The Lake Erie viticultural area was established in 1983, and is home to eight wineries. It is a multi-state AVA, covering the most western portion of New York State and extending into Pennsylvania and Ohio; the area covers 10 117 ha. The New York State portion includes the counties of Erie, Cattaraugus, and Chautauqua. Of the 8094 ha of vineyards, most (90%) are planted with Concord grapes. Lake Erie moderates the extremes of this northern region especially well within a 5 km-wide band between the lakeshore and the Allegheny Plateau. The lake's warming effect in winter and cooling effect in summer are trapped by the sudden rise in elevation to the plateau. The chances of late spring and early autumn frosts, thunderstorms, rainfall, and fog are also reduced. The result is a grape-growing season that extends to 200 days. Within this narrow band the soils are mainly gravelly loam. The dominant viniferas are Riesling and Chardonnay.

Finger Lakes

South of Lake Ontario and southeast of Rochester, this region centres on a group of deep, slender, finger-shaped lakes: Canandaigua, Keuka, and Seneca. The second largest grape-growing area in the state, its 5666 ha of vineyards are spread over the counties of Livingston, Monroe, Wayne, Seneca, Ontario, and Yates, as well as portions of Tompkins, Schuyler, Steuben, and Cayuga counties. The area has forty-five wineries, which produce 85% of New York's output; it is home to the Canandaigua Wine Company, the second-largest winery in the United States.

Wine production started in the 1820s, and it was here in 1953 that the first successful vinifera planting (Riesling) established itself. In 1961, the area produced the first botrytis wine made from vinifera grapes. The Finger Lakes viticultural area was formed in 1982.

The lakes moderate the local climate to give a 200-day growing season. In the autumn, dense cold air slides down from the steep slopes to the lakes, warms up, and rises to create a convection current. In the spring the process is reversed and the cool air retards budding until all danger of frost has passed. Above the lakes, the shallow topsoil on slopes of shale beds provides good drainage.

All types of wine are made in the area, including icewine and sparkling wine.

Cayuga Lake

East of the Finger Lakes AVA, the Cayuga Lake AVA is similar to the Finger Lakes in terms of the effect of the lake, topography, and soil. However, this area's lower elevations and deeper lake (133 m) create an even better situation for the growing of wine grapes.

Known in the past as the Cayuga Wine Trail, this area covers parts of the counties of Seneca, Tompkins, and Cayuga. The first wineries were not established until 1980; the viticultural area was established in 1988. It is now home to nine wineries with eighteen vineyards.

Hudson River

The Hudson River runs from upstate New York south to New York City. The AVA consists of approximately 9065 km^2 and encompasses all of Columbia, Dutchess, and Putnam counties, the eastern portions of Ulster and Sullivan counties, nearly all of Orange County, and the northern portions of Rockland and Westchester counties.

The wineries are located along the river and east toward the Connecticut border. Those along the Hudson River benefit from its moderating effect: the steep embankments create a funnel, channelling the warm ocean breezes from the Atlantic. The growing season averages 180 to 196 days. The geological formation is diverse: soils consist mainly of glacial deposits of shale, slate, schist, and limestone.

The Hudson River Region is the oldest wine-growing district in the United States. Wine has been made here continuously for over 300 years, since the Huguenots first settled at New Paltz in 1677. Brotherhood America's Oldest Winery, Ltd. (established 1839) is the oldest winery in the United States.

The Hudson River Region was formed in 1982. Today, twenty wineries operate 405 ha of vineyards. French-American hybrids, mainly Seyval Blanc, are the mainstay of the AVA. Chardonnay, Riesling, and Cabernet Franc are the key viniferas.

Long Island

Long Island has two AVAs, lying on the Atlantic coast northeast of New York City. Both are planted entirely with vinifera. Chardonnay, Merlot, and Cabernet Sauvignon are the predominant varieties. A few Riesling, Sauvignon Blanc, and sparkling wines (using the méthode Champenoise) are made. The two AVAs combined produce almost two million bottles of wine annually.

Long Island wines have lately displayed some remarkable and consistent qualities. As many as ten winemakers have shown an interest in establishing wineries in the future.

- North Fork of Long Island—this area covers 410 km^2 in Suffolk County and was established in 1986. Now twelve wineries cultivate 405 ha of vineyards in the townships of Riverhead, Shelter Island, and Southold. Ocean breezes moderate the conditions on this 10 km-wide peninsula, reducing the risk of early and late frosts. The Atlantic also moderates the average daily temperature and increases the amount of rainfall. The growing season lasts 233 days. Soils in North Fork have less silt and loam than in South Fork and require some irrigation, but are more fertile.
- The Hamptons, Long Island—located on a peninsula in Suffolk County and the townships of Southampton and East Hampton, the area covers 552 km^2. This viticultural area was established in 1985. The 22 ha of vineyards are shared by two wineries. As in North Fork, the climate is regulated by the Atlantic Ocean. Spring and autumn temperatures are lower than elsewhere on Long Island. The summer is very sunny with average precipitation. Spring fogs prevent early budding or damage from late spring frosts. The deep, fertile soil is composed of silt with a fine, sandy loam subsoil. This composition provides ample water retention and no irrigation is necessary. The growing season averages 182 days.

CANADA

As a wine producer, Canada is still young. Although a thousand years have passed since the first wine was made on Canadian soil, the industry looks back less than two centuries to the earliest commercial endeavours, and to the 1980s for the establishment of quality-driven, self-disciplinary measures. With the formation of national wine standards in 1996, international acceptance is on the horizon. Annual production is 250 000 hL, and per capita consumption is 11 L.

GEOGRAPHY

Canada is the second-largest country on earth, stretching more than 5500 km from east to west, and nearly 4600 km from near the North Pole to its southernmost point near the 42nd parallel at Pelee Island, Ontario. The United States borders it to the south and (with Alaska) to the northwest.

The vast majority of the country is unsuitable for the growing of grapes. Winter temperatures (Canada is also the second-coldest country on earth) dip below –30°C, guaranteeing winterkill of even the hardiest native varieties. In the Eastern Townships of Québec, for example, it gets so cold that growers must hill the soil over their vines every year to help them to survive.

Regions suitable for growing viniferas and hybrids exist in places moderated by large bodies of water: the southern coastal zones of the Atlantic and Pacific oceans; some inland zones around the Great Lakes; and near the deep Lake Okanagan in the British Columbia interior.

Ontario's main wine-growing regions have mostly heavy soils. Dense clays predominate, requiring the installation of extensive water-drainage systems. Exceptions are the sloped vineyards at the base of the Niagara Escarpment, and the many scattered pockets of loam, sand, or pebbles near the lakeshore.

British Columbia's Okanagan Valley, on the other hand, has soils that tend to be loose—predominantly sand and sandy loams. Here, what little rainfall there is drains away quickly, necessitating irrigation for all vines.

Wine Regions of Canada

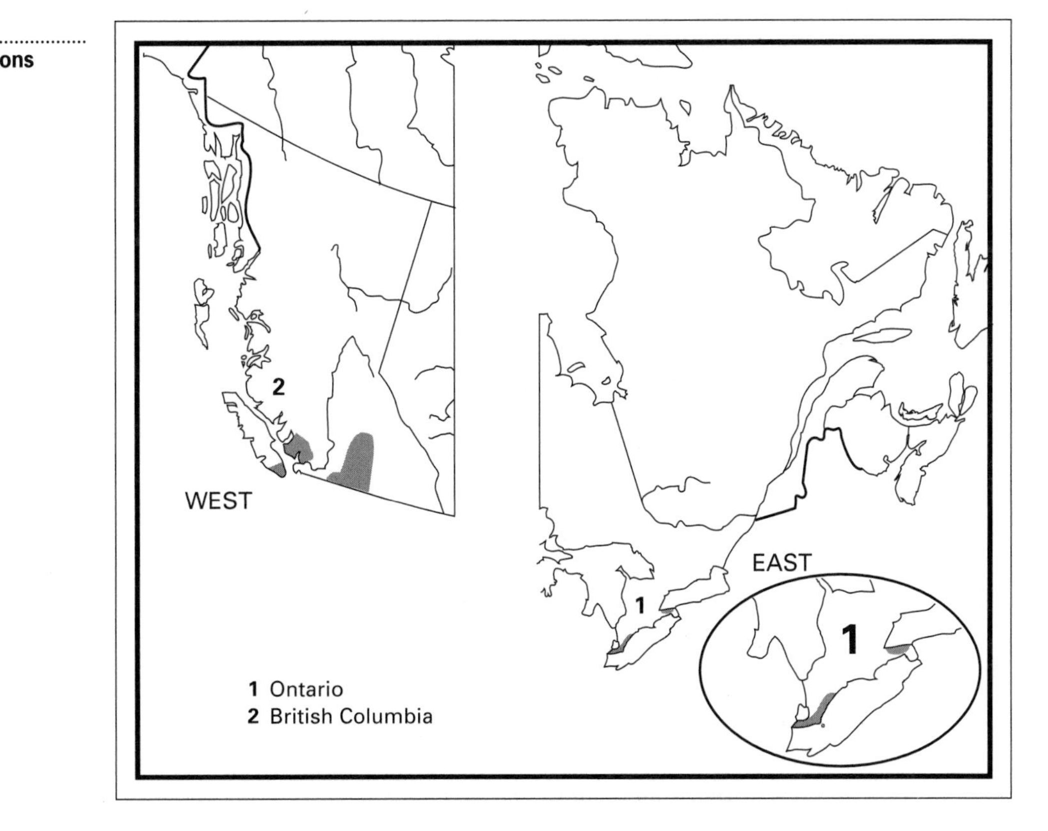

HISTORY

The first Canadian wine could have been made by none other than the Norse explorer Leif "The Lucky" Ericson. He and his crew may have landed in Newfoundland, around A.D. 1002. Supposedly, they discovered wild grapes growing nearby, and produced a batch of wine to help them survive the winter. They did survive, but left the following spring, never to return. Upon their return to Norway, Ericson described how his expedition had discovered "Vinland."

Five centuries later, French explorer Jacques Cartier reported finding grapes on Île d'Orleans in the St. Lawrence River. Originally, he considered naming it Île de Bacchus (prudent thinking prevailed).

Jesuit missionaries arrived with the early French settlers and they made the first sacramental wine from wild grapes. The flavour of these "holy wines" led to many attempts to cultivate vinifera vines, but harsh winter conditions ensured their failure. Grain was a better bet, so beer became the beverage of the pioneers.

Although vinifera varieties failed to flourish, some survived a winter or two and interbred with native species, creating the first generations of natural French-American hybrids.

By the turn of the nineteenth century, new endeavours using native grapes were undertaken. In 1811, Johann Schiller (a retired soldier) was the first to propagate the wild labrusca vine on a plot of land at the mouth of the Credit River. His commercial successes were notable but limited. The property sold in 1864 to J.M. de Courtenay, who promptly expanded the vineyard and became an outspoken promoter of Canadian wine (i.e., his own).

Foreign investment came in 1866, when a group of Kentucky businessmen bought land on Pelee Island in Lake Erie, and established Vin Villa Wines. Other small operations began to open in its tracks and the local wine industry was on its way.

Around the same time, the first British Columbia vineyard was coming to fruition. Father Charles Pandosy and his Oblate Fathers planted several acres at their mission in the Okanagan Valley, south of Kelowna. The grapes did well, but local farmers opted to focus their efforts on other fruits.

By the 1890s, more than forty wineries had been established in Canada. This did not sit well with everyone—a temperance movement quickly began to take hold.

In 1916, Ontario was the first province to pass a Temperance Act. Eventually, all the provinces passed legislation for some form of prohibition. Québec, for example, restricted access to distilled spirits, but continued to allow the sale and public consumption of wine and beer. In the 1920s, most temperance laws were repealed, at which time governments replaced them with provincial Liquor Control Boards.

In 1975, the licensing of Inniskillin—Canada's first "cottage" winery—in Ontario's Niagara Peninsula launched a revitalization of the industry that continues today. At first, just a handful of determined local grape farmers and passionate winemakers took up the challenge. But by the beginning of the 1980s, foreign money started trickling in.

For decades, experts had claimed that only labrusca and certain hybrid varieties could survive Canada's severe winters. In 1978, a young Paul Bosc proved them wrong when he established Château des Charmes and planted the first all-vinifera vineyard in the Niagara Peninsula. In Nova Scotia, Roger Dial created the Grand Pré Winery and proceeded to experiment with some little-known, winter-hardy Russian varieties.

Individual successes came slowly at first. There was a silver medal in Ljubljana, two bronzes in England, and a marketing coup when Bosc's 1982 "nouveau" landed in Paris a few days before the French could release theirs. Then, the new Pelee Island winery produced the first commercial icewine in 1983, while others introduced the first "serious" reds—single-vineyard Merlots and Pinot Noirs.

Few Canadians had the opportunity to taste these limited production wines, and broad domestic respect was still a long way off, but progress was unstoppable. Winemakers quickly mastered modern production techniques and developed skills in barrel fermentation and ageing. The industry took on the task of getting to better understand the country's mesoclimates as well as researching sites and clones best suited to cool-climate viticulture.

In 1988, a consensus of Ontario's estate winery owners established the Vintners' Quality Alliance (VQA). The organization designated three viticultural areas, banned the use of labrusca grapes, prohibited the addition of water, prescribed minimum sugar levels at harvest, and set out basic standards for winemaking and labelling. Two years later, British Columbia followed suit, accepting many of the Ontario guidelines and adding a few of its own.

Then, in 1991, Inniskillin stunned the wine world by winning a Grand Prix d'Honneur at VinExpo in Bordeaux, from among 4100 entries. The dam finally burst. Suddenly, the country's winemakers were being praised everywhere. More foreign money poured in to the industry. A French firm, Jaffelin, agreed to a joint venture with Inniskillin. Even Canadian drinkers began to believe in the quality of wines produced in their own backyard.

In 1996, the first oenology program focussing on cool-climate viticulture was established at Brock University in St. Catharines. In the same year, a national VQA committee agreed on National Wine Standards, providing the first comprehensive set of regulations and practices for quality wine production. Today, Canada's wines are sold in fine wine shops all over the world.

LAWS AND REGULATIONS

Under national VQA rules, geographical indications (appellations of origin) are divided into two categories: provincial and viticultural area.

Provincial

A provincial designation may be indicated on the label if:

- all grapes used in the production were grown in Canada.
- no less than 85% of the wine is from grapes grown in the province indicated.
- the wine meets all provincial VQA standards (which may be more, but not less stringent than national standards.
- the wine is approved by a provincial VQA certification process.

As of 1996, only British Columbia and Ontario were recognized by VQA Canada as provincial designations.

Viticultural Area

A Viticultural Area (VA) may be named if:

- all grapes used in the production process were grown in the province indicated.
- no less than 85% of the wine is from grapes grown in the named viticultural area.
- the wine meets all national and provincial VQA standards.
- the wine is approved by provincial VQA certification.

Other Regulations

Other regulations concerning, for example, sugar levels, allowable categories of wine, and labelling include the following:

- minimum sugar levels—set at harvest for various geographical designations and wine categories. Adjustments may be considered in exceptionally unfavourable years.

MINIMUM BRIX AT HARVEST

For wines to be labelled as indicated, grapes must achieve the minimum Brix levels shown.

Provincial designation	17.0
Viticultural Area designation	18.0
Vin de Curé	20.0 (but 32 Brix at time of pressing)
Late Harvest	22.0
Select Late Harvest	26.0
Botrytis Affected (BA)	26.0
Special Select Late Harvest	30.0
Totally Botrytis Affected (TBA)	34.0
Icewine	35.0

- chaptalization—permitted for all VQA wines except for vin de curé, late harvest, botrytized wines, and icewines. Wines with a provincial designation may be chaptalized by up to 60.5 g/L of sugar or 3.5% alcohol by volume. Wines listed as VA may receive no more than 42.5 g/L or up to 2.5% increase in alcohol by volume.
- sweet reserve—may be added to any VQA wines except for vin de curé, late harvest, botrytized wines, and icewines.
- vin de curé—must be produced exclusively from fresh grapes that are left to dry. Residual sugar and alcohol must result from the natural sugar of the grapes.
- late harvest, select late harvest, and special select late harvest wine—must be produced exclusively from fresh grapes naturally harvested from the vine. Special Select Late Harvest wine may further be labelled *Winter Wine* if harvested at –5°C or lower.

- liqueur wine—a category created to cover wines that are not fortified, but have a natural alcohol level between 14.9% and 20%.
- icewine—must be produced from grapes harvested naturally frozen on the vine and pressed in a continuous process while the air temperature is –8°C or lower. At no time can the grapes begin to thaw, or the wine will be downgraded to the Special Select Late Harvest category and may then be labelled *Winter Wine.*

ICEWINE

Considering that average winter temperatures often dip below –10°C (and for prolonged periods), it is no wonder that Canada has emerged as the world's leading producer of icewine. It is a blessing from nature.

To make icewine, grapes—Vidal Blanc, Riesling and, in some cases, Chenin Blanc or Gewürztraminer—are left on the vine well past normal harvest dates. During the cold months that follow, the grapes hang on the dormant leafless vines, protected from hungry birds with special netting. Once temperatures drop to –8°C (usually in the middle of a December night), the pickers are sent out to harvest the frozen fruit.

Pressing of the frozen grapes yields a very small amount of extremely concentrated juice, which can take several months to ferment to about 11% alcohol. It must retain a high natural residual sugar level of no less than 125 g/L.

As of 1997, the oldest Canadian icewines are not much more than a decade old, so it is not possible to conclude how long they will age. Some experts believe that Vidal icewines peak sooner than those produced from Riesling.

GRAPE VARIETIES

There are ninety-four varieties permitted by VQA National Wine Standards. Of these, sixty-four are Vitis vinifera and vinifera crosses. The other thirty are hybrids, scientifically referred to as interspecific crosses.

Interspecific refers to crosses of more than one species. Crosses of varieties of the same species are called intraspecific.

Vinifera Varieties

Chardonnay, Riesling, Gewürztraminer, Muscat, Pinot Blanc, and Pinot Gris are the primary white varieties cultivated. Cabernet Sauvignon, Cabernet Franc, Gamay, Merlot, Pinot Meunier, and Pinot Noir are the main reds. Whites make up the majority, although reds are growing in plantings.

Hybrid Varieties

Seyval Blanc and Vidal Blanc make up most of the hybrid planting of white varieties. Vidal is favoured for the production of late harvest wines and icewines. Seyval Blanc can be turned into a crisp, bone-dry, Sauvignon-Blanc-style wine but rarely is. Of the reds, Marechal Foch and Baco Noir are the most prevalent varieties grown. Many other red varieties are grown including De Chaunac, Chelois, Chambourcin, and (increasingly) Chancellor.

WINE REGIONS

At present, Canada recognizes seven VAs, of which three are in Ontario and four are in British Columbia.

ONTARIO

The province had almost 7000 ha of land under vine in 1995, with more than 800 growers. In total, Ontario produces 85% of the country's wine. Grapes are cultivated commercially only in areas that benefit from the winter-moderating effects of the Great Lakes. Three viticultural areas are recognized by the VQA, all located in the southern parts of the province.

Lake Erie North Shore

The area takes in the political boundaries of Essex, Kent, and Elgin counties, except for that part of Kent county to the north of the Thames River. Summers can be uncomfortably hot, and winters severe. From time to time, a prolonged cold spell can result in the complete freezing over of Lake Erie. When this happens, all moderating effect is lost and winterkill in the vineyards is widespread.

There are almost 100 ha of vineyards, mostly on Toledo clay, with patches of gravelly or sandy loam. The main grape varieties grown are Cabernet Franc, Cabernet Sauvignon, Chardonnay, Gamay, Gewürztraminer, Lemberger, Marechal Foch, Merlot, Pinot Blanc, Pinot Gris, Pinot Noir, Riesling, Seyval Blanc, and Vidal Blanc.

Niagara Peninsula

This large region consists of the land bounded by the south shore of Lake Ontario, from the Niagara River just below the Falls in the east, to Grimsby in the west, with the Niagara Escarpment naturally forming the southern boundary.

Sub-appellations to reflect the distinct nature of wines within the region are under discussion. Two being considered include the flatlands surrounding the town of Niagara-on-the-Lake, and the sloped vineyards on the "bench" of the Niagara Escarpment.

Ontario's fierce winters and blistering summers are here moderated by the protection of the Niagara Escarpment and the influence of two huge lakes, Ontario and Erie. These in combination generate air movement patterns that create a mesoclimate ideal for tender fruit (1426 degree days). Despite this, most vineyards experience some degree of winterkill each year. Soils are predominantly clay loam, with some sandy and silty areas.

The planting of Vitis vinifera varieties increases annually, although the majority of acreage is planted with hybrids (Seyval Blanc accounted for 10% of the crop in 1995). The main white varieties include Aligoté, Auxerrois, Chardonnay (including a rare Musqué clone), Gewürztraminer, Muscat, Pinot Blanc, Pinot Gris, Riesling, Vidal Blanc, and (most recently) Sauvignon Blanc. The most-cultivated red varieties are Baco Noir, Cabernet Franc, Cabernet Sauvignon, De Chaunac, Gamay, Marechal Foch, Merlot, Pinot Noir, and Zweigelt.

Pelee Island

This tiny island in Lake Erie, 3.2 km from the mainland, is the southernmost point in Canada (at 41.47° N) and has the longest growing season (1591 degree days). The island is relatively flat and the limestone outcropping that supports it is covered mainly with dense Toledo clay and a few patches of sandy loam. There are 200 ha under vine, devoted mainly to Cabernet Franc, Chardonnay, Gamay, Gewürztraminer, Merlot, Pinot Gris, Pinot Noir, Riesling, and Vidal Blanc.

BRITISH COLUMBIA

Canada's westernmost province has fewer than 1000 ha of grapes (about 120 registered vineyards) in commercial production. New plantings are almost exclusively vinifera varieties, although many hybrids are still grown. There are four recognized viticultural areas.

Fraser Valley

This viticultural area is situated to the south and east of Vancouver, and borders the United States. The climate is moderately cool (818 degree days), and the soil loamy over a clay base. There is one winery and a few tiny vineyards around Langley, growing Bacchus, Madeleine Angevine, Madeleine Sylvaner, Ortega, and Optima grapes.

Okanagan Valley

Officially, this viticultural area takes in all the land within the watershed of the Okanagan water basin. The region produces 97% of the province's wine, and 60% of that is grown in the area between Oliver and Osoyoos.

Temperatures fluctuate wildly between daytime highs and nighttime lows.Oliver has an average of 1460 degree days. With only 60 mm of rainfall from April through to the end of September, the valley is classified as Canada's only desert. Irrigation is mandatory for vineyard survival. The valley has mostly sandy soil with some glacial deposits in the higher altitudes.

The main (88%) production is white varieties such as Auxerrois, Bacchus, Chardonnay, Chasselas, Ehrenfelser, Gewürztraminer, Optima, Pinot Blanc, and Riesling. Reds grown include Baco Noir, Cabernet Franc, Chancellor, Marechal Foch, Merlot, Pinot Meunier, and Pinot Noir.

Similkameen Valley

The region consists of the watershed lands of the Similkameen River west of Osoyoos. It has 18 ha of vineyards. Summers are reasonably warm (1440 degree days), but fierce winds blowing up the valley in winter threaten European vines. Granitic sand and rocky glacial deposits make up the soil. There are two wineries located near Keremeos, growing mainly Auxerrois, Kerner, Merlot, Pinot Noir, and Riesling.

Vancouver Island

This island in the Pacific Ocean is 460 km long and 50 km–80 km wide. Its climate is understandably rainy, and cool (787 degree days) even when sunny. The soil is almost entirely heavy clay, with some pockets of sand and pebbles. One constant problem for growers is the wildlife population—birds, wasps, bears, etc. In 1994, one winery lost 60% of its crop to wasps.

OTHER WINE-GROWING AREAS

Québec

The Eastern Townships of Montréal are home to about 100 ha of vineyards shared by fifteen tiny estate wineries. The climate is so cold that even hardy hybrid vines must be buried to survive the icy winters. Production is almost exclusively white, with Seyval Blanc being the most-cultivated variety.

The Atlantic Provinces

In Nova Scotia, about forty growers supply almost exclusively hybrids to two wineries in the Annapolis Valley, and one located on the northeast shore. Grand Pré winery is the oldest, and has attracted attention with a wine made from the Russian varietal Michurinetz.

There is one grape and fruit winery overlooking the Northumberland Strait on the east coast of Prince Edward Island, and one berry winery near Markland in Newfoundland.

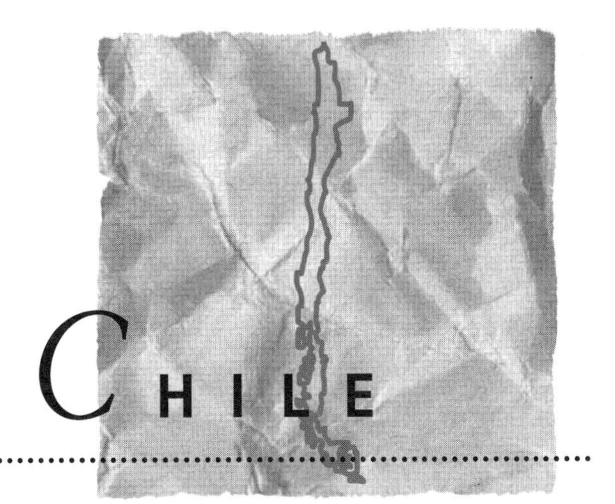

CHILE

This long, narrow country on the southwest coast of South America, tucked between the Andes and the Pacific Ocean, produces both table grapes and wine grapes. Its wine production region extends 1000 km down the length of the country and includes approximately 110 000 ha of vines, of which 35% is reckoned to be Pais grapes. The total wine production is approximately 3.6 million hL annually, and consumption is 29 L per capita. Isolated by the Andes, Chile is now the only country with vinifera varieties not grafted onto North American rootstocks.

GEOGRAPHY

Wine Regions of Chile

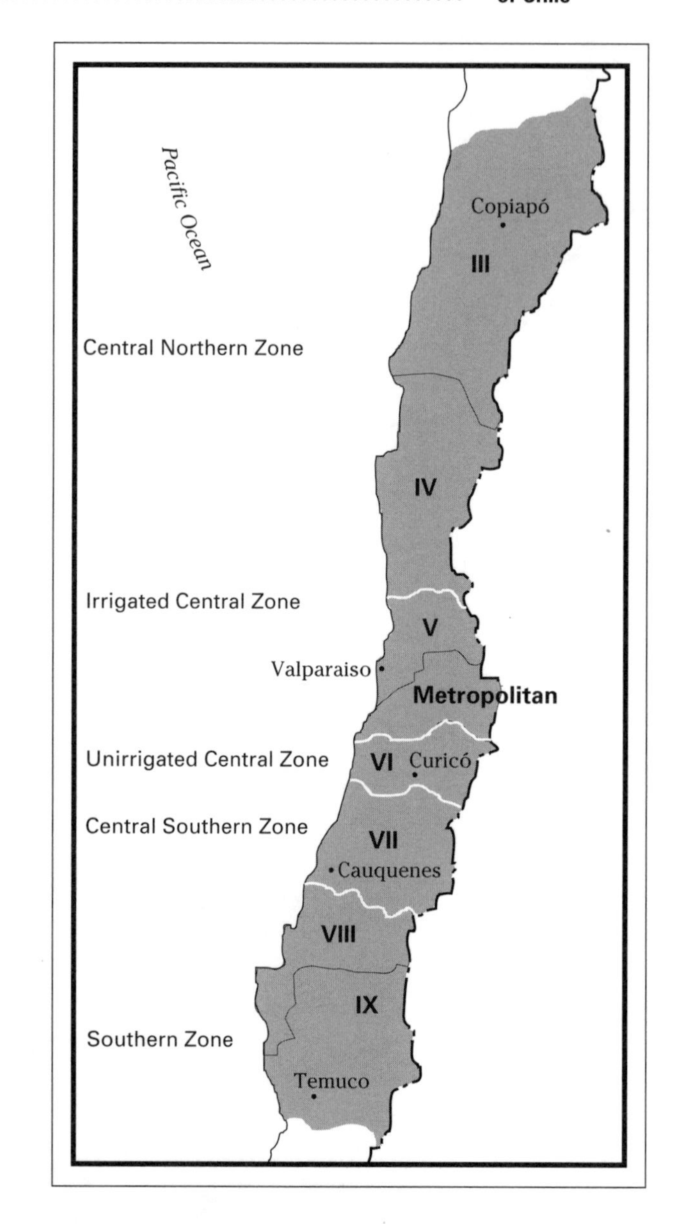

Bordered by Peru, Bolivia, and Argentina, Chile is 4630 km in length and has an average width of 160 km. From peaks in the Andes as high as 6870 m, the land cascades west to the Pacific Ocean. The climate varies dramatically from north to south, and from the coast to the eastern border. In the north, the arid region of the Atacama Desert receives rain every five to seven years. In the middle section, the climate is Mediterranean. Further south, the climate is colder and more humid. Considering that the country runs from about 18°S to 55°S, it is surprising that the yearly average temperature varies by only 13°C. In the north, the Peruvian (Humboldt) ocean current moderates the temperature, and in the south, west winds warm the coast in the winter.

Throughout Chile, the skies are rarely obscured by clouds during the growing season; the contrast of cool nights and hot, sunny days encourages the formation of tannins and pigments. The biggest factor in Chile's wine production is water—from precipitation, rivers, and irrigation.

HISTORY

Chile's history of wine is strongly influenced by the Spaniards, who first came to the land in the sixteenth century. Pizarro led the first group of conquistadores in 1530, but the Spaniards did not really gain a foothold until Pedro de Valdivia founded Santiago in 1541. The Catholic priests sent by Spain needed wine for the ritual of the Mass. The Abbot of Molina is reported to have found the Black Moscatel grape near Curicó, but the Jesuit Father Francisco de Carabantes is credited with having brought the first vine in 1548. This vine was the variety known as Pais in Chile, Criolla in Argentina, and Mission in California. It proliferated, becoming the main source of Chilean wine for the next four centuries.

As early as the seventeenth century, Spain tried to protect its export trade of oil and wine by prohibiting new plantings of vineyards and olive groves in its colonies. For the same reasons, the Spanish government ordered vines in the northern provinces be uprooted, in 1803. Only in 1821, after it had full independence from Spain, did Chile became a serious producer and exporter of wine.

By the mid-nineteenth century, Chile had a number of families who had become rich from the country's mineral deposits, and who toured Europe. When they returned to Chile, they brought back French winemakers and vines. Not only did these vines—Cabernet Sauvignon, Merlot, Cot (Malbec), Pinot Noir, Riesling, Sauvignon, and Sémillon—change the character of Chilean wines, they also became a unique collection, safe from the ravages of phylloxera.

While phylloxera devastated vineyards elsewhere in the world, international demand for Chilean wines skyrocketed. Wine output jumped from 51 400 hL in 1875 to 110 300 hL in 1883. Since then, Chile's wine trade has suffered (at various times) from high taxes and political, social, and economic upheaval. During the late 1980s and early 1990s, investments poured into the country's wine business. Today, Chilean wine has international appeal.

WINE PRODUCTION

Chile is in the midst of a spectacular technological revolution. Newcomers from France, Spain, and California have invested substantial amounts of capital and technology, and the old wineries are taking note and modernizing their equipment.

Thus, outmoded wine presses have been replaced by pneumatic ones, and the old-fashioned beechwood vats have been supplanted by French oak casks and stainless steel tanks. Sufficient cool storage is still a problem, but is improving. (One of the main drawbacks of Chilean white wines was their constant touch of oxidation.) Although the yield in the vineyards is relatively high, effort is being made to prune shorter and to select prime sites. The aim is wines with better character and structure. Each vintage seems to show improvements.

Overall, Chile provides very good fruity red wine in the Bordeaux style at an affordable price. The best wines are mostly varietals, and of these Cabernet Sauvignon, Merlot, Chardonnay, and Sauvignon seem to be the most successful. The Pais grape and table grapes still occupy large areas of vineyards. Chile exports only a small portion of its wine production. Brazil is the largest importer.

WINE REGIONS

The wine-producing region extends over from Copiapó to the River Cautín in the south. Chile does not have a distinctive classification for the various wine-producing regions similar to Europe's various appellation of origin systems. (The only formal classification that Chilean wines undergo is a taste test done by a group of experts on a yearly basis.) However, one can describe the wine-producing areas in terms of the administrative region, and by whether the areas are irrigated or unirrigated.

Chile is divided into thirteen administrative regions. To differentiate between these administrative regions and the wine regions, the following refers to administrative regions as regions, and wine regions as zones.

Region V, the Metropolitan Region of Santiago, Region VI, Region VII, and Region VIII have both irrigated and unirrigated areas. The unirrigated areas form a long, narrow stretch from Valparaiso to north of Concepción and along the Cordillera de la Costa (the mountain range closest to the coast). The larger irrigated areas run parallel for 320 km in the central valley and are watered by the rivers flowing down from the Andes. They produce better wines.

The best wines are found in the Metropolitan Region around Santiago, and in the two regions immediately to the south, Regions VI and VII.

CENTRAL NORTHERN ZONE

This area includes Regions III and IV—the vast provinces of Atacamá and Coquimbo respectively. In the Atacama province, two main areas of vineyards are located south of Copiapó and Vallenar. In Coquimbo province, the vineyards are located along the valleys of the Elqui, Limarí, and Choapo rivers. Because the climate is arid, the vineyards are irrigated. In addition to producing a large amount of table grapes, this zone produces wines from the Moscatel grape, with high alcoholic strength and low acidity; dessert and fortified wines; and Chile's famous brandy (called Pisco) from Moscatel.

IRRIGATED CENTRAL ZONE

Located in the basin tucked between the Andes and the Cordillera de la Costa, this is the main wine-production zone in Chile. It spans Regions V, VI, VII, and the Metropolitan Region. The main grape varieties planted are Merlot, Cabernet Sauvignon, Cabernet Franc, Cot (Malbec), and Petit Verdot.

- Region V—covers the old provinces of Aconcagua and Valparaiso. The summer can be arid and scorching with temperatures hovering around 30°C. The vineyards must be irrigated. Errazuriz is a major producer here.
- Region VI—consists of the ancient provinces O'Higgins, Colchagua, Curicó, and Talca. In Colchagua, the hotter temperatures produce full-flavoured red wines that are deep in colour. Some main producers are Los Vascos, Santa Emiliana, Santa Rita, and Undurraga. The area of Curicó and Talca extends south to the Maule River. Because of the cooler climate, these wines have a better balance of acidity. Some main producers are Miguel Torres, San Pedro, and Chile's largest producer, Concha y Toro.
- Region VII—stretches from Talca in the north, to Cauquenes in the south.
- The Metropolitan Region—located in the old province of Santiago. The best red wines (made from Cabernet Sauvignon as well as some Sémillon and Sauvignon) are produced in the Maipo Valley.

Unirrigated Central Zone

This area runs parallel to the Irrigated Central Zone, from Valparaiso to the Maule River and between the Cordillera de la Costa and the coast. The ocean moderates the climate—giving lower temperatures, more rainfall, and fog—so that the zone does not require irrigation. Some of the best Chardonnays are produced in the Casablanca Valley.

Central Southern Zone

The Central Southern Zone spans Regions VII and VIII, south of the Maule River and across the old provinces of Maule, Linares, and Ñuble. Extreme variations in temperature and abundant rainfall mean that the zone needs no irrigation and can produce white wines with good acidity, most notably from Sauvignon and Sémillon grapes. Excellent Riesling, Pinot Blanc, and Moscatel wines are produced along the coast.

Southern Zone

Wine production is dwindling in this colder zone, located in Regions VIII and IX, in the provinces of Concepción, Bío Bío, Arauco, Malleco, and Cautín. Most of the grapes are used to produce ordinary table wines, mainly from Pais and Muscat of Alexandria grapes.

South Africa

Whereas South Africa's wine trade was once best known for its famous sweet wine, Constantia, it now produces wines of a wider range and ranks eighth in wine production in the world. From its average annual crop of about 1 000 000 t of grapes grown in 116 149 ha of vineyards, it produces approximately 8 700 000 hL of wine. Consumption is 8 L per capita. South Africa has the potential to produce great wines.

SWEETNESS OF WINES

Still Wines	
extra dry	maximum 2.5 g/L
dry	maximum 4 g/L
semi-dry	4–12 g/L
semisweet	4–30 g/L
late harvest	20–30 g/L
special late harvest	a minimum of 22° Balling (i.e. Brix) in the must at harvest
sweet natural	minimum 30 g/L
noble late harvest	more than 50 g/L, as well as must with a minimum of 30 g sugar-free extract, and harvested at 28° Balling or more. These wines can reach 200 g/L.
Sparkling Wines	
extra dry	maximum 15 g/L
dry	15–35 g/L
semisweet	35–50 g/L

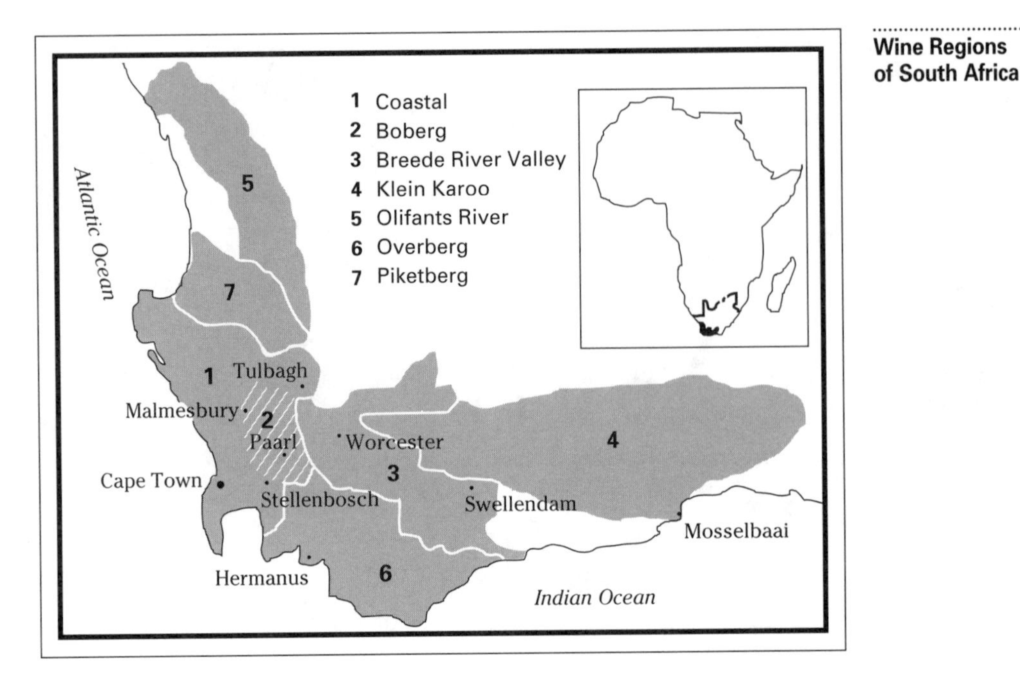

Wine Regions of South Africa

GEOGRAPHY

The Republic of South Africa covers an area of 1 221 037 km^2 at the tip of the continent. Its wine-growing regions are mostly located in the southwest, in Cape Province, between 33°S and 34°S. The Mediterranean-style climate in these areas is favourable to wine production. The growing season runs from September to May.

In terms of climate, wine grapes are grown in two distinct regions:

- Coastal Belt—west of the Drakenstein Mountains. This area has a good annual rainfall averaging 1000 mm, and is cooler than inland. Some loamy soils of medium fertility can produce wines of remarkable quality. Table Mountain sandstone, local granite, and Malmesbury shales are particularly good. Richer soils tend to yield a coarser wine of average calibre.
- Klein Karoo—further inland, between the Drakenstein Mountain range and the Swartberg Mountains, at higher altitudes than the Coastal Belt. This area has greater extremes of climate and an annual rainfall averaging 250 mm. Irrigation is necessary, and night harvesting is becoming a favoured practice to reduce heat stress in the grapes. Late frosts are not unusual in higher areas. The wines tend to be more robust, with a higher alcohol content. Large areas are planted for the production of brandy, sweet Sherry-style wines, and dessert wines.

HISTORY

Winemaking in South Africa began over three centuries ago at the Cape of Good Hope, when Jan van Riebeeck established a post for the Dutch East India Company. Believing that wine would be a good asset for trading, he planted Spaanse Druyven ("Spanish grapes") in the gardens of the settlement, in 1655. In 1661, he recorded in his diary that the first Cape wine had been made. Demand for the wine was high: the clergy used it for the Mass, and sailors drank it to prevent scurvy. Van Riebeeck planted over 1000 Bacchus vines on the slopes of Wynberg ("wine mountain") at Boschheuvel, which is known today as Bishopscourt. Simon van der Stel, the first governor of the new Dutch colony, followed Van Riebeeck's example by planting his estate in the Constantia valley with another 100 000 vines. Although the vines were productive, demand soon exceeded production and good profits could be made.

By 1688, vines were planted inland in the areas of Stellenbosch, Paarl, Tulbagh, and in Franschhoek, where French Huguenot refugees used their skills in grape growing and winemaking. The Huguenots' quest for freedom is still evident today in the names of the wine estates, e.g., Vredendal, Vredenburg, Vredenheim, Rust-en-Vreede. Later, the areas of Worcester, Robertson, Montagu, and Bonnievale in Klein ("little") Karoo were also planted with vines.

In the eighteenth and nineteenth centuries, South African wines were taken to Europe and became famous. On his deathbed, Napoleon Bonaparte requested Constantia wine made by the Cloete family. King Louis Phillippe of France (1773–1850), England's Queen Victoria (1819–1901), Sheridan (1751–1861), Bismarck (1815–1898), Baudelaire (1821–1867), and Longfellow (1807–1882) were among the other great figures who praised Constantia wine.

South Africa's once-famous Vin de Constance from Klein Constantia was probably a sweet, sometimes fortified Muscat de Frontignan and Pontac wine.

In 1806 the British occupied the Cape and encouraged wine export to Britain. Starting in 1825, they invested great sums of money in the flourishing wine industry. But in 1861, overseas tariffs collapsed the export trade.

In the years since, the South African wine industry has suffered phylloxera (1885), the South African (or Boer) War (1899–1902), prices falling due to overproduction (1900–1917), an imbalance in the industry (due to the merger of Oude Meester and Stellenbosch Farmers Winery to form the Cape Wine & Distillers handling 70% of all sales), and international sanctions during the period of apartheid.

On a more positive note, grape growers created their own co-operative, the Ko-operative Wynbouwe Vereniging (KWV), in 1918; Parliament granted the KWV certain controls in 1924; exports to Britain and the Netherlands resumed in 1926; the KWV acquired full control of the South African Wine Farmers Association (SAWFA) in 1950; and the Nederburg Auction began, as a very important showcase for national wines, in 1975. Fortunately the industry's tenacity and dedication has kept the quality of the wines high.

LAWS AND REGULATIONS

South Africa has a system of certification similar to the appellation of origin systems used in France and Germany, with a strong emphasis on grape cultivars. The Wine of Origin (WO) laws were introduced in 1973 and modified in 1992–1993.

Authorities regulate and supervise vineyards, winemaking practices, grape varieties, and vintages. They do not control yields, irrigation, or the use of fertilizers and pesticides. Acid correction is permitted; chaptalization is not. Sulphur is limited to 200 parts per million. Whereas the KWV used to control production and quotas, it no longer does; growers now decide which varieties and quantities will compete better in an international free market.

South Africa's Wine and Spirit Board performs analytical and taste tests of wine. The Board's seal on the bottle indicates that the wine has passed the required tests, and that the producer's claims on the wine label are sound. Prior to 1993, this seal was tricoloured and detailed (green gave the grape variety, red the vintage, blue the region of origin); now the seal is duller and shows only a serial number. Both official languages (English and Afrikaans) are often evident on the labels. All wines must state the alcohol content on the label, within a margin of half a percent for export wines, or one percent for wines to be consumed domestically

Labelling

Wines may be labelled as varietal wine if they have at least 75% of the named variety (for domestic sale), or 85% (for export). Labels on blends may list the grape varieties in descending order of proportion, but the label must give percentages for each variety that constitutes less than 20%.

The following are other labelling terms:

- Méthode Cap Classique (MCC)—indicating (since 1992) sparkling wine made by the méthode Champenoise.
- Grand Cru and Premier Grand Cru—not official statuses but the result of the wine producer's own assessment.
- Stein—semisweet white wine.
- Estate—similar to the French terms *château* and *domaine*. Wine labelled *Estate* may be wine made from fermented grape juice that is transported to a bottler or wholesaler to mature, barrel-age, fine, filter, and bottle. Or it may be wine made by one grower from grapes grown on different farms, provided these share comparable climate and soil.

Wines that have won the Veritas Show Award from the South African National Show have gold (16 out of 20 points) or double gold (17 out of 20 points) stickers affixed to them. (The double gold replaces the Superior gold sticker used prior to 1989.) Winning wine is of a very high standard: from an average 900 annual entries, only 15% receive the gold, and 8% receive the double gold.

GRAPE VARIETIES

Most European varietals are grown and vinified in South Africa. The most important grapes varieties according to total surface cultivation are Steen (31.2%), Sultana (9.9%), Colombard (8.8%), Palomino (6.3%), Hanepoot (6.3%), Cinsaut (5.3%), Cabernet Sauvignon (4.3%), Cape Riesling (3.9%), Sauvignon (3.7%), Clairette Blanche (2.5%), Hermitage/Pinotage (2%), Servin Blanc, locally called Raisin Blanc (1.6%), Muscadel (1.1%), Sémillion 1%, with Shiraz, Kanaän/Belies, and Ugni Blanc (all less than 1%). In addition, five white hybrids (Chenel, Weldra, Colomino, Grachen, and Follet) are used.

Some of these varieties have particularly distinct characters:

- Pinotage—a hybrid created in 1925, crossing Pinot Noir and Cinsaut (the latter being known locally as Hermitage). Pinotage can produce a flavourful wine if submitted to a hot fermentation and moderately aged in oak barrels. The wine has an acetone overtone. The best Pinotages need ageing to develop to a fine individual wine of length. Stellenbosch Farmers Winery commercially produced the first Pinotage in 1961.
- Shiraz—known elsewhere as Syrah, produces dense, flavourful reds that develop a smoky or chocolatey character with age

- Steen—known elsewhere as Chenin Blanc, the most widely-planted variety. It can achieve remarkable qualities with a hint of almond in its aroma. It is made in styles that range from dry table wines, to sweet botrytis-affected dessert wines, to sparkling wines, and as a base in rosés. It is also used for distilling brandy and spirits.
- Hanepoot—known elsewhere as Muscat of Alexandria, with an intense flowery bouquet and a deep, sumptuous, honey Muscat flavour.
- Cape Riesling or Riesling—not a true Riesling, but the French Crouchen Blanc. Since 1983, it cannot be exported to Europe labelled *Riesling*. The true Riesling is known here as Weisser Riesling or Rhine Riesling.
- Tinta Barocca—Portuguese Port grape used to make a good, earthy, varietal red wine, and also used in blends.

WINE PRODUCTION

South Africa is well known for its excellent wines in the Port and Sherry styles, the best of which rival the originals. But South Africa offers a great range of wines. Since the 1960s, South African winemakers have travelled abroad to master techniques used elsewhere—cold fermentation to produce crisp white wines, the use of French oak barrels, and how to handle European grape varieties. The results have been impressive: fruitier white wines, and more balanced (less oaky) red wines that express more finesse and varietal character. This applies particularly to Chardonnay, Pinot Noir, and Cabernet Sauvignon wines.

WINE REGIONS

Since 1973, the Cape Province has been divided into Wine of Origin regions (WO), districts, wards, and estates. At present, there are five regions containing twelve districts and thirty-nine wards. In total there are eighty-three estates, seventy co-operatives, and sixty private producers.

COASTAL

This region encompasses six of the best Cape districts (Constantia, Durbanville, Stellenbosch, Paarl, Swartland, and Tulbagh) and seven wards (Franschhoek Valley, Riebeekberg, Groenekloof, Simonsberg-Stellenbosch, Wellington, and Jonkershoek Valley).

- Constantia District—in the southern suburbs of Cape Town, with a Mediterranean climate, close proximity to the Atlantic, and average annual rainfall of 850 mm. This district produces (on the cool red granite slopes of the Constantia Mountain) a solid Chardonnay, Sauvignon (Fumé Blanc), Riesling, Steen (Chenin Blanc), sweet Muscat, Cinsaut, Pinotage, Cabernet Sauvignon, and Shiraz (Syrah). Some vineyards are located in lower-lying areas on sandy soils of Table Mountain sandstone origin.
- Durbanville District—located in the northern suburbs of Cape Town, 15 km from the cool Atlantic Ocean. The vineyards, situated on the red granite slopes of the Dorstberg Mountain, focus on red wines, particularly Shiraz (Syrah), Cinsaut, Cabernets, and Pinotage. Annual rainfall averages 350 mm.
- Stellenbosch District—surrounded by mountains in the east and northwest and rising to the west, where the dunes of the Cape Flats form a natural boundary along False Bay. The average annual rainfall is 500 mm. This fertile area has three types of soil: alluvial along the Eerste River, Table Mountain sandstone in the west (preferred for white wines), and granite slopes in the east (giving the best results for red wines). The area yields great Pinotage, Cabernet Sauvignon, Cabernet Franc, Merlot, Shiraz (Syrah), Weisser Riesling, Sauvignon, Clairette Blanche, and Chardonnay. Stellenbosch, also known as the "town of oaks," is the second oldest town in South Africa (1679) and the site of a university with a department in oenology and viticulture.
- Paarl District—60 km northeast of Cape Town in the fertile Berg Valley, surrounding the town of Paarl (so named because of the granite Paarl Mountains, which gleam like pearls when wet). This important wine district holds the headquarters of the KWV, on the outskirts of the town of Paarl. The district has wet winters averaging 650 mm of rainfall, and hot, dry summers. There are three types of soil: granite near Paarl, crumbling slate in the northeast, and sandy soil in the Berg Valley. The area is particularly known for Sémillion, Chardonnay, Palomino, Sauvignon, Shiraz (Syrah), and Pinotage.
- Franschhoek Ward—or "French Quarter," situated southeast of Paarl. Many of the 200 Huguenots who settled here were skilled winemakers in France. The climate is cool and wet, with 900 mm of annual rainfall. The area has a tradition of fine oak-aged Cabernet Sauvignon, Cabernet blends, Merlot, Sauvignon, Chardonnay, and Late Harvest wines, as well as Cap Classique sparkling wines.

- Swartland District—located east of the town of Darling and around Malmesbury, Moorreesburg, and Piketberg. The area has a hot dry climate and the vineyards require irrigation during the summer months. Apart from in the cool Saldanha area near the ocean, the rainfall is marginal (generally 450 mm–600 mm annual average, but as low as 250 mm near Malmesbury). Robust reds from the varieties Cinsaut, Tinta Barocca, Shiraz (Syrah), and Pinotage are produced with success. Fortified wines from Palomino and False Pedro (Pedro Luis), plus white wines from Sauvignon, Colombard, Riesling, Bukettraube, and Fernão Pires, are good and inexpensive.
- Tulbagh District—overlooked by the Witsenberg, Winterhoek, and Saronsberg mountains. This area is hot and dry, with only 350 mm annual rainfall, so requires irrigation of its sandy soil. Vineyards on the wetter slopes are less affected by heat and do not need irrigation. Many white blends of great quality—Gewürztraminer, Riesling, Chardonnay, Hanepoot (Muscat of Alexandria), Pinotage, Cinsaut, and Cabernet Sauvignon—are produced in this small area north of Worcester.

  Nicky Krone, of Twee Jonge Gezellen Estate, makes a fantastic méthode Champenoise Chardonnay/Pinot Noir in a bunkerlike structure built to withstand the area's frequent earth tremors. This same estate introduced cold fermentation in the 1960s and night picking in the 1980s.

BOBERG

This appellation is used for fortified wines made in the districts of Tulbagh and Paarl.

BREEDE RIVER VALLEY

This region includes the two districts of Worcester and Robertson, and part of the Swellendam district. There are seventeen wards: five (Aan-de-Doorns, Goudini, Nuy, Scherpenheuvel, Slanghoek) are in Worcester, eleven (Boesmansrivier, Bonnievale, Eilandia, Goree, Glerkliphoogte, Hooprivier, Klaaswoogds, Le Chasseur, McGregor, Riverside, Vinkrivier) are in Robertson, plus Buffeljags in the Swellendam district along the Breede River, attached to the region by the ward of Bonnievale.

- Worcester District—north of the Breede region and crossed by the many tributaries of the Breede River. Annual rainfall ranges from 1500 mm in the west to 210 mm in the east. Grape growers depend heavily on the water from the Branvlei dam for irrigation in the drier parts. This district produces a quarter of the total output of Cape Province wine. There are thirty-two co-operatives producing Weisser Riesling, Sauvignon, red and white Muscadel for dessert wines, and Colombard for distillation.
- Robertson District—southeast of the Worcester District on either side of the Breede River, with the rich lime soil of the valley, surrounded by high mountains. The hot and arid climate (340 mm annual rainfall) requires irrigation. This is the fourth-largest wine-producing district of the Cape Province. Estates and co-operatives produce an excellent Chardonnay and Shiraz (Syrah), as well as fortified wines including Palomino, Steen (Chenin Blanc), red Muscadel, Hanepoot (Muscat of Alexandria), Cinsaut, Colombard, Ugni Blanc, and Riesling.
- Swellendam District—located on the Breede River at the southern point of the Robertson District, Swellendam extends to the Indian Ocean. The average rainfall is 600 mm, and the average summer temperature rises to 21°C. Apart from alluvial sandy soils along the river banks, the soil is compact clay. This area mainly produces average bulk wine.

Klein Karoo

Klein Karoo is the largest region of the Cape Province, with the two wards of Montagu and Tradow. It stretches for over 250 km from Montagu in the west to Oudtshoorn in the east. The climate is dry to arid, with temperatures averaging 20°C–24°C and less than 280 mm of rainfall per year.

Because of inadequate supplies of water, the vineyards are confined to deep alluvial soils, which have a tendency to compact and so must be constantly irrigated.

The adaptable Steen (Chenin Blanc) thrives here, as well as the Muscadel and Muscats for which the area is famous. Boplaas Estate run by Danie Nel produces one of the best Port-style wines of South Africa. Among other fine wines from this estate are Cabernet Sauvignon and Sweet Muscatel.

Olifants River

This region north of Cape Town stretches north to south along the Olifants River, from Lutzville in the north to Porterville in the south. It includes the four wards of Spruitdrift, Koekenaap, Lutzville Valley, and Vredendal. The climate is hot and dry, with the lower rainfall near the ocean averaging an annual 300 mm. The vineyards are planted on terraces and are trellised to provide a cool canopy for the grapes. Within the region, three soil types favour grape growing: fertile soil near rivers; a red, sandy soil; and a lime-rich soil. This region produces good value in wines.

Independent Districts

- Piketberg District—north of Tulbagh and south of the Olifants River. This large area extending west to the Atlantic Ocean has high summer temperatures and meagre rainfall (175 mm annually). The irrigated land produces fortified wines and a small amount of Hermitage (Pinotage), Sauvignon, and sweet white blended wines.
- Overberg District—formerly known as the Caledon District, and located in the southernmost part of the Cape Province, it has the two wards of Elgin and Walker Bay. The annual rainfall is 750 mm. This cool district is well suited to Chardonnay, Sauvignon, and Pinot Noir. Walker Bay has the potential to produce some of Cape Province's best wines.
- Douglas District—about 100 km southeast of Kimberley, and originally part of the Orange River wine area, but given separate WO status in 1981. The hot and dry climate favours the production of fortified, dessert, and bulk wines. There are seven wards which stand on their own without relation to any other regions or districts. They are: Cederberg, Benede-Oranje, Ceres, Herbertsdale, Ruiterbosch, Swartberg, and Andalusia. These wards yield mostly co-operative bulk wines and table wines of good value.

CHAPTER 25 AUSTRALIA

This huge country—similar in size to the United States—is traditionally a beer-drinking nation and was once known as a supplier of cheap, sweet, fortified wines. Since the 1970s, however, Australia has proven again and again that it can produce great wines, and has become both a de facto wine-world power and a trendsetter. With approximately 73 000 ha under vine, it produces annually an average 5 900 000 hL of wine, of which 21% is exported to a receptive international market. Consumption is 18 L per capita.

MURRUMBIDGEE AND MURRAY

Murrumbidgee is an aboriginal word for "big water." In 1906, the Australian government authorized the construction of the Burrinjuck dam, which stores water from the Murrumbidgee River. This water is then diverted by a series of canals that irrigate 183 000 ha.

The Murray River is Australia's longest river and runs east for over 500 km to mark the border between North South Wales and Victoria.

Together the two river systems are the key to Australia's wine production.

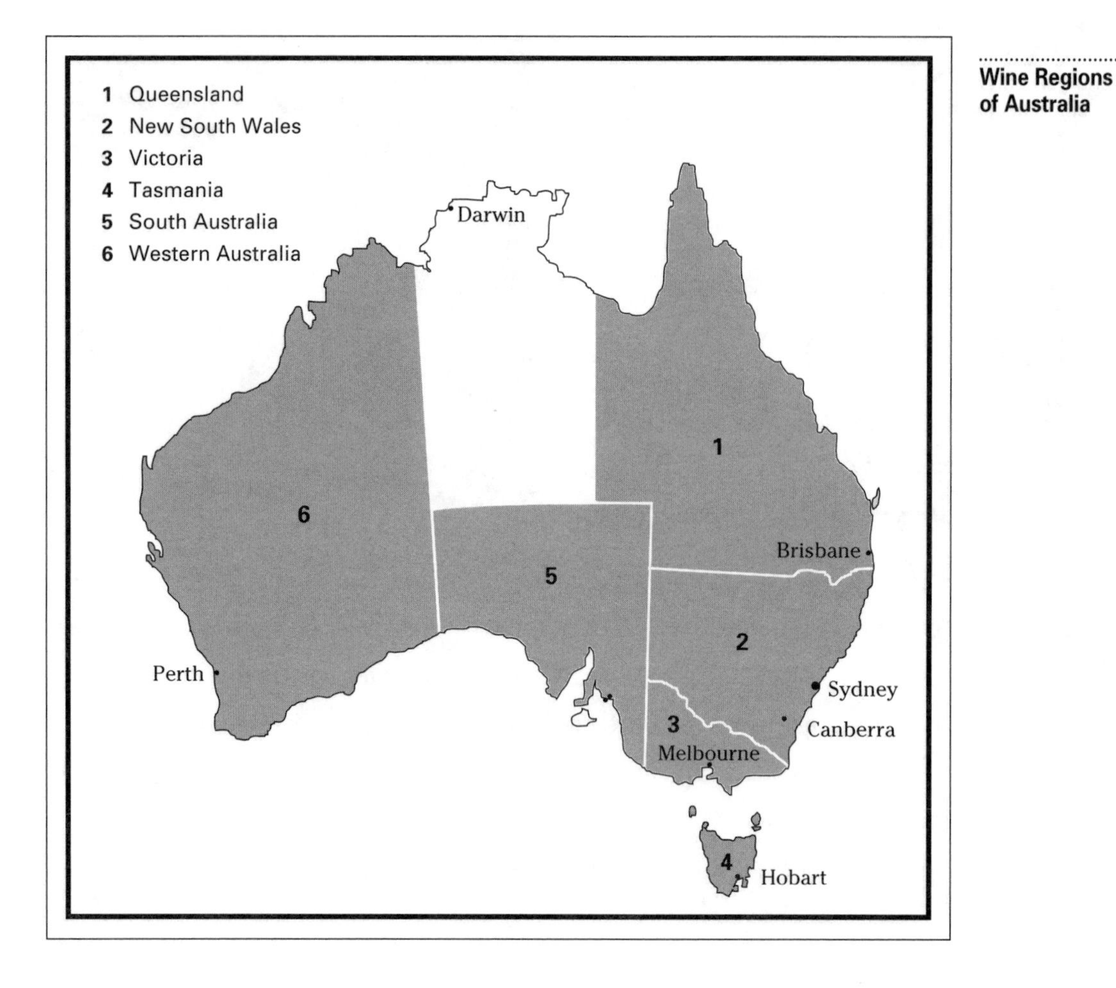

Wine Regions of Australia

GEOGRAPHY

Australia is a hot, sunny, and arid land of 7 686 810 km^2, between the Pacific and Indian Oceans. It is made up of the mainland states of Queensland, New South Wales, Victoria, South Australia, and Western Australia; the Capital Territory of Canberra; the Northern Territory; and the island state of Tasmania. Most of the mainland wine regions are in the southwest and southeast corners of the continent, with the best located between 34°S and 38°S. Tasmania lies further south, across the Bass Strait from Victoria. As a point of comparison to the northern hemisphere, Melbourne and Sydney in Australia's southeastern state of New South Wales are roughly equivalent in terms of latitude to the Californian cities of San Francisco and San Diego.

Temperatures in Australia vary greatly and the wine regions range as much as those in California. Western Australia, Tasmania, and the coastal vineyards of New South Wales (with the exception of Coonawarra) have an average rainfall of less than 600 mm. This low rainfall means that the vineyards are relatively free of diseases, but does mean that they must be irrigated. The Hunter Valley and the east coast of Australia get high summer precipitation and a greater incidence of downy mildew. This is controlled by spraying. Spring frosts occur in Tasmania, Victoria, and South Australia. The east coast of New South Wales suffers from damaging hailstorms during the mid-summer months (January and February), right up to harvesting time (March and April). In southern vineyards, hail is less damaging as it tends to occur during the months of October and November. All in all, Australia's climate is not kind: bush fires, droughts, floods, hail, frost, and pests can affect vintages.

THE RANGE OF CLIMATES

Using the California classification system developed by Professors Amerine and Winkler, Australia's wine areas span the range with, for example, Tasmania (2500 degree days) falling within Region 1 category and Pokolbin in New South Wales (4538 degree days) within Region 5. Of course, it is not only the average daily temperature, but also variations between day and night temperatures, that are reflected in the character of the wine.

In general, Australia's grape growers prefer either of the following soil types:

- in non-irrigated areas—well-drained, rich, red, loam soils with deep clay subsoils.
- in irrigated areas—sandy loam soils, with loam subsoils to permit the drainage of excess irrigation.

The vineyards regularly producing superior quality wines tend to be located on soils containing calcium carbonate or limestone. The red and black soils at Coonawarra (called terra rossa and rendzina, respectively), volcanic soils of Pokolbin, sandy loam soils of Reynella, and brick soils at Watervale all contain limestone. Heavy, rich, dark soils of alluvial origin yield rich crops suitable for the production of Port-style wines, dessert wines, and other fortified wines, but are not able to yield fine table wines.

HISTORY

Australia's wine industry had humble beginnings, and its evolution has sometimes been tortuous. There are no native grapes. The first vines were brought by Captain Arthur Phillip of the British Navy, when he arrived in New South Wales in 1788 to set up a penal colony. Because drunkenness became widespread in the new colony, he favoured having local production of beer and wine over the hard liquors drunk at the time. He planted vine cuttings from the Cape of Good Hope and from Rio de Janeiro.

Thus, Australia's first vineyard was planted at Farm Cove, near the site of the present-day Botanical Gardens in Sydney. These vines bore grapes, but soon died from the fungal disease called black spot. Phillip persisted and, in 1791, established a new vineyard near the Parramatta River. Unfortunately, he returned to England a year later and never sampled any local wine. However, he left his mark by encouraging the young colony's interest in wine. Over the next century, others were to try their hands at grape growing—with varying results.

From 1846 to 1852, southern Australia's vineyards quadrupled. Then, from 1856 to 1867, land under vine increased further—from 305 ha to 2500 ha. But the growth did not continue: a recession hit (1865–1885), phylloxera arrived in the Victoria area (1875) and spread; and when the major banks crashed in 1893, a depression, civil unrest, a saturated wine market, low prices, and tax barriers between states all diminished the health of the industry. Many vineyards were uprooted and coverage fell to 1750 ha.

The next century saw many ups and downs. For example, World War I (1914–1918) disrupted shipments to England, but a government policy brought returning soldiers to settlements along the Murray and Murrumbidgee rivers in South Australia, New South Wales, and Victoria. As a result, the irrigated lands were expanded. These inexperienced farmers failed, but entrepreneurial wine companies re-acquired these lands, and by 1920, South Australia was producing 75% of the country's wine, about 190 000 hL.

Production skyrocketed to about 606 000 hL from 22 000 ha of vines in 1927. But expansion was too rapid and once again prices plummeted due to the glut. The world depression of the 1930s and World War II (1939–1945) exacted their tolls.

At the end of the war, the wine industry started to undergo more change. Once again, former soldiers expanded irrigation. Wines previously sold in bulk started to be bottled with winery brand labels. Table wine production increased to 40% (to the detriment of fortified and sweet wines). Slowly, Australian wineries began to be noticed abroad. They won medals at international trade shows. The big boom finally arrived in the 1960s. Wine became fashionable with meals. This trend was enhanced by the media, aggressive marketing, the tourist trade, and the proliferation of wine and food societies.

Trailblazers

- Captain John MacArthur arrived in Sydney in 1790. After importing merino sheep, he turned to grape growing. He was the first to plant a vineyard for the purpose of making wine for sale. After a violent quarrel with the colony's Captain Bligh in 1809, MacArthur was exiled to England for eight years during which he made an extensive trip to France and to Switzerland to collect vines and learn about their cultivation. After his return to Australia in 1820, he planted first at Camden, and later on the Nepean River close to Penrith. In 1827, he made 75 708 L of wine.
- The explorer Gregory Blaxland purchased some "claret" vines from the Cape of Good Hope in 1816 and planted them on his 183-ha farm in the Parramatta River valley near the town of Eastwood. The vineyard was successful, and as early as 1822 he shipped a barrel of fortified wine to England. The wine was honoured with a silver medal. Extremely encouraged, he sent two more barrels and, in 1828 received the gold Ceres Medal. Blaxland also pioneered research to find varieties of vines resistant to the dreaded black spot.
- James Busby arrived in Australia in 1824 and the same year received a grant of 50 ha in the Hunter Valley. Soon his Kirkton Vineyard expanded and was to gain a reputation as finest in the region. As well, Busby taught viticulture at a school for orphaned boys near Liverpool, and in 1829 the vineyard cultivated by his students produced its first healthy crop. Busby continued to educate by writing increasingly accessible books about grape growing, by travelling to France and Spain and writing about what he learned, and by sharing his expertise with New Zealand. From his trip to Europe, Busby unselfishly donated 678 vine varieties to the Australian government to be planted in an experimental garden at Sydney. While these did not survive, cuttings sent to Kirkton, Camden, and Adelaide did. Busby himself was not an intensive grape grower and winemaker (his brother-in-law, Kelman, developed the property and planted vines as part of a mixed-crop farm), but he contributed immensely to the budding industry by educating others and by donating large collections of European vines.
- The first vineyards of Western Australia were planted in 1829 at Guildford by the botanist Thomas Waters. He established the Olive Farm, which today is Australia's oldest surviving winery.
- Captain John Septimus Roe planted a vineyard in Western Australia in 1840.
- The first Swan Valley commercial wine was made at Houghton in 1859 by a Dr. Ferguson.
- The Bassett family established a vineyard of some repute at Roma, Queensland, in 1863.

In recent years, the industry has won gold medals in all the most respected competitions. Roseworthy College near Adelaide (now a part of the University of Adelaide) has become one of the most notable wine institutions in the world. Australian-trained winemakers now lead some of the best wineries of France, Germany, Italy, Chile, and the United States. Not only are they contributing knowledge to their host wineries, but they will someday return to Australia to share their knowledge there.

LAWS AND REGULATIONS

The Australian Wine and Brandy Corporation (AWBC) regulates wine and brandy production and reports to the Federal Minister for Primary Industries and Energy. Australia has as yet no controlled appellation of origin system. The AWBC is working on a new set of laws and regulations, and (in accordance with the EU) has phased out use of the names Bordeaux, Burgundy, Champagne, and Chablis on Australian wine labels.

- Labels must give the winemaker's name and address.
- Labels may indicate a vintage, varietal, or geographic source; labels that do must contain a minimum of 85% from the indicated vintage, varietal, or geographic source.
- In the case of blends, varietals must be listed in descending order.

GRAPE VARIETIES

The number of grape varieties grown commercially in Australia was relatively small until the 1960s. Imports were subject to severe scrutiny to avoid infecting or infesting the vineyards.

Since then, a more efficient system for detecting pests and diseases (including a rigorous quarantine) has allowed the safe introduction of new varietals. Merlot, Gamay, Pinot Gris, Nebbiolo, Sangiovese, Zinfandel, Sylvaner, and Müller-Thurgau are expanding rapidly. Cabernet Sauvignon and, more recently, Chardonnay are gaining ground and are the most planted crops.

The most popular varieties are listed below.

Red Grapes

- Shiraz (85 000 t)—or Syrah, the primary red grape of Australia. Versatile, vigorous, and adaptable, it produces a wide range of wines, from fortified to table. The best ones have made Australia famous and demand now exceeds supply. In Rutherglen, Victoria, Shiraz (Syrah) produces fruity and full-bodied wines. From south and central Victoria the wines are spicier; Hunter Valley produces wines that are softer and heartier. Barossa's Penfolds Grange Hermitage (in South Australia) is the most celebrated.
- Cabernet Sauvignon (70 000 t)—increasing in importance in most regions. Large areas have been recently planted and the yield is expected to surpass that of Shiraz. More high performance clones of Cabernet Sauvignon have been planted to see if the quality is as good as the original. In Coonawarra it develops a typical black currant and minty aroma, and is regarded as Australia's best Cabernet Sauvignon. Southern Vales, Clare Valley, and Barossa Valley tend to be darker, with a more pronounced redcurrant nose, and a softer finish. In Hunter Valley it seems to do well, because the loose clusters of berries are generally resistant to rain damage during harvest. In Victoria, Cabernet Sauvignon tends to be herbaceous and have a distinct bell-pepper nose. It is produced both as a single varietal and blended with Merlot (or more typically, with Shiraz) because of limited availability. Cabernet-Shiraz blends give a textural sensation in the middle palate that is often missing in Cabernet Sauvignon on its own.
- Grenache (21 000 t)—a heavy producer. It yields wines that mature at an early age. Because the must is rich in sugar, it is a favourite for producing fortified wines. It is also suited to making good commercial dry reds and rosés. The dry reds are best when young since they lack colour, tannin, and personality—weaknesses that are accentuated when grown in hot, irrigated areas. The best ones are made in the Adelaide area, in the Barossa Valley, and in Southern Vales in South Australia.

- Pinot Noir (14 000 t)—an early-maturing variety good for growing in cool regions. It is one of the key grapes used in the production of sparkling wines. Until the 1970s, the use of Pinot Noir was impeded by a lack of reliable clones. Considerable genetic variation within the variety made it difficult to find the suitable clone for high performance, reasonable yield, and fine character. It is planted mainly in southern Victoria and the Adelaide Hills, and with very promising results in Tasmania.
- Mataro (8500 t)—known elsewhere as Mourvèdre. This variety used to be the fourth most important red grape planted in Australia, but is less used now. It is still used in the blending of Port-style wines, and is occasionally combined with Shiraz (Syrah). It is very dark, with a neutral flavour and a harsh finish.

Other significant varieties grown in lesser amounts include Malbec, Carignan, Cinsaut, Zinfandel, and Merlot.

White Grapes

- Chardonnay (95 000 t)—all the rage since the 1970s and planted at an incredible pace. This grape prefers cool, sunny areas with chalky soil. Excellent results have been achieved in Tasmania, southern Victoria, and South Australia. Winemakers improved their use of oak to retain more elegance and better structure without sacrificing nut and peach aromas.
- Muscat Gordo Blanco (67 900 t)—a versatile grape known elsewhere as Muscat of Alexandria. This sweet, spicy, all-purpose grape is eaten fresh, dried into raisins, and used in winemaking. It is one of the main constituents of Australian cream "sherries", of many dessert wines, fruity and aromatic table wines, and sparklers. Plantings are concentrated in the Murrumbidgee and Murray irrigation areas and in the Swan Valley of Western Australia. Many other Muscat varieties, with darker skins, are also used in Australia.
- Sémillon (55 000 t)—before Chardonnay took off, the most popular grape in Australia for making dry table wine. It was Australia's answer to white Burgundy. It established the reputation of Hunter Valley whites and was very popular in New South Wales. When made without oak, it has a distinct similarity to Riesling, but without the spiciness. Today, Sémillon plantings rank third to those of Chardonnay and Muscat, but are on the increase. The best Sémillon hails from Hunter Valley. Locally, it is sometimes known as Hunter River Riesling, Green Grape, or White Madeira.

- Riesling (45 000 t)—responsible for producing some of Australia's best whites, before 1970. Produced in either of the dry Alsatian or sweet German styles, it can achieve superb results in the Barossa Valley, Clare Valley, and Tasmanian grape-growing regions. The very rare examples affected by noble rot are sumptuous.
- Sauvignon (10 800 t)—plays an important role in the hotter regions because of its high natural acidity. In cooler areas, its varietal character is more pronounced, but still lacks typical grassy aromas. This problem is compounded by the use of oak. It is known also as Surin.

Other significant varieties grown in lesser amounts include Doradillo, Pedro Ximénez, Palomino, Clare Riesling, Trebbiano, Tokay or Hárslevelü, Albillo, Sercial, Chasselas, Clairette, Verdelho, Marsanne, and Aucerot.

WINE PRODUCTION

Australia has approximately 900 wineries. Only 37% of their wines are sold in bottles; almost a third are sold in flagons (2-L screwtop), roughly the same again in casks (4 L bag-in-a-box style), and the remainder in bulk. Dry, varietal, table wines of the highest calibre dominate Australia's wines, but Cabernet Sauvignon and Shiraz (Syrah) blends have become a trademark. As well, dry Rieslings have supplanted sweet ones, Chardonnay and Cabernet Sauvignon wines have increased, and (by 1985) one-third of the wines were made from premium vinifera grape varieties. The wine market continues to boom, and new wine areas are being developed throughout the country.

WINE REGIONS

As of 1995, Australia had thirty-four Wine Zones (WZ) and Regions. Grouped according to states, they are as follows:

New South Wales

Central Western New South Wales, Far Western New South Wales, Greater Canberra, Holiday Coast, Hunter, Illawarra, Murray, New England, Orana, Riverina, Sydney

Victoria

Central Victoria, Geelong, Gippsland, Melbourne, Mornington Peninsula, North-Eastern Victoria, North-Western Victoria, Western Victoria, Yarra Valley

South Australia

Central South Australia, Eyre Peninsula, Far North, Kangaroo Island, South-Eastern South Australia, Murray Mallee, Yorke Peninsula.

Western Australia

Darling Ranges, Margaret River, Mount Barker-Frankland, North Perth, South West Coastal, Swan Valley, Warren-Blackwood

The following are the most important areas, grouped by states.

QUEENSLAND

Queensland's grape-growing areas are around Roma and on the Granite Belt close to the New South Wales border. The climate is hot and appropriate for the growing of table grape varieties, with the exception of the Granite Belt area.

- Granite Belt—situated at altitudes of 750 m–900 m. The climate is cool with threatening spring frosts. The summer is hot enough to bring grapes to maturity. Rain is abundant during the summer and lasts until harvest time. The small wine production is sold locally and consists mainly of Sémillon and Shiraz (Syrah), plus some Chardonnay and Cabernet Sauvignon.

NEW SOUTH WALES

New South Wales is the cradle of the wine industry in Australia. About 28% of Australian wines come from this state.

- Hunter Valley (Hunter WZ)—one of the best wine areas of Australia. It is divided in two: Lower Hunter Valley (which is the older district around Pokolbin and Rothbury), and the Upper Hunter area, which is west of Muswellbrook. The average annual grape production is 18 000 t.

  Temperatures range from 4°C in July to 21°C in January. Rain during the summer, heavy rains at harvest time, and hail are threats. Birds do their share of damage, too. The geological formation of the vineyards varies and consists mainly of shale, tuffs, sandstone, and conglomerate resting on a lava bed. There are some areas around Pokolbin made of red, volcanic clay loam, especially suitable for growing red grapes. Shiraz (Syrah) and Sémillon here produce some of Australia's best wines, but their importance is challenged by the latest favourites of this area—Chardonnay and Cabernet Sauvignon. Thus far, Hunter Valley is one of the few places in Australia to have escaped the ravages of phylloxera.

- Mudgee (Orana WZ)—a tiny area, located 260 km northwest of Sydney and close to the Hunter River. Mudgee's vineyards are the highest in Australia (520 m–610 m). The climate is moderate, but subject to frost and flood. Annual average rainfall is 600 mm–1300 mm. Mudgee's soils consist of red loam containing ironstone and limestone, over a bed of heavy clay. The area produces rich and deep red Shiraz (Syrah) and Cabernet Sauvignon wines, as well as opulent Chardonnays.
- Corowa (Murray WZ)—on the northwest bank of the Murray River, at the border of New South Wales and Victoria. The area produces mainly dessert and fortified wines.
- Riverina (Riverina WZ)—the old Murrumbidgee Irrigation Area located in the southwest of New South Wales, around the towns of Griffith, Leeton, and Yenda. This dry area (400 mm annually) relies on irrigation which, in combination with steady sunshine, results in bountiful yields. Its average annual production is the highest in New South Wales and one-sixth of Australia's total. The temperatures range from averages of 3°C in July to 32°C in February. Soils vary from heavy clay to light sandy loam, with patches of sandy soil on clay. Most (70%) of the wines are white, of which half are made from Sémillon and Trebbiano. Half the reds are produced from Shiraz (Syrah). The star of the area is the botrytized Sémillon. The rest is of average quality.
- Cowra (Central Western New South Wales WZ)—located near the town of Orange. This warm area makes rich and buttery Chardonnays.
- Young (Greater Canberra WZ)—near the town of Orange. The area produces rich Shiraz (Syrah) and Sémillon wines.
- The Capital Territory of Canberra has a small but important production of Cabernet Sauvignon, Chardonnay, and Riesling.

Victoria

- North-East Victoria (North-Eastern Victoria WZ)—encompassing Rutherglen, Wangaratta, and part of Corowa. Warm days, cool nights, and 640 mm annual rainfall are characteristic of the area. Wine production is focussed on big reds and very good dessert and fortified wines.

- Murray River Valley (North-Western Victoria WZ)—around the town of Swan Hill and south of the fork of the Murrumbidgee River. This area is parched, so must be irrigated to yield any crop, but these conditions push the grapes to higher sugar levels. Many Sherry-style wines, dessert wines, and brandies are produced. Shiraz (Syrah), Sémillon, and some Cabernet Sauvignon are used for table wines.
- Mildura (North-Western Victoria WZ)—south of the fork of the Darling and Murray rivers, 160 km upstream from the Murray River valley vineyards. The arid land relies on irrigation diverted from the Murray River. Red sand lies over a layer of clay and limestone. The area is phylloxera-free, though spraying is necessary to control pests and mildew. The district produces a wide assortment of grapes for table and fortified wine, as well as for brandy and fortifying spirits.
- Goulburn Valley (Central Victoria WZ)—120 km north of Melbourne, near the town of Seymour. Temperatures are moderate in general, but heatwaves can send them to 43°C in the summer months. The annual precipitation of 580 mm is supplemented with irrigation. Downy mildew and spring frosts are the big threats. The soils vary from sandy loam to grey alluvial outwash near the river bed. High quality wines are made from Shiraz (Syrah), Cabernet Sauvignon, Chardonnay, Marsanne, Riesling, Chasselas, and Muscat. The area's grapes tend to have high sugar and low acidity.
- Bendigo (Central Victoria WZ)—a modest production area west of the Goulburn River. Sparse vineyards yield dark-coloured, full-bodied red wines with lots of character. Cabernet Sauvignon and Shiraz (Syrah) are blended to produce the best wines of the area.
- Macedon (Central Victoria WZ)—a cool, windswept area northwest of Melbourne, on the granitic Macedon range. It can yield very good Cabernet Sauvignon, Pinot Noir, Shiraz (Syrah), and Chardonnay when the weather is kind.
- Grampians (Western Victoria WZ)—a hilly area associated mostly with sparkling wines. It is located 220 km west of Melbourne near Stawell, on the western slopes of the Dividing Range. At 350 m above sea level and with the poorest soil of Australia, spring frosts, and rainfall in the winter but little in the growing season, it is a difficult area for grape growing. The soil is primarily of volcanic origin and varies from grey granitic sand to red-brown gravelly loam, on a bed of clay and granite.

Against these odds, a summer average of 20°C (which allows the fruit to mature and retain good acid levels) and drip irrigation help produce some superb wines. Cabernet Sauvignon, Shiraz (Syrah), Cinsaut, Malbec, Mataro, Pinot Noir, Chardonnay, Sémillon, Chasselas, Folle Blanche, Tokay, and Riesling are the main varieties.

- Pyrenees (Western Victoria WZ)—between the Grampians and Bendigo districts. The eight wineries produce rich, spicy red wines, pleasant white wine, and a fine Sauvignon labelled Fumé Blanc.
- Geelong (Geelong WZ)—west of Melbourne and the Corio Bay, around the town of Geelong, settled by Swiss. It was here that phylloxera first appeared in 1875. The area was devastated, and most growers turned to other crops. Viticulture did not revive here until the mid 1960s. The cool, dry climate produces very fine, scented Pinot Noirs and Chardonnays as well as racy Shiraz (Syrah), Cabernet Sauvignon, Sauvignon, and Riesling wines.
- Yarra Valley (Yarra Valley WZ)—just east of Melbourne. This very cool and dry area produces austere, distinctive, elegant wines with light body and good acidity. Good sparkling wines are also produced, but it is the Pinot Noir that is most responsible for this district's fame. Excellent Cabernet Sauvignon, graceful Chardonnay, spicy Shiraz (Syrah), and rich Merlot are also attracting attention.
- Mornington Peninsula (Mornington Peninsula WZ)—a relatively new area, 40 km south of Melbourne. The cool maritime climate is well suited to delicate Chardonnays, elegant Cabernet Sauvignons, and scented Pinot Noir, Shiraz (Syrah), and Merlot wines. Even Viognier and Pinot Gris have given excellent results. Most of the wines are sold at small "cottage" wineries.

Tasmania

Early attempts to grow grapes failed when mildew and vine diseases attacked in the island's cool damp climate. Today, these problems are easily controlled and a major revival of island wine production is taking place, with particularly strong results from Chardonnay and Pinot Noir. New plantings of Riesling and Gewürztraminer could surprise us with their quality very soon. The total production is small but generally of good quality.

South Australia

South Australia supplies a wide array of wines from Sherry-style to sparkling, as well as a good range of varietal red and white table wines. More than half of all Australian wine is made in just seven of the districts in this state. Most of the wines are of excellent quality and are among Australia's finest.

- Riverland (Murray Mallee WZ)—extends 120 km along the Murray River between the towns of Renmark and Morgan. The climate is hot and dry with summer temperatures hovering around 32°C and average rainfall below 200 mm. This rainfall is supplemented by irrigation from the Murray River. The soils are either red sand or alluvial sandy loam, usually over a clay subsoil with patches of limestone. Initially, only fortified wines, dessert wines, and spirits were made here. Today, with an efficient irrigation system, modern refrigeration, and cold fermentation equipment, the area produces large amounts of respectable table wine. Chardonnay, Chenin Blanc, Colombard (Columbard), Shiraz (Syrah), Palomino, Doradillo, Riesling, and Grenache are the dominant varietals.
- Clare Valley (Central South Australia WZ)—140 km north of Adelaide and the northernmost area in South Australia. The vineyards are planted on rolling hills between 305 m–610 m. The climate is continental, with hot summers and an annual rainfall of 610 mm (little need to irrigate). Heavy loam predominates, with frequent patches of limestone and sandstone. Wines from this region are concentrated and robust yet manage to retain a certain elegance. Good fruity Rieslings are made in both dry or semi-dry styles. Shiraz (Syrah), Cabernet Sauvignon, and Malbec are deeply coloured and full of extract. Sémillon produces well, but Chardonnay and Pinot Noir usually end up flabby.
- Barossa (Central South Australia WZ)—arguably Australia's best known wine district. It is 55 km northeast of Adelaide and extends for 55 km between the towns of Nuriootpa and Lyndoch. The vineyards cover the rolling hillsides of the valley and climb up the craggy slopes of the Barossa Range (on the eastern perimeter) to altitudes of 180 m–270 m. The summers are dry and hot, with temperatures up to 28°C.

Annual rainfall ranges from 500 mm at the lower elevations to 800 mm on the hillsides, and falls mainly during the winter and spring. A great variety of soils include red-brown heavy loam, patches of limestone, terra rossa, and sand. Traditionally, the area is known for the Sherry- and Port-style wines and dessert wines that are produced in the lower and hotter part of the valley. The cool hills produce wines that are more delicate. The hills yield good Shiraz (Syrah), Sémillon, and Grenache with some Riesling, Cabernet Sauvignon, Chardonnay, Gewürztraminer, Sauvignon Blanc, Mataro (Mourvèdre), and Malbec. A small amount of sparkling wine is produced. This area is free from phylloxera.

- Eden Valley (Central South Australia WZ)—an extension of the Barossa Valley. The difference is a cooler climate, which yields somewhat fruitier wines.
- Adelaide Metropolitan (Central South Australia WZ)—encompassing the Adelaide Hills as well as the Adelaide Plains (also called Angle Vale). Vine plantings are under intense pressure from urban development around the city of Adelaide. The area ranges from the flat land of the heavy loam plains to the steep inclines of the Lofty Range. The temperature on the plains is moderate and the annual precipitation is 500 mm. In the cooler foothills, with an average rainfall of 640 mm, black currant scented Cabernet Sauvignon, crisp, fruity Riesling, flavourful Sauvignon Blanc, Pinot Noir, and Chardonnay are produced.
- Southern Vales (Central South Australia WZ)—south of Adelaide. Because Southern Vales is often confused with the Southern Vales Co-op, it has been suggested that the region change its name to Reynella-McLaren Vale. The climate is mild with rainfall averaging 600 mm and no frost. The soils vary from pure sand, sandy loam, and heavy loam, to limestone, rich alluvial soil, and heavy red clay. The area is free of pests and produces good Shiraz (Syrah), Cabernet Sauvignon, Merlot, Chardonnay, and Sauvignon.
- Coonawarra (South-Eastern South Australia WZ)—the southernmost major wine district of Australia. Coonawarra is an aboriginal word meaning "wild honeysuckle." The land is relatively flat and consists of terra rossa, and the black soil known as rendzina. The limestone subsoil is rich in subterranean water not far below ground level—a situation that encourages vigorous growth of the vines.

The climate is cool with an average rainfall of about 640 mm. Ripening slowly and late into the season, the grapes develop good varietal character, soft tannin, and fine acid balance. Cabernet Sauvignon, Shiraz (Syrah), Pinot Noir, Riesling, Sauvignon, and Chardonnay are prestigious and in great demand both domestically and internationally.

- Padthaway (South-Eastern South Australia WZ)—directly north of Coonawarra. Padthaway was created when land prices in Coonawarra rose too high for the big companies that wanted to increase their vineyard holdings. The climate is warmer than in Coonawarra. The red wines are slightly richer and the Chardonnays have a more pronounced citrus aroma.

WESTERN AUSTRALIA

Western Australia is the latest state to experience a wine boom. The climatic conditions range from scorching in the Swan Valley, to cool in the Mount Barker and Frankland River areas.

- Swan Valley (Swan Valley WZ)—the centre of the state's wine production, near Perth. The vineyards are perched on the banks of the Swan and Canning rivers. Summer temperatures are blistering at 40°C throughout the growing season. The rainfall averages 900 mm, but falls only during the winter. Irrigation is not possible because the Swan River is tidal and the water brackish. The district has a fertile and deep, alluvial, sandy loam soil known as Marri land. The combination of fertile soil and hot climate produces grapes high in sugar and low in acid. The wines are full bodied and rich, with a tendency to high alcohol levels. White wines make up two-thirds of the production, most of which is Chenin Blanc. The rest is made from Chardonnay, Verdelho, Sémillon, and Muscat. Grenache and Cabernet Sauvignon dominate the red varieties. Excellent dessert wines and Sherry-style wines are also produced.
- Margaret River (Margaret River WZ)—south of the town of Bunbury, 250 km south of Perth. The Indian Ocean moderates the climate to one very suitable for the production of elegant Cabernet Sauvignon, Merlot, Shiraz (Syrah), Zinfandel, Chenin Blanc, and Chardonnay wines. This small area is relatively new, but the quality and demand for the wines are bound to bring expansion.

- Mount Barker-Frankland River (Mount Barker-Frankland WZ)—north of Albany. The cool maritime climate is one of the best in Australia for the production of classic wines. The first vineyard was planted in 1966 and the results have been so successful that the demand surpasses the supply. Cabernet Sauvignon, Pinot Noir, Shiraz (Syrah), Riesling, Chardonnay, Sauvignon Blanc, Sémillon, and Chenin Blanc are all noted for their elegance and distinction. This area is in full development.
- Moondah Brook—located north of Perth, near the Swan Valley. It is best known for its Chenin Blanc and Verdelho.

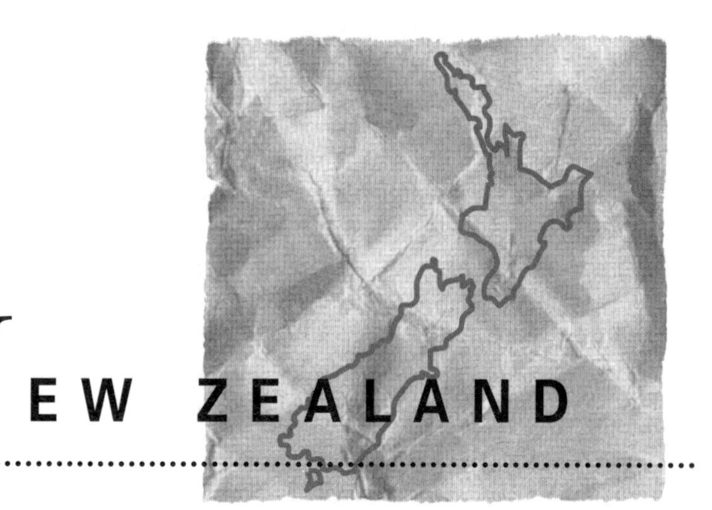

CHAPTER 26 NEW ZEALAND

Although New Zealand has been making wine for over a hundred years, it is only in the last quarter of the twentieth century that the industry really took off. From 179 ha of vineyards in 1923, to 2000 ha in 1973, the number increased dramatically to 6070 ha by the mid 1990s—and they continue to expand. New Zealand's annual production is around 405 000 hL, and consumption is 8 L per capita.

Wine Regions of New Zealand

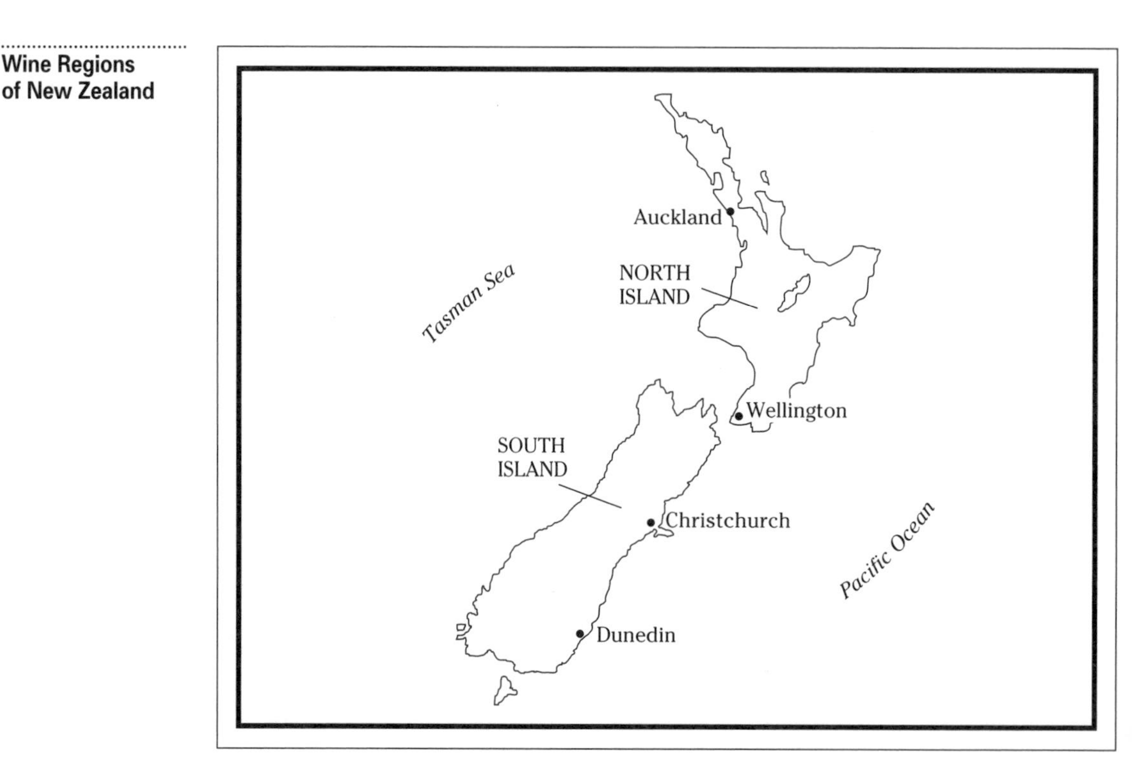

GEOGRAPHY

Located in the South Pacific Ocean, southeast of Australia, New Zealand consists of two main islands. North Island and South Island, which lie between 34°S and 47°S, are separated by Cook Strait. Both islands are volcanic in origin and quite mountainous. North Island still experiences volcanic activity, as well as earthquakes. On South Island, the Southern Alps create a spine that runs the length of the island, with Mount Cook the highest peak (3764 m). Glaciers are found at elevations as low as 300 m.

The Pacific Ocean makes the climate generally mild and humid, and weather can change dramatically in the course of a day. West winds often blow forcefully and bring abundant rainfall. The South Island's Alps can register 5700 mm annual precipitation. Average summer temperatures are about 20°C at Auckland, 17°C at Wellington, and 14°C at Dunedin. Average winter temperatures are about 11°C at Auckland, 8°C at Wellington, and 5°C at Dunedin. In many areas, New Zealand has the right soil and climate to produce excellent wines.

HISTORY

New Zealand's history of viticulture is entwined with the histories of the Maori people, the New Zealand settlers (particularly the British and French, who brought vines to cultivate), and Australia.

The first vines arrived in 1819 under the British flag and the patronage of an Anglican missionary, Reverend Samuel Marsden, Chief Chaplain of the Government of New South Wales in Australia. Although Marsden never resided in New Zealand, he made seven visits from 1814 to 1837 and established the first mission. In 1819, Marsden reported 100 vines had been planted at Kerikeri in 1819. In 1835, Darwin landed at Kerikeri and remarked that the vines tended by the Maoris there were thriving.

James Busby was the next person to bring vine cuttings to the country. Arriving in 1833, he became the first British Agent in New Zealand under the New South Wales government. He established his first vineyard at Waitangi, and his light, white wines were later praised by a visiting French Navy commander, Dumont D'Urville.

Other French citizens were soon to follow. Settlers landed at Akaroa, where they planted vines from their native France; these vineyards eventually succumbed to powdery mildew. In 1838, Bishop Pompallier landed at Hokianga, where he established a mission and planted vines in order to make sacramental wine. These French vineyards took hold, particularly around Napier. The Mission Vineyard at Greenmeadows is the oldest vineyard in New Zealand, and remains under the administration of the Society of Mary.

The second half of the nineteenth century brought even more interest in producing wine. For example, Joseph Soler planted vines at Wanganui in 1865, and William Beetham returned to New Zealand from a trip to France with a strong desire to grow grapes at Masterton. It was not only individuals who were interested in grape growing, but also the government. Romeo Bragato, who held a Diploma of Oenology from Conegliano in Italy was "borrowed" from Australia, ostensibly to establish a Department of Agriculture. His report after touring the area in 1895 was generally favourable: various areas could grow Vitis vinifera. He also said that phylloxera had reached certain vineyards, and American rootstocks should be imported to avoid devastation. Unfortunately, this recommendation was not followed, and soon most of New Zealand's vineyards fell to either phylloxera or powdery mildew.

Bragato was recalled to New Zealand in 1901 by the government, and was offered a position as the head of the Department of Viticulture. He ran an experimental vineyard and nursery at Te Kauwhata, grafted vinifera on American rootstocks, built a winery, and made wine from the grafted vines. In 1908, a wine made from the grafted Cabernet Sauvignon at Te Kauwhata received a gold medal at the Franco-British Exhibition in London and was described as "a wine approaching the Bordeaux clarets in lightness and delicacy."

Still, farmers were not convinced that vinifera grapes grafted onto American rootstocks would produce wines of quality. As well, the prohibitionist crusade registered its first victory in 1909: in Masterton, 7000 Pinot Noir vines were uprooted. Apathy, a temperance movement, bad winemaking practices (notably extreme chaptalization and watering wines to reduce acidity), and the difficulty in obtaining cuttings of the right varieties meant that many New Zealanders opted for American hybrids. Thus, many farmers planted table grapes (which eventually could be made into fortified wine) and the industry stagnated.

World War II (1939–1945) increased the demand for alcoholic beverages and vineyards expanded. But, again, most of the vines were the American hybrids unsuitable for New Zealand wine production.

As though to make up for lost time, New Zealand's wine industry has made a complete turnaround. To quote Michael Broadbent's *Pocket Guide to Wine Vintages*, "Prior to 1970 New Zealand was noted for sheep, butter, and beautiful, uncrowded countryside. Then vines were planted and, throughout the 1980s, dramatic advances were made." Fortified and sweet wine production has dwindled and the selection of exciting table wines continues to increase. Even though the domestic market still favours beer over wine (in 1990, 144 L versus 13 L per capita annually), supermarkets have been permitted to sell wine only since 1990, and bar hours remain restricted, New Zealand's wine industry seems to be coming into its own.

LAWS AND REGULATIONS

Because of the youth of the industry, New Zealand has had few regulations: varietal wines must contain 75% of the named variety, and blends must list the grape varieties in descending order.

Traditionally, labels have provided information required for export to other countries, and regional names on labels have indicated that the grapes are predominantly from the named area. As of 1996, the exception is the Wairarapa-Martinborough area, where producers may comply with the guidelines that wines labelled Wairarapa-Martinborough are produced from grapes entirely of that area.

As of the mid 1990s, however, New Zealand is developing a system to certify the origins of its wines.

GRAPE VARIETIES

The cool climate of New Zealand has led to whites dominating the vineyards. Chardonnay is the most prominent and successful. It is blended with Pinot Noir and bottle fermented to meet the demand for premium sparkling wines. Some red varieties are growing in popularity. Cabernet Sauvignon is the most planted red grape. It is produced as a varietal, as well as blended with Merlot, Malbec, and/or Cabernet Franc. The leading varieties are as follows.

White Grapes

- Chardonnay—the most cultivated white variety in New Zealand. The best examples are to be found in Gisborne, Marlborough, Hawke's Bay, and Nelson. They are usually harmonious, with good acidity.
- Sauvignon Blanc—the second most important grape in terms of hectares of vines, but the best in terms of quality. New Zealand's very good Sauvignon Blancs have a grassy character, with crisp acidity and gooseberry aromas similar to those in the best Sancerres. The Marlborough area tends to produce the best examples, followed closely by Auckland and Hawke's Bay.
- Riesling—not yet widely planted but on the increase. Good dry versions are found in Hawke's Bay, Nelson, and Marlborough. Some rare but excellent botrytized Rieslings appear from time to time.
- Gewürztraminer—grown in minute quantities. They can display the characteristic litchi and faded-rose-petal nuances. From Gisborne they can be very pleasing, when made into the dry style of Alsace.

Red Grapes

- Cabernet Sauvignon—the country's most-planted red, and the grape that first brought fame to New Zealand. The best wines come from Auckland, either as varietals or blended with Merlot. These are inclined to display a herbaceous character. Hawke's Bay and Marlborough also produce good Cabernet Sauvignons that tend to be laden with fruit flavours.
- Pinot Noir—inconsistent in quality thus far, but showing potential for very good wines. Notable examples have come from Central Otago, Canterbury, and Wairarapa-Martinborough.
- Merlot—grown by the French settlers, but ignored for some time after. Merlot is often blended with Cabernet Sauvignon to give body and texture.

Some varietals are produced with good results. It achieves its best results in Auckland, Marlborough, and Hawke's Bay.

Other white varieties are Müller-Thurgau, Sémillon, Chenin Blanc, Chasselas, and Palomino. The other red varieties are Cabernet Franc, Pinotage, Malbec, and Gamay.

WINE PRODUCTION

Far from the days of producing sweet and fortified wines only, New Zealand wines are often very good. Today, the Sauvignon Blancs from Marlborough and Hawke's Bay count among the best in the world. As well, other European varieties—particularly Chardonnay and Pinot Noir—are making a great impression internationally. Winemakers have learned to put their wines through malolactic fermentation to reduce acidity, and to make better use of French casks. Wines in the styles of Port and Sherry are still produced, as well as some good sparkling and dessert wines.

WINE REGIONS

Roughly from north to south, the eleven wine regions are as follows.

NORTH ISLAND

- Northland/Matakana—a small production zone located at the tip of the country, northeast of the town of Whangarei from Waihopo to Kaiwaka.
- Auckland—from Warkworth to Pukekohe around the city of Aukland and including Henderson, Kumeu, Huapai, Waiheke Island, and Great Barrier Island. The climate is warm and humid, and the soil is primarily clay. The area is relatively small and dominated by red grapes.
- Waikato—named for the Waikato River. The area's two districts are the Hauraki Plains north of Huntly near Hamilton, and a smaller area in the lower valley of the Waihou River south of Thames. Waikato includes Te Kauwhata, known for its first great Cabernet Sauvignons. This district is humid and relatively warm. Chardonnay, Chenin Blanc, Sauvignon, Müller-Thurgau, and Riesling share the majority of the white crop; Cabernet Sauvignon is the primary red grape.
- Bay of Plenty—small vineyard holdings located near the bay east of Edgecumbe. The main production is white wines.

- Gisborne—tucked between the Raukumara Range and the east coast. The main area is located on rich, alluvial loam soils around the town of Gisborne, from Ormond to Manutuke. This area specializes in Chardonnay, Müller-Thurgau, Sauvignon, Gewürztraminer, and Muscat.
- Hawke's Bay—an important area on the east coast between Napier and Havelock North, including Hastings. This sunny area is drier than Gisborne, with cooler springs and warm autumns. The well-drained coastal plain produces some of the best Chardonnay, Cabernet Sauvignon, Merlot, Sauvignon Blanc, and Müller-Thurgau.
- Wairarapa-Martinborough—the southernmost vine-growing district on the North Island and near to Wellington, the capital. The main area is located on a stony plateau south of the town of Martinborough and southeast of the Tararua Range. The climate is sunny and dry but very windy. The area produces good Pinot Noir, Chardonnay, and Sauvignon.

South Island

- Nelson—a small area situated on north-facing hills, south of the town of Nelson. The climate is the warmest of the South Island and produces very good Chardonnay, Riesling, and some Pinot Noir wines.
- Marlborough—in the Wairau Plains facing Cloudy Bay and east of the town of Blenheim. The area is dry, temperate, and very sunny. The area is most successful for white varieties such as Sauvignon Blanc, Chardonnay, Müller-Thurgau, and Riesling. Some interesting Pinot Noirs and herbaceous Cabernet Sauvignons are also produced.
- Canterbury—located near the town of Christchurch in the lowest part of the Waimakariri River valley. This area is quite cool, with a long ripening season. The best wines are made from Chardonnay, Riesling, and Pinot Noir. The northern subregion of Wairapa is recognized as a separate area.
- Otago—a tiny area of 150 ha and the coolest and the most southerly region. It is located west of the town of Alexandra, where the Kawarau and Clutha rivers meet. Sunny, dry summers and long autumns are ideal for making excellent Chardonnay, Riesling, and Pinot Noir wines.

INDEX OF WINES

GLOSSARY

aroma: the scent of the grape variety; more generally, the scent of the wine

back blending: a method used to sweeten wines with grape juice

Baumé: the French scale for measuring the concentration of grape sugars in must

blend: a wine made by combining the wines of different grape varieties or origin

bloom: the whitish covering on the grape skin

botrytized wines: sweet wines made from grapes affected by noble rot (Botrytis cinerea)

bouquet: the combination of scents in a wine that are created during the winemaking process; more generally, the scent of the wine

Brix: the North American scale for measuring the concentration of grape sugars in must

budbreak: a stage in vine development in which small shoots emerge from vine buds, in Spring

carbonation method: a method of making sparkling wine by injecting tanks of wine with carbon dioxide

carbonic maceration: in making red wine, a process of two fermentations involving carbon dioxide and the softening of the grapes

chaptalization: the addition of sugar to grape juice or must, either before or during fermentation to increase alcohol

charmat method: a method of making sparkling wine in a sealed tank; also called cuve close

clarification: the process of removing suspended particles and insoluble materials from grape juice or new wine

clone: an entity derived from a single cell or from a single parent plant

cultivar: a cultivated variety of vine

decant: to gently pour wine from its bottle into another vessel, leaving the sediment behind

delimited: defined for the precise limits of the production, area, vinification methods, etc.

demarcated: marked for the limits of a region

dormancy: a period of rest or inactivity (i.e., minimal metabolic activity) in the winter

downy mildew: a fungal disease that attacks green parts of the vine and looks like white cotton; also called peronospera

extract: the sum of all the non-volatile solids in wine, that is, the sugars, non-volatile acids, minerals, etc.

filtering: the process of straining solid particles from wine using filters

fining: the addition of a coagulant to make suspended particles sink so that they can be removed from grape juice or new wine

flavonoids: the compounds (e.g., anthocyanins, flavones) that give each grape variety its distinctive aromatic and colouring qualities

free-run: the juice that runs from crushed grapes without them having been pressed; also the wine that flows from a fermentation tank, without the must having been pressed

fructose: one of the two principal sugars in grapes (the other is glucose); also known as levulose

fruit set: a stage in vine development in which the flower becomes a grape berry

fusel oils: a term for a group of alcohols in wine that contribute to its complex aromas

glucose: one of the two principal sugars in grapes (the other is fructose)

hybrid: a cross of two varieties of vines

lactic acid: a mild acid found also in milk, yogurt, etc.

lees: the sediment or dregs in winemaking, consisting of dead yeast cells and grape particles

legs: drops of wine running down the inside of the glass

malic acid: an apple-flavoured acid, one of the principal acids in grapes

malolactic fermentation: a secondary fermentation that converts malic acid to the softer lactic acid

méthode Champenoise: the traditional method of making Champagne and other sparkling wines, involving still wine undergoing a second fermentation in the bottle

must: the mixture (grape juice, stem fragments, grape skins, seeds, and pulp) after the grapes are crushed and de-stemmed

noble rot: the vine disease caused by Botrytis cinerea, which can have beneficial results in some situations

nose: the smell of the wine, a combination of aroma and bouquet

Oechsle: the German scale for measuring the concentration of grape sugars in must

pectins: a constituent of grapes that tends to cause cloudiness in wine; this cloudiness can be eliminated with pectic enzymes

phenolic compounds: chemical compounds (natural colour pigments, tannins, and flavour compounds) found in wine

photosynthesis: a biochemical reaction that converts the energy of the sun to form sugars in the vine

phylloxera: a pest that attacks vine roots

Pierce's disease: a deadly bacterial disease found particularly in hot and/or arid climates

pigeage: the process of punching the cap of skins, seeds, and stems in making red wine

pomace: the pulp remaining in the primary fermentation tank when the free-run juice has been drawn off

powdery mildew: a fungal disease that attacks the green parts of the vine and looks powdery; also called oidium

press wine: wine made from the must after it is pressed, rather than free-run wine

racking: removing clear wine from the settled sediment

raisining: drying harvested grapes to dehydrate them and concentrate their sugars

remontage: circulating the must over the cap of skins, seeds, and stems in making red wine

süssreserve: unfermented grape juice used to sweeten must or wine

tannins: a group of chemicals that give astringency to wine

tartaric acid: a harsh-tasting acid; one of the principal acids in grapes

tastevin: a small, shallow, bumpy, silver cup traditionally used by winemakers, cellar masters, and wine service professionals

terroir: the combination of climate, soil, and landscape in a vine-growing site

transfer method: a method of making sparkling wine, similar to the méthode Champenoise, but which transfers the wine for its second fermentation

ullage: the space in a container after some of the wine has evaporated; also, the process of this evaporation

varietal: a wine named for the single grape variety it is made from; also, the grape variety itself

veraison: an intermediate stage in the grape's development, from hard green berries to soft and coloured grapes

Vitis vinifera: species of European vine that produces most of the wines around the world

INDEX